Quantum Optics and Fundamentals of Physics

## Fundamental Theories of Physics

*An International Book Series on The Fundamental Theories of Physics: Their Clarification, Development and Application*




Volume 63

# Quantum Optics and Fundamentals of Physics

*by*

Jan Peřina

Zdeněk Hradil

and

Branislav Jurčo

*Department of Optics,*
*Palacký University,*
*Olomouc, Czech Republic*

KLUWER ACADEMIC PUBLISHERS
DORDRECHT / BOSTON / LONDON

A C.I.P. Catalogue record for this book is available from the Library of Congress.

ISBN 0-7923-3000-5

---

Published by Kluwer Academic Publishers,
P.O. Box 17, 3300 AA Dordrecht, The Netherlands.

Kluwer Academic Publishers incorporates
the publishing programmes of
D. Reidel, Martinus Nijhoff, Dr W. Junk and MTP Press.

Sold and distributed in the U.S.A. and Canada
by Kluwer Academic Publishers,
101 Philip Drive, Norwell, MA 02061, U.S.A.

In all other countries, sold and distributed
by Kluwer Academic Publishers Group,
P.O. Box 322, 3300 AH Dordrecht, The Netherlands.

*Printed on acid-free paper*

# Table of contents

# Preface

In last years increasing attention has been again devoted to interpretations of quantum theory. In the same time interesting quantum optical experiments have been performed using nonlinear optical processes, in particular frequency down conversion, which provided new information about nature of a photon on the basis of interference and correlation (coincidence) phenomena. Such single-photon and twin-photon effects of quantum optics provide new point of view of interpretations of quantum theory and new tests of its principles.

The purpose of this book is to discuss these questions. To follow this goal we give brief reviews of principles of quantum theory and of quantum theory of measurement. As a fundamental theoretical tool the coherent state technique is adopted based on a general algebraic treatment, including the description of interaction of radiation and matter. Typical quantum behaviour of physical systems is exhibited by nonclassical optical phenomena, which can be examined using photon interferences and correlations. These phenomena are closely related to violation of various classical inequalities and Bell's inequalities. The most important part of this book discusses quantum optical experiments supporting quantum theory.

This book may be considered as a continuation of previous monographs by one of the authors on Coherence of Light (Van Nostrand Reinhold, London 1972, second edition D. Reidel, Dordrecht 1985) and on Quantum Statistics of Linear and Nonlinear Optical Phenomena (D. Reidel, Dordrecht 1984, second edition Kluwer, Dordrecht 1991), which may serve as a preparation for reading this book.

The book will be useful to researchers in the fields of quantum optics, quantum theory, measurement theory and optical communications as well as to students of physics, optics, optical engineering, optoelectonics and photonics.

The text of the book has been worked out as follows: Chapters 2 and 4 have been written by B. Jurčo excepting Secs. 2.9 and 4.6–4.8. Chapter 3 has been prepared by Z. Hradil. All the other text is by J. Peřina.

We would like to thank Professors L. Mandel, J. G. Rarity and P. R. Tapster, S. Reynaud, F. Selleri, M. C. Teich, D. F. Walls, H. Walther, E. Wolf and

A. Zeilinger for kind sending us reprints and preprints of their papers related to the content of this book. One of us (J. P.) thanks Prof. A. Zeilinger for kind possibility to visit the Innsbruck University and to see interesting experiments on photon interferences and coincidences and on "nonlocal" effects in quantum theory (in the whole book we use the word nonlocal in the inverted commas to distinguish nonlocality related to quantum correlations from other kinds of nonlocality, e.g. in teleportation [Popescu (1994)]). Another of us (B. J.) would like to thank Professors J. Tolar and W. Lücke for discussions, A. von Humboldt Foundation and CERN for support and Prof. H. D. Doebner for hospitality in Clausthal. The support by Professors M. Bertolotti, M. C. Teich and E. Wolf was considerable for the progress in this work. Many discussions with our colleagues, particularly with Prof. V. Peřinová, Prof. R. Horák, Dr. J. Bajer and Dr. A. Lukš were useful for the text of the book. We are obliged to MSc. Jan Peřina, Jr. for careful preparation of the manuscript by means of computer in the camera-ready form and for his comments. Permission to reproduce figures is acknowledged to Professors P. Grangier, H. J. Kimble, L. Mandel and Y. Yamamoto and to the American Physical Society, Elsevier Science Publishers, Institute of Physics of Polish Academy of Sciences and Masson Publishing House.

Olomouc, March 1994

Jan Peřina
Zdeněk Hradil
Branislav Jurčo

# Chapter 1

# Introduction

Classical optics stimulated and strongly influenced the origin and development of modern physical disciplines, such as theory of relativity and quantum mechanics. However, it was considered to be closed as a classical scientific discipline investigating generation, manipulation and detection of optical radiation until the laser was discovered. The discovery of the laser as a handmade source producing new quality of light permitted to create new fields of physics and optics, such as holography, representing new imaging process conserving phase information owing to the use of a reference beam, nonlinear optics, providing new nonlinear optical phenomena determined by the dependence of quantities of the medium on the incident laser intensity, etc. Many laser applications are available in science and techniques. High coherence and strong intensity (light degeneration, i.e. large mean number of photons present in the coherence volume) of laser light manifest themselves as the most important properties of laser beams.

Such a progress, together with new detection methods developed in nuclear physics enabling to detect single photons, stimulated R. J. Glauber to formulate quantum theory of coherence in 1963, giving basis for quantum optics. The so-called coherent states, introduced by E. Schrödinger in 1927, were the basic tool further developed by R. J. Glauber and other authors in connection with quantum theory of coherence and quantum optics. Such quantum states, representing an infinite sum of number occupation (Fock) states, are minimum uncertainty states suitable for the description of boson cooperative phenomena, such as optical coherence. These states are closest to classical states as it is permitted by quantum theory – they include only quantum fluctuations of physical vacuum. The development of the coherent-state technique made it possible to obtain a complete classical limit of quantum electrodynamics and to discover new aspects of quantum theory related to phase-space formulations and correlation measurements. Thus optics was again able to influence quantum theory.

The generalized coherent-state technique is useful for treatments in quantum theory and in quantum optics with a particular emphasis on algebraic properties. Although these rather sophisticated methods are typical for general quantum field theory, they are also useful in more practical quantum optics because they represent simple qualitative methods for obtaining various closed-form solutions of quantum optics.

Using phase-space methods, quantum optics was able to predict many nonclassical phenomena, such as squeezing of vacuum fluctuations in field quadratures, antibunching of photons, sub-Poisson photon statistics exhibiting reduced photon-number fluctuations below the Poisson level of ideal laser light, quantum oscillations in photocount distributions, violation of various classical statistical inequalities, collapses and revivals of atomic inversions, etc. New quantum features of interferences, wave-particle dualism, complementarity principle and Einstein-Podolsky-Rosen (EPR) "paradox" were obtained which are described by strongly quantum correlated (entangled) states exhibiting various "nonlocal" properties. Interesting methods were developed for measurements of phase-space quasidistributions. Many of these quantum effects can be observed in micromasers and in cavity quantum optics and cavity quantum electrodynamics dealing with the interaction of single atoms with light fields. Interesting direction of contemporary research is devoted to investigations and measurements of quantum phase.

For verifications of these quantum properties interference and correlation measurement techniques were developed in quantum optics for low-intensity fields involving higher-order correlations and coincidences combined with the use of nonlinear effects. The optical parametric process of frequency down conversion, producing quantum entangled states, has been proved to be very fruitful in examing the quantum phanomena mentioned above.

Such quantum effects have been observed in interference and correlation experiments in full agreement with principles and predictions of quantum theory. The correlation experiments by L. Mandel and co-workers at the University of Rochester and by other groups were performed with resonance fluorescence light of an atom in 1977 and 1978. In these experiments, representing the first direct probe to quantum operation of the atom, photon antibunching was observed. The interference and coincidence experiments performed later on included polarization and momentum transfer properties, feedback and feedforward configurations, nonlinear processes, particularly frequency down conversion, squeezed and sub-Poisson light generation in nonlinear processes clearly demonstrating quantum aspects of light, atoms and their interaction, wave-particle dualism of a single photon, strong violation of classical inequalities including Bell's inequalities and providing new interpretations of EPR "paradox" and "nonlocal" quantum effects. Some of them related to quantum intereference may serve as a basis for future techniques of quantum cryptography for safe conservation and transfer of information. All these experiments are in full agreement with quantum theory and they contradict any classical theory. In these investigations realizations of quantum superposition principle are available and various quantum nondemolition measurements can be developed. Especially experiments in cavity quantum optics and with induced coherence in cascaded frequency down converters are able to provide new information about the nature of a photon.

Quantum optical methods may effectively be applied to other fields of physics and science, such as solid-state physics, particle physics (field theories, statistical interpretations of high-energy collisions of particles in colliders), in biophysics (statistical interpretations of human vision, cooperative operation of biological cells), etc.

The structure of the book is as follows. In Chapters 2 and 3 we present the basic ideas and methods of quantum theory, i.e. vectors and operators in Hilbert space are introduced and analyzed together with statistical operator (density matrix), quantum states, observables and measurements, spectral representations, uncertainty relations, phase-space representations, symmetries, etc. Particular attention is devoted to quantum systems and dissipations, quantum measurements, quantum optical measurements, quantum nondemolition measurements, high-precision measurements and quantum phase measurements.

Chapter 4 represents a brief review of the coherent-state technique, including the coherent states of a harmonic oscillator and their representations, squeezed states, atomic coherent states of angular momenta, generalized coherent states introduced in different ways as minimum uncertainty states, displaced states and annihilation operator eigenstates. Various field operator orderings are considered in relation to possibilities of measurements of quasidistributions. Algebraic properties of coherent states can provide a unified approach to the coherent-state techniques. The quantum dynamical and statistical properties are treated in terms of Heisenberg-Langevin equations and quantum characteristic functions or alternatively in terms of the master equation and generalized Fokker-Planck equation for quasidistributions. In many treatments of quantum statistics in nonlinear phenomena the so-called generalized superposition of coherent fields and quantum noise, extending the classical superposition of signal and noise by including nonclassical effects in relation to two-photon coherent states, can be applied at least approximately.

Such quantum optical theoretical methods represent effective tools for investigations of the above mentioned nonclassical optical phenomena and their relations, particularly they are suitable for examining quadrature and intensity squeezing, photon antibunching, sub-Poisson behaviour of photon statistics, oscillations of photon-number distributions, etc., which represents the content of Chapter 5. Moreover, new interesting suggestions for quantum phase measurements and for measurements of quasidistributions are also included.

Theoretical and experimental methods of photon interferences and correlations of second as well as higher orders involving linear as well as nonlinear interactions are stressed in Chapter 6, with a particular attention to single photon interferences, principle of complementarity and wave-particle dualism for a single photon. For such investigations the frequency down-conversion process and various feedback and feedforward techniques are useful or other three- and four-wave interactions and anharmonic oscillators involving the Kerr nonlinearity can be employed.

There are many classical statistical inequalities having interesting physical interpretation violated by quantum optical processes. Moreover, in relation to the EPR "paradox" and "nonlocal" quantum phenomena, Bell's inequalities can be formulated, which are in agreement with classical theories and in contradiction with quantum theory. Such discussions are contained in Chapter 7, including stronger formulations using triple correlations with the help of three detectors.

Chapter 8 is devoted to a thorough analysis of quantum optical experiments for verification of quantum theory, including single-photon interference experiments, correlation experiments with polarized photons, influence of various feedback mechanisms for the control of statistics of light, also with the application of down-converted photons in entangled states, resonance fluorescence experiments indicating photon antibunching, cavity experiments with a single atom and experiments with squeezed light and sub-Poisson light, generated in nonlinear optical processes, such as optical parametric processes, four-wave mixing on atoms and in optical fibres, phase conjugation, etc. In particular, experiments with pairs of down-converted photons in a single nonlinear crystal or in a cascade of nonlinear crystals connected by an idler beam can give new information about the nature of a photon and its properties. All these experiments represent an excellent verification of validity of the principles, predictions and consequencies of quantum theory and fundamentals of quantum physics.

We believe that methods, ideas, theoretical and experimental results presented in this book support quantum theory representing a correct picture of physical laws in Nature, in full agreement with all observations performed until now.

# Chapter 2

# Fundamentals of quantum theory

In this chapter we present basic ideas of quantum theory.

## 2.1 Axioms of quantum mechanics

In this section we will collect the basic principles of quantum mechanics. They will be presented in the form of axioms as they were formulated by von Neumann (1932). The physical considerations leading to these axioms are discussed in details in all standard textbooks, for instance by Messiah (1961) and Jauch (1968), and we will not repeat them here. For the reader's convenience, the necessary notions of functional analysis are summarized in the Appendix to this chapter. The reader who is not interested in mathematical details can understand all formally, without worrying about domains, topology etc. and he can consider only the formulation of the **Axiom 2** valid for operators with pure point spectrum. Very good references for this chapter are books by Berezin and Shubin (1991) and Blank, Exner and Havlíček (1993).

We start with the basic notion of a *state* of a system (with a finite number of degrees of freedom). Let us first recall its meaning in the classical theory. The time evolution of a system is given by a solution of the equations of motion and is completely determined by the initial condition (a point of the phase space of the system). Also the time evolution of any classical *observable* (function on the phase space) is completely determined by the initial condition. So the states of the classical systems appear to be completely described by the points of their phase spaces. Similar situation is met in the classical statistical mechanics where we should assume a more general notion of a state. There the state of a system is completely described by the probability distribution on the phase space. If the probability distribution is supported in one point, then the state is called *pure state.*

From the point o view of physical measurement of an observable in a state, the state is considered as a result of the preparation. The latter becomes indispensable in the quantum case where the interaction of the macroscopic measuring apparatus with the microscopic quantum object is to be incorporated. We have the formal definition [Jauch (1968)]:

**Definition 1** *A state is the result of a series of physical manipulations on the system which constitute the preparation of the state. Two states are identical if the relevant conditions in the preparation of the state are identical.*

Let us note that this definition contains the notion of a system without any specification. This may lead to situations in which some system may be understood as a state or set of states of some more general systems.

We also have a similar definition of the notion of an observable [Jauch (1968)] according to which

**Definition 2** *An observable is measured on a system through the interaction with a suitable macroscopic apparatus (the measuring device) and the eventual reading (or recording) of a scale.*

After these preliminary definitions we can formulate:

**Axiom 1** *With any quantum mechanical system a complex separable Hilbert space $\mathcal{H}$ can be associated. The states of the system are then represented by the statistical operators on $\mathcal{H}$, i.e. positive Hermitian operators $\hat{\varrho}$ with unit trace* $\mathrm{Tr}(\hat{\varrho}) = 1$. *The observables are represented by self-adjoint operators on $\mathcal{H}$.*

We shall sometimes distinguish between observables and representative operators using lower case letter $a$ for the observable represented by the operator $\hat{A}$.

The statistical operator $\hat{\varrho}$ is also called the *density operator* or the *density matrix.*

A particular example of a statistical operator is the projector $\hat{E}_\psi$ on the *ray* (one-dimensional subspace) in $\mathcal{H}$ generated by the unit vector $|\psi\rangle$, defined as

$$\hat{E}_\psi|\varphi\rangle = \langle\psi|\varphi\rangle\,|\psi\rangle\,, \qquad |\varphi\rangle \in \mathcal{H}. \tag{2.1}$$

This definition suggests a very convenient shorthand notation

$$\hat{E}_\psi = |\psi\rangle\langle\psi|\,.$$

The states represented by such one-dimensional projectors are called *pure states.* All other states are called *mixed.*

Let us assume a statistical operator $\hat{\varrho}$ of the form

$$\hat{\varrho} = \sum_i c_i |\psi_i\rangle\langle\psi_i| , \tag{2.2}$$

where $\{|\psi_i\rangle\}$ is some set of normalized states in $\mathcal{H}$ and the sum of the real positive numbers $c_i$ is equal to one. In this respect we are referring to the statistical operator $\hat{\varrho}$ as to the *incoherent superposition* of the pure states $|\psi_i\rangle$ to distinguish it from the normal *coherent superposition* of the states $|\psi_i\rangle$, which is a pure state of the form $|\psi\rangle = \sum_i \alpha_i|\psi_i\rangle$, where the complex numbers $\alpha_i$ satisfy $\sum_i |\alpha_i|^2 = 1$.

The pure states of the quantum system are represented by rays in the Hilbert space. The difference between a ray and a unit vector generating it is usually omitted. The pure states can be equivalently characterized as states represented by the statistical operators $\hat{\varrho}$ fulfilling

$$\hat{\varrho}^2 = \hat{\varrho} . \tag{2.3}$$

This corresponds to the terminology mentioned in the connection with the classical statistical systems as will be seen later.

In order to formulate the second axiom, let us first introduce the following definition:

**Definition 3** *The observables $a_1, a_2, \ldots, a_n$ are said to be compatible if arbitrarily accurate measurement of their simultaneous values can be carried out in one experiment.*

The probability that the measured quantities of the observables $a_1, a_2, \ldots, a_n$ in a state given by the statistical operator $\hat{\varrho}$ do not exceed the values $\lambda_1, \lambda_2, \ldots, \lambda_n$, respectively will be denoted as $P_{\hat{\varrho}}(\lambda_1, \lambda_2, \ldots, \lambda_n)$ (*joint probability distribution*).

**Axiom 2** *The observables $a_1, a_2, \ldots, a_n$ are compatible if and only if their corresponding representative self-adjoint operators $\hat{A}_1, \hat{A}_2, \ldots, \hat{A}_n$ commute. The joint probability distribution $P_{\hat{\varrho}}$ is then given by*

$$P_{\hat{\varrho}}(\lambda_1, \lambda_2, \ldots, \lambda_n) = \mathrm{Tr}\left(\hat{E}^{(1)}_{\lambda_1}\hat{E}^{(2)}_{\lambda_2}\ldots\hat{E}^{(n)}_{\lambda_n}\hat{\varrho}\right) , \tag{2.4}$$

*where $\hat{E}^{(i)}_{\lambda_i}$ are the projectors belonging to the spectral families of the operators $\hat{A}_i$.*

In the case of a pure state $|\psi\rangle$ one readily obtaines

$$P_{\psi}(\lambda_1, \lambda_2, \ldots, \lambda_n) = \|\hat{E}^{(1)}_{\lambda_1}\hat{E}^{(2)}_{\lambda_2}\ldots\hat{E}^{(n)}_{\lambda_n}\psi\|^2 .$$

The physical meaning of the **Axiom 2** is clear: we can simultaneously measure only the values of observables which are represented by commuting operators. Futhermore, the result of the measurement is a random quantity which cannot be predicted exactly, with the probability distribution given by (2.4).

According to the property of the spectrum, the projector $\hat{E}(\Delta) = 0$ for every open interval $\Delta \in R$ which does not contain any point of the spectrum. So it follows from **Axiom 2** that the probability that observable $a$ will take a value $\lambda$ not belonging to the spectrum of $\hat{A}$, is zero. Thus:
*the observable $a$ can take only these values belonging to spectrum of $\hat{A}$.*

For operators with pure point spectrum the conclusion of **Axiom 2** can be reformulated in a simpler way:
*in an experiment prepared for the measurement of $n$ compatible observables $a_1, a_2, \ldots, a_n$ in the state $\hat{\varrho}$, the probability $p_{\hat{\varrho}}(\mathbf{\Delta})$ that the values of these observables will belong to the subsets $\Delta_1, \Delta_2, \ldots, \Delta_n$ of $R$, respectively, is given by*

$$p_{\hat{\varrho}}(\mathbf{\Delta}) = \mathrm{Tr}\left[\hat{E}^{(1)}(\Delta_1)\hat{E}^{(2)}(\Delta_2)\ldots\hat{E}^{(n)}(\Delta_n)\hat{\varrho}\right], \tag{2.5}$$

*where $\hat{E}(\Delta_i), i = 1, 2, \ldots, n$ is the projector on the closed subspace of the Hilbert space $\mathcal{H}$, which is a direct sum of the characteristic subspaces corresponding to the eigenvalues of the operator $\hat{A}_i$ belonging to the set $\Delta_i$.*

Up to now we did not mention the dynamics (time evolution) of the quantum system. In the classical case the time evolution is governed by the observable called hamiltonian, which has the meaning of the total energy of the system. The time evolution can then be understood as a one-parameter group of canonical transformations (with parameter being the time $t$) of the classical phase space.

The total energy of the quantum system is represented by the self-adjoit operator $\hat{H}$ called *hamiltonian* of the system. The basic dynamical axiom is (in the non-dissipative case):

**Axiom 3** *If the state of the quantum system at the moment $t = 0$ is described by the statistical operator $\hat{\varrho}(0)$, then at the moment $t$ it is given by the statistical operator $\hat{\varrho}(t)$*

$$\hat{\varrho}(t) = \hat{U}(t)\hat{\varrho}(0)\hat{U}^{-1}(t), \tag{2.6}$$

*where $\hat{U}(t)$ is a unitary operator called evolution operator. If at least in the moment $t = 0$, $\hat{\varrho}(0)D_{\hat{H}} \subset D_{\hat{H}}$ holds, then there exists the operator $d\hat{\varrho}/dt$ defined on the domain $D_{\hat{H}}$ and*

$$i\hbar\frac{d\hat{\varrho}}{dt} = [\hat{H}, \hat{\varrho}] \tag{2.7}$$

*is valid with $\hbar$ being the Planck constant $h$ divided by $2\pi$.*

*In the case of a conservative system, the operator $\hat{H}$ is time-independent and the operators $\{\hat{U}(t)\}_{t\in R}$ form a strongly continuous one-parameter group of unitary operators generated by $-(1/\hbar)\,\hat{H}$:*

$$\hat{U}(t) = \exp\left(-\frac{i}{\hbar}\hat{H}t\right)\,; \qquad \hat{U}^{\dagger}(t) = \hat{U}^{-1}(t)\,. \tag{2.8}$$

We see that if the system is at $t = 0$ in the pure state $|\psi\rangle$, then, at the arbitrary moment $t$, it is described by the pure state $\hat{U}(t)|\psi\rangle$; if $|\psi\rangle$ is at least at the moment $t = 0$ contained in $D_{\hat{H}}$, then its time evolution is described by the *Schrödinger equation*

$$i\hbar\frac{d|\psi\rangle}{dt} = \hat{H}|\psi\rangle\,. \tag{2.9}$$

The next axiom is not generally valid for all systems. The systems for which it is assumed to be valid are sometimes called *coherent systems.*

**Axiom 4** *Any statistical operator describes some state of the (coherent) system and any self-adjoint operator represents its observable.*

This axiom leads to the *superposition principle*, stating that the sum of two vectors $\psi_1$ and $\psi_2$ corresponding to two states of the system corresponds also to a state of the system. There are systems for which this axiom does not hold. Their state space is a direct sum of *coherent subspaces* (*superselection sectors*). Any vector belonging to a coherent subspace corresponds to some state of the system. However, sums of vectors from different coherent subspaces do not describe any states of the system. All operators corresponding to the observables should then commute with projectors on the coherent subspaces so that these are invariant under the action of the corresponding operators (this condition is only necessary; whether a given operator with the above property describes some observable still depends on the physics of the system). In this case we are speaking of *superselection rules.* If some operator from the class described above commutes with all operators from this class, then it is a linear combination of the projectors on the coherent subspaces. Such an operator is called *superselection operator.*

The statistical operator in the case of superselection rules should also commute with projectors on the coherent subspaces. This easily follows from the fact that the projectors on the pure states appearing in the decomposition of the statistical operator should also commute with the projectors on the coherent subspaces, because the corresponding pure states generating them represent the possible states of the system.

Let us formulate the last axiom

**Axiom 5** *The state of a quantum mechanical system is symmetric under the permutation of identical bosons and antisymmetric under the permutations of identical fermions.*

We will postpone the explanation of this axiom to Section 2.7.

## 2.2 More on quantum mechanics: states, observables and yes-no experiments

Now we shall formulate some consequences of the axioms given in Section 2.1. In this chapter we assume only coherent systems. All considerations can be easily extended to the more general case of incoherent systems.

It follows from the axioms that all informations about a quantum system are encoded in the state of the system as described by the statistical operator $\hat{\varrho}$. Then we can predict the expectation value $\langle a_{\hat{\varrho}} \rangle$ of the measurement of any observable $a$ (represented by an operator $\hat{A}$) in the state described by $\hat{\varrho}$. As is immediately seen from **Axiom 2**, using the spectral decomposition of the operator $\hat{A}$ this *expectation value* is given by

$$\langle a_{\hat{\varrho}} \rangle = \mathrm{Tr}(\hat{A}\hat{\varrho}), \tag{2.10}$$

if the trace exists.

Let the self-adjoint operator $\hat{A}$ correspond to the observable $a$ and $f$ be such real function of a real variable that the self-adjoint operator $f(\hat{A})$ is well defined. Let us assume that the observable $f(a)$ has the following physical interpretation: it is measured with the same measuring device as the observable $a$, but the measuring apparatus has now a new scale. The value $y$ of the new scale corresponds to the set of values $f^{(-1)}(y)$ (inverse image of $y$ under $f$) of the old one. According to the rules of the probability theory, it holds that

$$\langle f(a)_{\hat{\varrho}} \rangle = \int f(\lambda)\, dP_{\hat{\varrho}}(\lambda)\,,$$

where $P_{\hat{\varrho}}(\lambda)$ is the probability distribution corresponding to the measurement of observable $a$. Using the definition of the operator $f(\hat{A})$, it is easy to show that $\langle f(a)_{\hat{\varrho}} \rangle = \mathrm{Tr}[f(\hat{A})\hat{\varrho}]$ (provided the trace exists). As a consequence of this fact it can be shown that the operator corresponding to the observable $f(a)$ and the operator $f(\hat{A})$ have the same spectral decompositions, i.e. they are identical.

Similar considerations can be made for the set of commuting self-adjoint operators. The result is that the functional relations between the observables are the same as functional relations between the corresponding self-adjoint operators. The corresponding relation between the spectra (possible values of the observables) is straightforward (see Appendix to this chapter).

Probability theory tells us that a random quantity takes its particular value $\lambda$ with the probability one if and only if $\lambda$ is its expectation value and its mean square deviation is zero. It follows from Sec. 2.1 that mean square deviation $\langle(\Delta a)^2_{\hat{\varrho}}\rangle$ for the measurement of the observable $a$ in the state $\hat{\varrho}$ is given by

$$\langle(\Delta a)^2_{\hat{\varrho}}\rangle = \mathrm{Tr}\left[(\hat{A} - \langle a_{\hat{\varrho}}\rangle)^2 \hat{\varrho}\right] . \tag{2.11}$$

For a pure state $|\psi\rangle$ this reduces to

$$\|\hat{A}|\psi\rangle - \langle a_\psi\rangle|\psi\rangle\|^2 .$$

As an immediate consequence of this we have that the observable $a$ in the pure state $|\psi\rangle$ takes (in repeated measurements) the same value $\lambda$ if and only if $|\psi\rangle$ is the eigenstate of the operator $\hat{A}$ with eigenvalue $\lambda$. Using the decomposition of the statistical operator $\hat{\varrho}$ we also immediately conclude that in this case the "if and only if" condition is that the statistical operator is an incoherent superposition of (several) eigenstates of $\hat{A}$ corresponding to the single eigenvalue $\lambda$.

Let $\hat{E}_\lambda$ be the spectral family of projectors corresponding to a self-adjoint operator $\hat{A}$ and let $\hat{E}(\Delta) = \int_\Delta d\hat{E}_\lambda$ be the projector introduced with the help of the characteristic functions of Borel subsets $\Delta$ of the real line (see Appendix to this section). According to **Axiom 2** the probability, that in an experiment prepared for the measurement of the observable $a$ in the state $\hat{\varrho}$ the measured value will belong to the subset $\Delta$, is

$$P_{\hat{\varrho}}(\Delta) = \int_\Delta dP_{\hat{\varrho}}(\lambda) = \mathrm{Tr}\left[\hat{E}(\Delta)\hat{\varrho}\right] .$$

At least in principle, every measurement of the observable $a$ can be reduced to some sequence of experiments in which we are asking whether the value of the observable $a$ belongs to some subset $\Delta$ (interval) or not. The results of such partial measurement take only values "yes" or "no" (*yes-no experiment*). According to the last equality, this yes-no experiment can be connected with the measurement of the observable corresponding to the projector $\hat{E}(\Delta)$. The observable $a$ takes its values only from the $\Delta$ with probability 1 if and only if the state of the system given by $\hat{\varrho}$ is described as an incoherent superposition of the pure states belonging to the subspace $\hat{E}(\Delta)\mathcal{H}$. If $\Delta = \{\lambda\}$ and $\lambda$ is an eigenvalue of $\hat{A}$, then $\hat{E}(\{\lambda\})\mathcal{H}$ is the characteristic subspace $M_\lambda$ corresponding to this eigenvalue. If $\lambda$ belongs to the continuous spectrum, there is no state of the system in which the observable $a$ takes the value $\lambda$ with probability one. Nevertheless, there are states in which the observable takes values from any neighbourhood $\Delta$ of $\lambda$. Such states can be constructed from the complete system of generalized eigenvectors of $\hat{A}$ (see Appendix to this chapter) as

$$|\psi\rangle = \int_{\Delta} \tilde{\psi}(m)\Phi^*(m)\, d\mu\,.$$

Then it is clear that such $|\psi\rangle \in \hat{E}(\Delta)\mathcal{H}$ and values which the observable $a$ can take on in this state are close to $\lambda$. In this sense the generalized eigenstate of $\hat{A}$ with eigenvalue $\lambda$ is often referred to as the state of the system in which the observable $a$ takes on the value $\lambda$ with probability one.

If the operator $\hat{A}$ has pure point spectrum (let us remember that in this case the spectrum $\sigma(\hat{A})$ of $\hat{A}$ is the closure of the set of all eigenvalues), then, according to the properties of the continuous spectrum, $E(\lambda) = 0$ for the point $\lambda$ of $\sigma(\hat{A})$ which is not an eigenvalue. So the observable with the pure point spectrum can take on only its eigenvalues as possible values. This is in agreement with the obvious fact that no measurement can distinguish among the points of the set and the points of its closure.

Let $\hat{E}$ be any projector in $\mathcal{H}$. As already discussed, the measurement of the corresponding observable is a yes-no experiment (the projector has only two eigenvalues 0 and 1). In any state given by $\hat{\varrho}$ the probability that this observable takes on value 1 is $\mathrm{Tr}(\hat{E}\hat{\varrho})$ and the probability of taking 0 is $\mathrm{Tr}[(\hat{1} - \hat{E})\hat{\varrho}]$ ($\hat{1}$ being the identity operator). In particular, if the projector $\hat{E} = \hat{E}_\psi$ corresponds to the pure state $|\psi\rangle$, we have the probability $\langle\psi|\hat{\varrho}|\psi\rangle$ that the corresponding observable takes on the value 1 in the arbitrary state given by $\hat{\varrho}$ and the probability $1 - \langle\psi|\hat{\varrho}|\psi\rangle$ that the observable takes on the value 0. In this sense we can interpret the state described by the statistical operator $\hat{\varrho} = \sum_i c_i|\psi_i\rangle\langle\psi_i|$ as a statistical mixture of the pure states $|\psi_i\rangle$ with the corresponding probabilities $c_i$. In fact we know that we can verify the statistical predictions of the quantum mechanics experimentally if we make the corresponding measurement on a large number of identical copies of the system. It is often technically impossible to prepare all these copies in the same pure state. So we are forced to use the description of the state by the statistical operator. Another situation, when such a description is necessary, takes place if we want to describe measurement on some sub-system of a larger system.

Let us assume for simplicity a system composed of two (non-identical) sub-systems. If the sub-systems themselves have the Hilbert spaces $\mathcal{H}_1$ and $\mathcal{H}_2$, then the states of the composite system belong to the tensor product $\mathcal{H} = \mathcal{H}_1 \otimes \mathcal{H}_2$. In general the pure states $|\psi\rangle$ of the composite system need not be of the form $|\psi_1\rangle \otimes |\psi_2\rangle$. If this happens, i.e. if the composite system is described by the statistical operator $\hat{\varrho}_\psi$ corresponding to the pure state $|\psi\rangle$ which is not expressible as a tensor product of pure states of the sub-systems, then both *reduced states* (*reduced statistical operators*) $\mathrm{Tr}_1\hat{\varrho}$ and $\mathrm{Tr}_2\hat{\varrho}$ of the sub-systems are mixed. We have used the symbols $\mathrm{Tr}_i$, $i = 1, 2$, for the traces taken with respect to the i-th component of the tensor product $\mathcal{H}_1 \otimes \mathcal{H}_2$.

We have already noted that every measurement on the system can be reduced to a sequence of yes-no experiments and that such partial experiments are related to the measurement of an observable described by some projector. The set of all projectors in $\mathcal{H}$ is a particular example of a more general notion of the so-called *proposition system* [Jauch (1968)]. Nevertheless it turns out that all proposition systems connected with a quantum mechanical system are (essentially) of this form; in general we have to investigate the superselection rules, and also, the possibilities of Hilbert spaces being not only complex but also real or quaternionic. Some known examples of such non-standard realizations of quantum mechanics suggest their equivalence with the standard case over complex numbers.

The state of the system can be experimentally determined with the help of the probabilities $P(\hat{E})$ (for every projector $\hat{E}$) that the corresponding observable described by $\hat{E}$ takes the value 1 in this state. The probability distribution $P(\hat{E})$ over closed subspaces of $\mathcal{H}$ has the following properties:
i) $0 \leq P(\hat{E}) \leq 1$.
ii) $P(0) = 0, \qquad P(\hat{1}) = 1$.
iii) For each sequence of mutually orthogonal projectors $\hat{E}_i$ $(\hat{E}_i\hat{E}_j = 0)$, $P(\sum_i \hat{E}_i) = \sum_i P(\hat{E}_i)$ holds.
iv) If $\{\hat{E}_i\}$ is an arbitrary sequence of projectors such that $P(\hat{E}_i) = 1$ for all $i$, then $P(\cap\hat{E}_i) = 1$, where $\hat{E}\cap\hat{E}'$ is the projector on the subspace $\hat{E}\mathcal{H}\cap\hat{E}'\mathcal{H}$.
v) For every pair of different projectors $\hat{E} \neq \hat{E}'$ there is a state such that $P(\hat{E}) \neq P(\hat{E}')$.
Conversely (for the Hilbert spaces of $\dim\mathcal{H} \geq 3$), *Gleason's theorem* holds: If the function $P(\hat{E})$ fulfils the above conditions i)–iii), then there exists a unique statistical operator $\hat{\varrho}$ such that $P(\hat{E}) = \mathrm{Tr}(\hat{E}\hat{\varrho})$.

Let us now consider the dispersion $\sigma_{\hat{\varrho}}(\hat{E})$ of the observable represented by the projector $\hat{E}$ $(\hat{E}^2 = \hat{E})$ in the state given by $\hat{\varrho}$:

$$\sigma_{\hat{\varrho}}(\hat{E}) = \mathrm{Tr}(\hat{E}\hat{\varrho}) - \left[\mathrm{Tr}(\hat{E}\hat{\varrho})\right]^2 . \tag{2.12}$$

The *total dispersion* $\sigma_{\hat{\varrho}}$ of the state given by $\hat{\varrho}$ is then defined as the upper limit of the set $\{\sigma_{\hat{\varrho}}(\hat{E})\}$, where $\hat{E}$ runs over the set of all projectors in $\mathcal{H}$. In general this is a number from the interval $[0, 1/4]$. The states for which $\sigma_{\hat{\varrho}} = 0$ are called *dispersion-free*. Every dispersion-free state is pure. The converse is not true in general.

In classical mechanics, the role of the projectors is taken over by the Borel sub-sets of the phase space. The probability that the pure state — point of the phase space — belongs to some Borel set can take on only two values 0 or 1 depending on the fact whether the point belongs to it or not. Thus all pure states of the classical system are dispersion-free. So the situations in classical and quantum mechanics are different. The theories with *hidden variables* represent an attempt to describe quantum probabilities by classical probability distributions. They suppose that the pure states of a quantum mechanical system are dispersion-free but that for some reasons, we are able to prepare only their mixtures. Namely, they assume the situation that there exists a space $X$ with some finite measure $\varrho$ (normalized so that $\varrho(X) = 1$) such that for every state described by the probabilities $P(\hat{E})$ which are expressed as a mixture

$$P(\hat{E}) = \int_X P_\xi(\hat{E})\, d\varrho(\xi)$$

of dispersion-free states. The states represented by probabilities $P_\xi(\hat{E})$ are not necessarily prepared by known physical equipment. For this reason the variables $\xi$ are called hidden. However, it is well known that if this really happens, the set of all projectors in $\mathcal{H}$ can contain only two projectors $\hat{0}$ and $\hat{1}$, and consequently the quantum mechanical system which admits hidden variables should be trivial. A similar theorem holds also in the more general situation of a proposition system involving superselection rules [Jauch (1968)].

## 2.3 Time evolution

From the **Axiom 3** of Section 2.1 we know how the time evolution of a state given by $\hat{\varrho}$ of a quantum mechanical system is described. We know also that the only possible way how to determine the state is a measurement of the mean values of the observables. The time evolution of the mean value of an observable $a$ is easily obtained from the axioms. We will assume only conservative systems in this section, if not stated otherwise.

The starting point of all considerations of this section will be the expression for the time dependent mean value $\langle a_{\hat{\varrho}}\rangle(t) = \mathrm{Tr}[\hat{A}\hat{\varrho}(t)]$ of the observable $a$. The property of the trace $\mathrm{Tr}(\hat{A}\hat{B}) = \mathrm{Tr}(\hat{B}\hat{A})$ ($\hat{A}$ is trace class operator, $\hat{B}$ is bounded) gives

$$\langle a_{\hat{\varrho}}\rangle(t) = \mathrm{Tr}\left[\hat{A}\hat{U}(t)\hat{\varrho}(0)\hat{U}(-t)\right] = \mathrm{Tr}\left[\hat{U}(-t)\hat{A}\hat{U}(t)\hat{\varrho}(0)\right] . \tag{2.13}$$

The two expressions (2.13) lead to the two possible interpretations of the time evolution of the quantum system called the *Schrödinger* and *Heisenberg pictures*, respectively. In the Schrödinger picture the operators corresponding to the observables are assumed to be time independent and the time evolution of the system is described by the time dependence of the states. In the Heisenberg picture it is assumed on the contrary that the states of the system are time independent and that the time evolution of the system is given by the time evolution of the operators under the assumption that if the observable

$a$ is at the moment $t = 0$ described by the operator $\hat{A}$, then at arbitrary time $t$ it is described by the operator

$$\hat{A}(t) = \hat{U}(-t)\hat{A}\hat{U}(t). \tag{2.14}$$

In the case of bounded $\hat{A}$ also the time derivative of (2.14) is well defined:

$$i\hbar\frac{d\hat{A}}{dt} = [\hat{A}, \hat{H}]. \tag{2.15}$$

The last equation is formally used also for the unbounded operators.

Both pictures are of course equivalent, because according to (2.13) they lead to the same experimental results. The formulation of the **Axiom 3** corresponds to Schrödinger picture. We should also note that so-called *Dirac's* or *interaction picture*, which is an intermediate between the above mentioned pictures, can be defined. It is often used in many practical calculations. It assumes that the hamiltonian $\hat{H}$ consists of the "free" and "interaction" parts $\hat{H} = \hat{H}_0 + \hat{V}$. If this happens and we denote the evolution operators corresponding to the free hamiltonian $\hat{H}_0$ and to the full hamiltonian $\hat{H}$ of the system by $\hat{U}_0(t) = \exp[-(i/\hbar)\hat{H}_0 t]$ and $\hat{U}(t) = \exp[-(i/\hbar)\hat{H}t]$, respectively, then the time evolution of the operators is given by the free hamiltonian $\hat{H}_0$ according to the rule

$$\hat{A}(t) = \hat{U}_0(-t)\hat{A}\hat{U}_0(t) \tag{2.16}$$

and the time evolution of the states is assumed to be governed by the rule

$$\hat{\varrho}(t) = \hat{U}_0(-t)\hat{U}(t)\hat{\varrho}\hat{U}(-t)\hat{U}_0(t). \tag{2.17}$$

If the free and interaction parts commute, then, as is easily seen, the time evolution of the states is governed by the evolution operator $\exp[-(i/\hbar)\hat{V}t]$. The time evolution is given in such a way that the mean value of an observable $a$ at the time $t$ is $\mathrm{Tr}[\hat{A}(t)\hat{\varrho}(t)]$. The corresponding differential equations are

$$i\hbar\frac{d\hat{A}(t)}{dt} = [\hat{A}(t), \hat{H}_0] \tag{2.18}$$

and

$$i\hbar\frac{d\hat{\varrho}(t)}{dt} = [\hat{V}(t), \hat{\varrho}(t)], \tag{2.19}$$

where $\hat{V}(t) = \hat{U}_0(-t)\hat{V}\hat{U}_0(t)$ is the interaction part of the hamiltonian in the Dirac picture. We will not discuss the precise assumptions for getting these differential equations from their integral forms.

Let the operator function $\hat{\varrho}(t)$ describe the evolution (in the Schrödinger picture) of a conservative quantum system described by the hamiltonian $\hat{H}$. Let $\hat{A}$ correspond to an observable and let the eigenstates $|e_i(t)\rangle$ of $\hat{\varrho}(t) =$

$\sum_i c_i|e_i(t)\rangle\langle e_i(t)|$ be such that the sum $\sum_i c_i\langle e_i(t)|(i/\hbar)(\hat{H}\hat{A}-\hat{A}\hat{H})|e_i(t)\rangle$ converges uniformly on some open set $G \in R$. Let us suppose also that for $t \in G$ the operators $\hat{A}\hat{\varrho}(t)$, $\hat{H}\hat{A}\hat{\varrho}(t)$, $\hat{A}\hat{H}\hat{\varrho}(t)$ be trace class and that there is a real number $c$ such that for each eigenvector $|e_i(t)\rangle$, $\|\hat{A}|e_i(t)\rangle\| \leq c$ holds. Under these assumptions the function $\langle a_{\hat{\varrho}}\rangle(t) = \mathrm{Tr}[\hat{A}\hat{\varrho}(t)]$ has the derivative at all points $t \in G$ and we have

$$\frac{d\langle a_{\hat{\varrho}}\rangle(t)}{dt} = \frac{i}{\hbar}\mathrm{Tr}\left[(\hat{H}\hat{A} - \hat{A}\hat{H})\hat{\varrho}(t)\right]. \tag{2.20}$$

Let us note that if the operator $\hat{A}$ is bounded, then it is sufficient to assume that the operators $\hat{H}\hat{A}\hat{\varrho}(t)$ and $\hat{H}\hat{\varrho}(t)$ are trace class for $t \in G$. Then the above derivative is continuous.

If the operator $\hat{B} = (i/\hbar)(\hat{H}\hat{A} - \hat{A}\hat{H})$ is essentially self-adjoint (its closure is self-adjoint), then we will denote its closure $\hat{B}$ by $(i/\hbar)[\hat{H}, \hat{A}]$, too. In this case we can write $d\langle a_{\hat{\varrho}}\rangle(t)/dt = \langle b_{\hat{\varrho}}\rangle(t)$. Defining the operator $\hat{A}^\bullet$ corresponding to the observable $da/dt$ as

$$\hat{A}^\bullet = \frac{i}{\hbar}[\hat{H}, \hat{A}],$$

we have

$$\frac{d}{dt}\mathrm{Tr}\left[\hat{A}\hat{\varrho}(t)\right] = \mathrm{Tr}\left[\hat{A}^\bullet\hat{\varrho}(t)\right]. \tag{2.21}$$

Formally we can get from the axioms the same equations in the case of a non-conservative system when the hamiltonian $\hat{H}$ is an explicit operator function $\hat{H}(t)$ of the time (e.g. if the system is interacting with a time-dependent external field). The precise formulation of analogous statements heavily depends on the properties of the hamiltonian $\hat{H}(t)$ and may be highly non-trivial in this case. For the practical purposes the differential equations of the time evolution of the statistical operator or of the operators are taken as the starting point in the case of a non-conservative system.

It is in accordance with our experience and with all we have already said to introduce *stationary states* as the states for which the values $\langle a_{\hat{\varrho}}\rangle$ do not depend on time for any observable $a$. Such states can be characterized by the property equivalent to the statement that the corresponding statistical operator $\hat{\varrho}$ commutes with the hamiltonian $\hat{H}$, i.e. $[\hat{H}, \hat{\varrho}] = \hat{0}$ in $D_{\hat{H}}$. If the system is described by a pure state $|\psi\rangle$, then the "if and only if" condition means that the state is an eigenstate of the hamiltonian $\hat{H}$

$$\hat{H}|\psi\rangle = \lambda|\psi\rangle. \tag{2.22}$$

The last equation is often called the *time independent Schrödinger equation.* A simple consequence is that if the hamiltonian has pure continuous spectrum, the corresponding quantum system has no stationary state.

Similarly we say that the observable $a$ is an *integral of motion* if the result of all yes-no experiments connected with this observable (the quantities $\mathrm{Tr}[\hat{E}(\Delta)\hat{\varrho}]$ for any Borel subset $\Delta \subset R$) does not depend on time for each state of the system. The observable $a$ is an integral of motion if and only if the corresponding operator $\hat{A}$ commutes with the hamiltonian $\hat{H}$. So e.g. the energy of the conservative quantum system is an integral of motion. All above statements concerning the stationary states as well as the integrals of motion can be very easily obtained formally using the formulas for the time derivatives of $\hat{A}$ and $\hat{\varrho}$. The precise proofs should of course include a little of the functional analysis explained in the Appendix to this chapter.

The eigenstates and generalized eigenstates of the hamiltonian play a very important role in quantum mechanics. Using the spectral decomposition

$$\hat{H} = \int \lambda \, dE_\lambda$$

of the hamiltonian $\hat{H}$ we have immediately the following expression for the spectral decomposition of $\hat{U}(t)$

$$\hat{U}(t) = \int \exp\left(-\frac{i}{\hbar}\lambda t\right) d\hat{E}_\lambda \,. \tag{2.23}$$

It means that the time evolution of the system is contained completely in the spectral decomposition of $\hat{H}$. As follows from the facts collected in the Appendix to this chapter this decomposition can explicitly be obtained by utilizing the generalized eigenstates.

## 2.4 Uncertainty relations

Let us return to the mean square deviation $(\delta a)^2_{\hat{\varrho}} \equiv \langle(\Delta a)^2_{\hat{\varrho}}\rangle$ for measuring the observable $a$ in the state given by $\hat{\varrho}$. In the following we will omit the subscript $\hat{\varrho}$ for convenience. According to the definition we have

$$(\delta a)^2 = \mathrm{Tr}\left[(\hat{A} - \langle a\rangle)^2 \hat{\varrho}\right] \,. \tag{2.24}$$

Let us now consider two self-adjoint operators $\hat{A}$, $\hat{B}$ (not necessarily commuting) corresponding to the observables $a$ and $b$, respectively. In order to avoid inessential complications with the definition domains we shall assume the operators $\hat{A}$ and $\hat{B}$ to be bounded. The operator

$$\left(\frac{\hat{A} - \langle a\rangle}{\delta a} - i\frac{\hat{B} - \langle b\rangle}{\delta b}\right)\left(\frac{\hat{A} - \langle a\rangle}{\delta a} + i\frac{\hat{B} - \langle b\rangle}{\delta b}\right)$$

is then automatically positive. Similar consideration is valid if we exchange the operators $\hat{A}$ and $\hat{B}$ in this expression. So the expectation values in the state given by $\hat{\varrho}$ of the operators obtained in this way are greater or equal to zero. From this observation we immediately get the so-called *uncertainty relation*

$$\delta a \delta b \geq \frac{1}{2} \left| \mathrm{Tr} \left[ i(\hat{A}\hat{B} - \hat{B}\hat{A})\hat{\varrho} \right] \right| . \tag{2.25}$$

In the general case of unbounded operators we should suppose for the validity of the above inequality that the operators $\hat{A}\hat{\varrho}$, $\hat{B}\hat{\varrho}$, $\hat{A}\hat{B}\hat{\varrho}$ and $\hat{B}\hat{A}\hat{\varrho}$ are trace class operators. If $\hat{\varrho}$ is the statistical operator of a pure state corresponding to the vector $|\psi\rangle \in \mathcal{H}$, then it is sufficient to assume that $|\psi\rangle \in D_{\hat{A}\hat{B}-\hat{B}\hat{A}}$. If we know in this general case that the operator $\hat{C} = i(\hat{A}\hat{B} - \hat{B}\hat{A})$ is a restriction of some self-adjoint operator (in the case of $\hat{A}$ and $\hat{B}$ bounded, $\hat{C}$ is of course self-adjoint) which is again denoted as $\hat{C}$, then we can write the uncertainty relation

$$\delta a \delta b \geq \frac{1}{2} |\langle c \rangle| .$$

The physical consequence of this uncertainty principle is the following. An experiment measuring any observable $a$ can in principle be organized so that the system is prepared in a state for which the value of $\delta a$ is arbitrarily small. This can be attained e.g. in the situation of a yes-no experiment corresponding to a very small interval of values of the observable $a$. If the result of the measurement is "yes", then, immediately after the measurement, we have the system in the considered state. If the observables $a$ and $b$ are measurable simultaneously (compatible), then one can prepare in one experiment a state with both values $\delta a$ and $\delta b$ arbitrarily small.

From the uncertainty relation it follows that in general it is not possible for a pair of non-compatible observables to measure them simultaneously with an arbitrary accuracy. Possible values of $\delta a$ and $\delta b$ should satisfy the above inequality. For instance it may happen, as it will be seen later in the case of position and momentum operators, that the right-hand side of the uncertainty inequality is strictly positive for any state. For such pairs of observables there is no state of the system in which the values of $\delta a$ and $\delta b$ are simultaneously arbitrarily small. So there is principally no experiment in which the values of the corresponding observables can be measured with arbitrary precision. The uncertainty relation is seen to support the choice of **Axiom 2** in such a way .

We should mention that the behavior of the system under measurement was already incorporated in the proposition system of quantum mechanics (see Sec. 2.2) which is empirically given. From this point of view it is no surprise that the uncertainty principle is a purely mathematical consequence of the properties of the observables. The physical considerations leading to the uncertainty principle and its practical consequences are discussed in detail in all textbooks [Messiah (1961)].

Sometimes it is useful to use a more precise inequality instead of the standard uncertainty relations, which reads ($\hat{A}$, $\hat{B}$ self-adjoint)

$$(\delta a)^2(\delta b)^2 \geq \frac{1}{4}\left\{\left(\mathrm{Tr}\left[i(\hat{B}\hat{A}-\hat{A}\hat{B})\hat{\varrho}\right]\right)^2 + \left(\mathrm{Tr}\left[(\Delta\hat{A}\Delta\hat{B}+\Delta\hat{B}\Delta\hat{A})\hat{\varrho}\right]\right)^2\right\}. \tag{2.26}$$

It is again supposed that all terms in the inequality are well defined. For $\hat{A}$ and $\hat{B}$ bounded this inequality is a consequence of the Schwarz inequality

$$\left[\mathrm{Tr}(\hat{A}_1\hat{A}_2)\right]^2 \leq \mathrm{Tr}(\hat{A}_1^\dagger\hat{A}_1)\mathrm{Tr}(\hat{A}_2^\dagger\hat{A}_2),$$

($\dagger$ denotes the Hermitian conjugation) which holds for Hilbert-Schmidt operators $\hat{A}_1$ and $\hat{A}_2$. The uncertainty principle as given by (2.25) is then a consequence of (2.26) if we neglect the term containing the anticommutator $(\Delta\hat{A}\Delta\hat{B}+\Delta\hat{B}\Delta\hat{A})$.

## 2.5 Symmetries

Let us remind again the fundamental role of the proposition system formed by all projectors $\hat{E}$ (or equivalently by corresponding subspaces $\hat{E}\mathcal{H}$) in the Hilbert space $\mathcal{H}$. We know that all measurements on the quantum system under consideration can be reduced to a sequence of "yes-no" experiments connected with the measurement of the observables corresponding to projectors $\hat{E}$. For simplicity let us assume a coherent system with $\dim(\mathcal{H}) \geq 3$. From the above point of view it is natural to understand the *symmetry* (*symmetry transformation*) of the system as a bijection (one-to-one map) $\alpha$ of the set of all subspaces $\hat{E}\mathcal{H}$ onto itself, which preserves the structure of the proposition system. In general it is sufficient to assume that
i) $\hat{E}_1\mathcal{H} \subseteq \hat{E}_2\mathcal{H} \Rightarrow \alpha(\hat{E}_1\mathcal{H}) \subseteq \alpha(\hat{E}_2\mathcal{H})$,
ii) $\alpha((\hat{E}\mathcal{H})^\perp) = (\alpha(\hat{E}\mathcal{H}))^\perp$,
where $\perp$ has been used to denote the orthogonal complement in $\mathcal{H}$. Let us note that the above requirements have a straightforward physical interpretation.

The symmetry $\alpha$ induces then, by Gleason's theorem, a natural transformation $\alpha_*$ on the states given by $\hat{\varrho}$, defined as

$$(\alpha_*\hat{\varrho})(\hat{E}) = \hat{\varrho}\left[\alpha^{-1}(\hat{E})\right],$$

such that the predictions of "yes-no" experiments (and consequently of all measurements on the system) remain unchanged.

The fundamental *Wigner's theorem* says that to any symmetry $\alpha$ of the system under consideration there corresponds a unique transformation $\hat{U}$ on the corresponding Hilbert space $\mathcal{H}$ which is unitary or antiunitary ($\langle \hat{U}\psi|\hat{U}\varphi\rangle = \langle\psi|\varphi\rangle$) and vice versa.

Only some discrete symmetries are represented by antiunitary operators, e.g. the time reversal.

## 2.6 Canonical quantization

Up to now we did not mention any particular example of a quantum system. The basic example is the one-dimensional particle without inner degrees of freedom (spin, etc.) and without constraints. In this case we use the Hilbert space $\mathcal{H} = L^2(R, dx)$ of square integrable functions as the state Hilbert space of the considered system. Let us remind that the inner product $\langle\psi|\varphi\rangle$ of two vectors $|\psi\rangle$ and $|\varphi\rangle$ corresponding to two functions $\psi(x)$ and $\varphi(x)$ is defined as

$$\langle\psi|\varphi\rangle = \int_R \psi^*(x)\varphi(x)\, dx\,,$$

$*$ denotes complex conjugation. The operator $\hat{x}$ representing the position observable (we assume the Schrödinger picture) is taken in this case as

$$(\hat{x}\psi)(x) = x\psi(x) \tag{2.27}$$

with the obvious domain. The operator $\hat{x}$ is self-adjoint.

Let us assume further the displacement of the particle by $\tau$. It is obvious that this represents a symmetry of the system in the sense of Sec. 2.5. The corresponding unitary transformation $\hat{U}(-\tau)$ is given by

$$[\hat{U}(-\tau)\psi](x) = \psi(x-\tau)\,. \tag{2.28}$$

Moreover, as easily seen the operators $\{\hat{U}(\tau)\}_{\tau\in R}$ form a strongly continuous one-parameter group of unitary transformations. According to the Stone's theorem, there exists unique self-adjoint operator $\hat{p}$ such that

$$\hat{U}(\tau) = \exp\left(\frac{i}{\hbar}\hat{p}\tau\right)\,. \tag{2.29}$$

The operator $\hat{p}$ is interpreted as the operator corresponding to the momentum observable of the particle. This is in accord with the classical mechanics where in the Hamilton formalism the generator of the space translations of the system is the momentum $p$. For sufficiently regular function $\psi$ we can write its Taylor expansion in the form

$$\psi(x+\tau) = \exp\left(\tau\frac{d}{dx}\right)\psi(x)$$

which utilizes the momentum operator

$$(\hat{p}\psi)(x) = \frac{\hbar}{i}\frac{d}{dx}\psi(x), \tag{2.30}$$

again with the obvious domain. One can show that this operator is self-adjoint, too. The functions

$$\varphi_n(x) = \frac{1}{\left(\sqrt{\hbar\pi}\,2^n n!\right)^{1/2}} \exp\left(-\frac{x^2}{2\hbar}\right) H_n\left(\frac{x}{\sqrt{\hbar}}\right), \qquad n = 0, 1, \ldots, \tag{2.31}$$

where $H_n(x)$ are Hermite polynomials, form an orthonormal basis in $L^2(R, dx)$. The finite linear combinations of the $\varphi_n(x)$ form a common dense domain for the position and momentum operators $\hat{x}$ and $\hat{p}$ which is invariant under their actions. In this domain the so-called *canonical commutation relation* holds

$$[\hat{x}, \hat{p}] = i\hbar\hat{1}, \tag{2.32}$$

where $\hat{1}$ is again the identity operator. The commutator replaces the Poisson brackets from the classical mechanics in such a way after suitable normalization. From this point of view it is natural to refer to the procedure of replacing the basic classical observables by the operators and their Poisson brackets by their commutator as to the *canonical quantization*. The operators $\hat{x}$ and $\hat{p}$ are examples of operators with the pure continuous spectra. The complete set of generalized eigenvectors of the operator $\hat{x}$ is formed by the set of Dirac's delta functions $\{\delta(x - x_0);\ x_0 \in R\}$. In the sense explained in the Appendix to this chapter we may introduce the dual bra vectors $\langle x_0|$ and write $\psi(x_0) = \langle x_0|\psi\rangle$. In the case of the momentum operator $\hat{p}$ the complete set of generalized eigenvectors is formed by the set of functions $\{(1/\sqrt{2\pi})\exp(ikx/\hbar);\ k \in R\}$ and the corresponding unitary isometry (see Appendix to this chapter) between $L^2(R, dx)$ and $L^2(R, dk)$ is just the ordinary Fourier transform. The generalized eigenvectors of the operator $\hat{p}$ are thus ordinary smooth functions. Nevertheless they do not belong to the Hilbert space $L^2(R, dx)$. The situation is similar for the case of generic linear differential operator in $L^2((a, b), dx)$ where $a, b \in R \cup \{\pm\infty\}$.

The fundamental relation (2.32) is often replaced by the so-called *Weyl relation*, which is the relation fulfilled by the bounded operators $\hat{U}(\tau)$ given by (2.29) and $\hat{V}(s) = \exp(i\hat{x}s)$

$$\hat{U}(\tau)\hat{V}(s) = \exp(i\tau s)\hat{V}(s)\hat{U}(\tau). \tag{2.33}$$

This is made to avoid the domain problems. The representation of the Weyl relation by the operators $\hat{U}(\tau) = \exp(\tau d/dx)$ and $\hat{V}(s) = \exp(i\hat{x}s)$ is the so-called *Schrödinger* representation. This representation is *regular* in the sense that the operators $\hat{U}(\tau)$ and $\hat{V}(s)$ are strongly continuous operator functions of the parameters $\tau$ and $s$ respectively and that for the value 0 of these parameters they coincide with the identity operator $\hat{1}$.

The well-known *Stone-von Neumann theorem* claims:
any regular representation of the Weyl relations (2.33) in separable Hilbert space $\mathcal{H}$ is unitarily equivalent to at most countable direct sum of Schrödinger representations.

To formulate an analogous statement for the canonical commutation relation (2.32) we need more complicated assumptions. We will return to this point after discussing the second fundamental system: the harmonic oscillator. We set $\hbar = 1$ for simplicity and assume the unit mass $m$ and frequency $\omega$ of the oscillator. The Hilbert space is again $L^2(R, dx)$ and the position and momentum operators are introduced in the same way as was done above. The operator $\hat{H}$ corresponding to the energy of the system is now taken as the closure of the formal differential operator

$$\hat{H} = \frac{1}{2}(\hat{p}^2 + \hat{x}^2). \tag{2.34}$$

It is convenient to introduce now the so-called *creation* and *annihilation* operators $\hat{a}^\dagger$ and $\hat{a}$ respectively by formulas

$$\hat{a}^\dagger = \frac{1}{\sqrt{2}}(\hat{x} - i\hat{p}), \qquad \hat{a} = \frac{1}{\sqrt{2}}(\hat{x} + i\hat{p}). \tag{2.35}$$

They are adjoints of each other on the domain

$$D = \{|\psi\rangle ; \sum_n |\langle\psi|\varphi_n\rangle|^2 n < +\infty\} \tag{2.36}$$

and they satisfy

$$\begin{aligned} \hat{a}|\varphi_0\rangle &= 0, \\ \hat{a}|\varphi_n\rangle &= \sqrt{n}\,|\varphi_{n-1}\rangle, \qquad n = 1, 2, \ldots, \\ \hat{a}^\dagger|\varphi_n\rangle &= \sqrt{n+1}\,|\varphi_{n+1}\rangle, \qquad n = 0, 1, \ldots, \end{aligned} \tag{2.37}$$

where $|\varphi_n\rangle$ are the kets introduced in (2.31). It follows that on a dense domain they satisfy the commutation relation

$$[\hat{a}, \hat{a}^\dagger] = \hat{1}. \tag{2.38}$$

The hamiltonian (2.34) is then expressed in terms of $\hat{a}$ and $\hat{a}^\dagger$ as

$$\hat{H} = \hat{a}^\dagger\hat{a} + \frac{1}{2}\hat{1}. \tag{2.39}$$

Simple algebraic manipulations based on (2.38) lead to the following statements:

- The vectors $|\varphi_n\rangle$ given by (2.31) are eigenvectors of the operator $\hat{H}$ with eigenvalues $n + 1/2$ and the operator $\hat{H}$ is self-adjoint in the domain (2.36). Let us denote $|\varphi_n\rangle = |n\rangle$, hence $\varphi_n(x) = \langle x|n\rangle$.
- The Fock state (the occupation number state $|n\rangle$) equals

$$|n\rangle = \frac{1}{\sqrt{n!}}(a^\dagger)^n|0\rangle\,. \tag{2.40}$$

- Using the spectral decomposition of $\hat{H}$ we can describe the harmonic oscillator in terms of the Hilbert space $l^2$ of sequences $|c\rangle = \{c_1, c_2, \ldots\}$ with finite sum $\sum_i |c_i|^2$ (see Appendix to this chapter). The state of the harmonic oscillator is then represented by its coefficients expressing it as a linear combination of the elements of the base (2.31).

Let us now assume arbitrary Hilbert space $\mathcal{H}$ such that we have on it an action of the operators $\hat{a}$ and $\hat{a}^\dagger$ fulfilling the commutation relation (2.38). Let $\mathcal{H}$ also contain a normalized state $|0\rangle$ such that $\hat{a}|0\rangle = |0\rangle$ and let the set of vectors $(1/\sqrt{n!}\,)(a^\dagger)^n|0\rangle$ form a basis in $\mathcal{H}$. Such Hilbert space is called *Fock space.* All Fock spaces appear to be unitarily isomorphic. The unitary isomorphism sends a vacuum into a vacuum and transforms the actions of the creation (annihilation) operators correspondingly. In fact it is uniquely defined by the identification of the corresponding bases (2.40). In the above sense we are speaking also of the *Fock representation.* The Fock representation is of course unitarily equivalent to the Schrödinger representation.

Now we can formulate another version of the Stone-von Neumann theorem (uniqueness of quantization). Let us assume that there is such dense domain $D$ in a Hilbert space $\mathcal{H}$ contained in domains of operators $\hat{x}$ and $\hat{p}$ that these operators leave $D$ invariant and on $D$ the canonical commutation relation (2.32) is valid. Let us also suppose that the symmetric operator $\hat{x}^2 + \hat{p}^2$ defined on $D$ is essentially self-adjoint (its closure is self-adjoint). So we have a representation of the canonical commutation relation (2.32) and it follows that any such representation is unitarily equivalent to some direct sum of Schrödinger (Fock) representations.

The generalization of all said above to the $n$-dimensional case ($n$ being finite) is straightforward and can be easily reformulated to this case. The substantial fact is that in this more general case we have $\mathcal{H} = L^2(R^n, dx_1, dx_2, \ldots, dx_n)$ and $n$ pairs of position and momentum operators $\hat{x}_i$, $\hat{p}_i$, $i = 1, 2, \ldots, n$ fulfil the relations

$$[\hat{x}_i, \hat{p}_j] = i\hbar\delta_{ij}\hat{1}\,, \qquad [\hat{x}_i, \hat{x}_j] = \hat{0}\,, \qquad [\hat{p}_i, \hat{p}_j] = \hat{0}\,. \tag{2.41}$$

Similarly for the pairs of creation and annihilation operators it holds that

$$[\hat{a}_i, \hat{a}_j^\dagger] = \delta_{ij}\hat{1}\,, \qquad [\hat{a}_i, \hat{a}_j] = \hat{0}\,, \qquad [\hat{a}_i^\dagger, \hat{a}_j^\dagger] = \hat{0}\,. \tag{2.42}$$

A direct generalization of both the versions of the Stone-von Neumann theorem holds also in this case. The case $n = 3$ corresponds to the more realistic system of a particle in the space of 3 spatial dimensions.

## 2.7 Identical particles and the second quantization

Let us now assume a particle in three dimensions with possibly additional inner degrees of freedom as e.g. spin. So the Hilbert space $\mathcal{H}$ of the particle is a tensor product $L^2(R^3) \otimes \mathcal{L}$, where the Hilbert space $\mathcal{L}$ describes the inner degrees of freedom. Using a suitable complete set of commuting operators we can represent the Hilbert space $\mathcal{H}$ as $L^2(X, d\mu)$, where the set $X$ corresponds to the values of the considered complete set of operators.

The quantum mechanical description of a system consisting of several identical particles is based on the **Axiom 5** of Sec. 2.1. Let us start with a system containing two identical particles. The appropriate Hilbert space $\mathcal{H}$ is now a tensor product $L^2(X, d\mu) \otimes L^2(X, d\mu)$ with $L^2(X, d\mu)$ being the single particle Hilbert space. Two identical particles are assumed to be indistinguishable in the sense that their exchange does not lead to the change of the state of the complete system. It means particularly that pure states $\psi(\xi_1, \xi_2)$ and $\psi(\xi_2, \xi_1)$ may differ only by a factor $c$ with unit norm $|c| = 1$. Moreover exchanging the arguments of the function $\psi$ twice we see that $c^2 = 1$ and the function $\psi$ describing the system is symmetric if $c = 1$ and antisymmetric if $c = -1$.

In more general case of $n$ identical particles the situation is similar. The function of $n$ variables $\psi(\xi_1, \xi_2, \ldots, \xi_n)$ describing a system of $n$ identical particles should be either totally symmetric or totally antisymmetric in variables $\xi_1, \xi_2, \ldots, \xi_n$. This is in accordance with the real behaviour of all known particles. In the symmetric case we are speaking of the *Bose-Einstein statistics* and in the antisymmetric one of the *Fermi-Dirac statistics*. It is a deep result of relativistic quantum field theory that the particles with integer spin obey the Bose-Einstein statistics and those with half-integer spin obey the Fermi-Dirac statistics.

Let now $P$ denote the permutation of $n$ particles and $\hat{U}_P$ denote the corresponding operator defined as

$$\hat{U}_P \, \psi(\xi_1, \xi_2, \ldots, \xi_n) = \psi(\xi_{P(1)}, \xi_{P(2)}, \ldots, \xi_{P(n)}) \,. \tag{2.43}$$

As is easily seen this operator is unitary. Let us now define operators

$$\Pi_+ = \frac{1}{n!} \sum_P \hat{U}_P \tag{2.44}$$

and

$$\Pi_- = \frac{1}{n!}\sum_P \delta_P \hat{U}_P\,, \tag{2.45}$$

where the summations run over all permutations of $n$ particles and $\delta_P$ is $+1$ or $-1$ according to whether the permutation is even or odd. The operators $\Pi_+$ and $\Pi_-$ are orthogonal projectors from $\mathcal{H} = (L^2(X, d\mu))^{\otimes n}$ to the symmetrical or antisymmetrical subspaces $\mathcal{H}^+$ and $\mathcal{H}^-$ of $\mathcal{H}$. According to the above discussion these subspaces correspond to systems of identical bosons or identical fermions, respectively.

All said above does not, of course, depend on the particular realization $L^2(X, d\mu)$ of the single particle Hilbert space.

In the following we assume only systems consisting of bosons. The case of the system containing fermions can be treated similarly, however this case is not so important for problems of quantum optics. Let us now consider a system of $n$ non-interacting bosons. Such a system is sometimes called also *free Bose gas.* Let the single particle Hilbert space be $\mathcal{H}_1$. According to the above discussion the Hilbert space $\mathcal{H}_n$ of the entire system is $\mathcal{H}_n = \Pi_+(\mathcal{H}_1^{\otimes n})$. If the vectors $|r\rangle$ form a basis of $\mathcal{H}_1$, then the vectors $|r_1 r_2 \ldots r_n\rangle$ of the form

$$|r_1 r_2 \ldots r_n\rangle = \Pi_+(|r_1\rangle \otimes |r_2\rangle \otimes \ldots \otimes |r_n\rangle) \tag{2.46}$$

are still orthogonal and complete in $\mathcal{H}_n$ but no more normalized. To normalize them it is convenient to introduce the so-called *occupation number representation.* Let $n_i$ denotes the number of indices among the $r_1$, $r_2$, ..., $r_n$ which are equal to $i$, $i = 1, 2, \ldots$, so that $n = \sum_i n_i$. It follows that two vectors $|r_1 r_2 \ldots r_n\rangle$ are equal if and only if the corresponding occupation numbers $n_1$, $n_2$, ... are equal. The norm of $|r_1 r_2 \ldots r_n\rangle$ is then easily expressed it terms of the occupation numbers $n_i$ as

$$\big\| |r_1 r_2 \ldots r_n\rangle \big\|^2 = \frac{n_1! n_2! \ldots}{n!}\,.$$

Hence, we have the basis $|n_1 n_2 \ldots\rangle$ of $\mathcal{H}_n$ of the form

$$|n_1 n_2 \ldots\rangle = \sqrt{\frac{n!}{n_1! n_2! \ldots}}\,|r_1 r_2 \ldots r_n\rangle\,. \tag{2.47}$$

The vectors of the basis depend exclusively only on the numbers $n_i$.

In the most general situation of the quantum field theory it is necessary to drop the constrain fixing the total number of particles under consideration and assume a more general system with an undetermined number of particles. The Hilbert space $\mathcal{H}$ of the system is now taken as a direct sum $\mathcal{H} = \oplus_{n=0}^{\infty}\mathcal{H}_n$ of the Hilbert spaces $\mathcal{H}_n$ corresponding to the $n$ particle systems. According to the definition, the elements $|\psi\rangle$ of $\mathcal{H}$ are sequences of vectors $|\psi_n\rangle \in \mathcal{H}_n$

fulfilling $\| |\psi\rangle \|^2 = \sum_n \| |\psi_n\rangle \|^2 < \infty$. The scalar product in $\mathcal{H}$ is defined with the help of the scalar products in $\mathcal{H}_n$

$$\langle\varphi|\psi\rangle = \sum_n \langle\varphi_n|\psi_n\rangle \,.$$

Addition and multiplication with a scalar are of course the component-wise operations. The space $\mathcal{H}_0$ is assumed to be the one-dimensional Hilbert space representing the state $|0\rangle$ of the system with no particle. Collecting together the bases of all Hilbert spaces $\mathcal{H}_n$ in a clear manner, we get a basis of $\mathcal{H}$. We can now introduce a pair of creation and annihilation operators $\hat{a}_r^\dagger$ and $\hat{a}_r$ for any $r$ acting in $\mathcal{H}$ as

$$\begin{aligned} \hat{a}_r|n_1 n_2 \ldots n_r \ldots\rangle &= \sqrt{n_r}\,|n_1 n_2 \ldots n_r - 1 \ldots\rangle \,, \\ \hat{a}_r^\dagger|n_1 n_2 \ldots n_r \ldots\rangle &= \sqrt{n_r + 1}\,|n_1 n_2 \ldots n_r + 1 \ldots\rangle \,. \end{aligned} \tag{2.48}$$

Particularly it holds that (the vacuum stability condition)

$$\hat{a}_r|0\rangle = 0 \,.$$

The operator $\hat{a}_r^\dagger$ is the adjoint of $\hat{a}_r$ and it holds that

$$[\hat{a}_r, \hat{a}_s^\dagger] = \delta_{rs}\hat{1} \,, \qquad [\hat{a}_r, \hat{a}_s] = \hat{0} \,, \qquad [\hat{a}_r^\dagger, \hat{a}_s^\dagger] = \hat{0} \,. \tag{2.49}$$

The operator $\hat{N}_r = \hat{a}_r^\dagger \hat{a}_r$ has the meaning of the particle number operator for the state $r$,

$$\hat{N}_r|n_1 n_2 \ldots n_r \ldots\rangle = n_r|n_1 n_2 \ldots n_r \ldots\rangle \,. \tag{2.50}$$

The total number operator $\hat{N}$ is defined as the sum $\hat{N} = \sum_r \hat{N}_r$. To avoid the infinity of the energy of the vacuum, the energy operator $\hat{E}_r$ for the state $r$ is defined as $\hbar\omega_r \hat{N}_r$ and the total energy operator as $\hat{E} = \sum_r \hbar\omega_r \hat{N}_r$ with $\omega_r \in R$. We can write

$$|n_1 n_2 \ldots\rangle = \frac{1}{\sqrt{n_1! n_2! \ldots}} (\hat{a}_1^\dagger)^{n_1} (\hat{a}_2^\dagger)^{n_2} \ldots |0\rangle \,, \tag{2.51}$$

hence we have the generalization of the Fock space of the previous section to the case of infinitely many pairs of creation and annihilation operators. From the above discussion it follows also that the free Bose gas with an undetermined number of particles is indistinguishable from a collection of distinct harmonic oscillators with frequencies $\omega_r$.

Now we discuss briefly the connection of the Bose gas with an undetermined number of particles with the quantized real scalar field. We assume again $\hbar = 1$ and also the unit light velocity $c$. The classical real scalar field $A(x)$ is the solution of the *Klein-Gordon* equation

$$(\Box + m^2)A(x) = 0 \,, \tag{2.52}$$

where as usually

$$\Box = \frac{\partial^2}{\partial t^2} - \sum_{i=1}^{3} \frac{\partial^2}{\partial x_i^2}.$$

From the point of view of quantum optics the most important case is that of the zero mass $m = 0$ and we further restrict ourselves to this case. It is well known that we can write

$$A(x) = A^{(+)}(x) + A^{(-)}(x) = \frac{1}{(2\pi)^{3/2}} \int \frac{d\mathbf{k}}{2|\mathbf{k}|} a(\mathbf{k}) \exp(-ikx) + \text{c.c.}, \quad (2.53)$$

where $A^{(-)}(x) = [A^{(+)}(x)]^*$, $k = (\mathbf{k}, |\mathbf{k}|)$ denotes the momentum four-vector and $x = (\mathbf{x}, t)$ the point in Minkowski space. The scalar product $kx$ of the four-vectors $k$ and $x$ is taken as $kx = |\mathbf{k}|t - \mathbf{k} \cdot \mathbf{x}$. Let us assume that the function $a(\mathbf{k})$ belongs to the Hilbert space $\mathcal{L}_1 = L^2\left(R^3, d\mathbf{k}/(2|\mathbf{k}|)\right)$ and let us choose some basis $f_r(\mathbf{k})$ in this space so that we can decompose the Fourier components $a(\mathbf{k})$ and $a^*(\mathbf{k})$ of the $A(x)$ as

$$a(\mathbf{k}) = \sum_r A_r f_r(\mathbf{k}), \qquad a^*(\mathbf{k}) = \sum_r A_r^* f_r^*(\mathbf{k}) \quad (2.54)$$

with

$$A_r = \langle f_r | a \rangle, \qquad A_r^* = \langle a | f_r \rangle. \quad (2.55)$$

The (*second*) *quantization* of the real scalar field $A(x)$ consist of replacing the coefficients $A_r^*$ and $A_r$ by the creation and annihilation operators $\hat{a}_r^\dagger$ and $\hat{a}_r$ fulfilling (2.49) and acting on the Fock space $\mathcal{H}$ corresponding to the Bose gas with an undetermined number of particles. The single particle Hilbert space $\mathcal{H}_1$ is assumed to be identical to the above introduced Hilbert space $\mathcal{L}_1$ with $f_r = |r\rangle$. According to (2.55) we can interpret, in the classical case, the functions $a(\mathbf{k})$ and $a^*(\mathbf{k})$ as linear functionals on the $\mathcal{L}_1$. This interpretation leads after quantization naturally to the *operator-valued generalized functions* $\hat{a}(\mathbf{k})$ and $\hat{a}^\dagger(\mathbf{k})$. To explain this notion more precisely let us introduce the *operator-valued linear functional* $\hat{B}$ on $\mathcal{L}_1$ as a mapping from the entire $\mathcal{L}_1$ into the space of all linear operators on the Fock space $\mathcal{H}$, such that

- the operators $\hat{B}(f)$ for all $f \in \mathcal{L}_1$ are defined on a common dense domain $D \subset \mathcal{H}$,
- $\hat{B}(\alpha_1 f_1 + \alpha_2 f_2) = \alpha_1 \hat{B}(f_1) + \alpha_2 \hat{B}(f_2)$ for any complex $\alpha_1$ and $\alpha_2$.

The linear operator-valued functional $\hat{B}^\dagger$ is the *adjoint* of the operator-valued functional $\hat{B}$ if for all $f \in \mathcal{L}_1$, $\psi \in D$ and $\varphi \in D^*$ it holds that

$$\langle \varphi | \hat{B}(f) \psi \rangle = \langle \varphi \hat{B}^\dagger(f^*) | \psi \rangle,$$

where $D^*$ is the common dense domain of all operators $\hat{B}^\dagger(f)$. In our case $\mathcal{L}_1 = L^2(R^3, d\mathbf{k}/(2|\mathbf{k}|))$ and it is natural to write symbolically

$$\hat{B}(f) = \int \frac{d\mathbf{k}}{|\mathbf{k}|} \hat{B}(\mathbf{k}) f(\mathbf{k})$$

and $\hat{B}(\mathbf{k})$ is called the *operator-valued generalized function.* The operator-valued generalized functions $\hat{a}(\mathbf{k})$ and $\hat{a}^\dagger(\mathbf{k})$ should be understood in this sense. On the common dense domains $D$ and $D^*$

$$D = D^* = \{|\psi\rangle \in \mathcal{H};\ \langle\psi|\hat{N}|\psi\rangle < \infty\}, \tag{2.56}$$

where $\hat{N}$ is the total number particle operator in the Fock space, we can write formally

$$\begin{aligned}
(\hat{a}(\mathbf{k})\psi_n)(\mathbf{k}_1, \mathbf{k}_2, \ldots, \mathbf{k}_{n-1}) &= \sqrt{n}\,\psi_n(\mathbf{k}, \mathbf{k}_1, \ldots, \mathbf{k}_{n-1}), \\
(\hat{a}^\dagger(\mathbf{k})\psi_n)(\mathbf{k}_1, \mathbf{k}_2, \ldots, \mathbf{k}_{n+1}) &= \frac{1}{\sqrt{n+1}} \sum_{i=1}^{n+1} 2|\mathbf{k}|\delta(\mathbf{k} - \mathbf{k}_i) \\
&\quad \times \psi_n(\mathbf{k}_1, \ldots, \check{\mathbf{k}}_i, \ldots, \mathbf{k}_{n+1}), \\
\psi_n &\in \mathcal{H}_n,
\end{aligned} \tag{2.57}$$

where the symbol $\check{\mathbf{k}}_i$ means that the corresponding variable is omitted. Hence the operators $\hat{a}(\mathbf{k})$ and $\hat{a}^\dagger(\mathbf{k})$ represent respectively the annihilation and creation operators of the particle with momentum $\mathbf{k}$. As a simple consequence we have the following commutation relations

$$\begin{aligned}
[\hat{a}(\mathbf{k}), \hat{a}(\mathbf{k}')] = [\hat{a}^\dagger(\mathbf{k}), \hat{a}^\dagger(\mathbf{k}')] &= \hat{0}, \\
[\hat{a}(\mathbf{k}), \hat{a}^\dagger(\mathbf{k}')] &= 2|\mathbf{k}|\delta(\mathbf{k} - \mathbf{k}').
\end{aligned} \tag{2.58}$$

It is easy to check that $\hat{a}(f_r^*)$ and $\hat{a}^\dagger(f_r)$ really coincide with $\hat{a}_r$ and $\hat{a}_r^\dagger$ as introduced in the discussion of the Bose gas ($|r\rangle \equiv f_r$).

The quantization of the real scalar field $\hat{A}(x)$ is then given by the Fourier transform (2.53), which should be understood on a proper class of operator-valued generalized functions to make a good sense. The resulting operator $\hat{A}(x)$ is also properly understood as an operator-valued generalized function in the $x$-representation. We finish this section with the commutation relations for the field operators $\hat{A}(x)$ and $\hat{A}(y)$

$$[\hat{A}(x), \hat{A}(y)] = \frac{1}{(2\pi)^3} \int \frac{d\mathbf{k}}{2|\mathbf{k}|} \{\exp[-ik(x-y)] - \exp[ik(x-y)]\}\,. \tag{2.59}$$

The positive frequency part $\hat{A}^{(+)}(x)$ and negative frequency part $\hat{A}^{(-)}(x) = (\hat{A}^{(+)})^\dagger(x)$ are then interpreted as annihilation and creation operators at the space-time point $x$, respectively.

Let us also note that a variant of the Stone-von Neumann theorem can be shown also in the framework of the second quantization [Berezin (1966)].

## 2.8 Quantization of the free electromagnetic field

Here we set again $c = \hbar = 1$. Let us assume the classical radiation field described by the (four-)vector potential $(\mathbf{A}, A^0)$ in the *Coulomb gauge*

$$A^0 = 0, \qquad \mathrm{div}\mathbf{A} = 0. \tag{2.60}$$

It is well known that in this case we can write

$$\mathbf{A} = \mathbf{A}^{(+)} + (\mathbf{A}^{(+)})^*,$$

$$\mathbf{A}^{(+)}(x) = \frac{1}{(2\pi)^{2/3}} \int \frac{d\mathbf{k}}{2|\mathbf{k}|} \mathbf{a}(\mathbf{k}) \exp(-ikx),$$

$$\mathbf{k} \cdot \mathbf{a}(\mathbf{k}) = 0. \tag{2.61}$$

The vector function $\mathbf{A}$ solves the wave equation

$$\Box \mathbf{A} = 0 \tag{2.62}$$

and the intensities $\mathbf{E}(x)$ and $\mathbf{B}(x)$ of the electric and magnetic fields are expressed in terms of the vector potential $\mathbf{A}(x)$ as

$$\mathbf{B} = \mathrm{rot}\mathbf{A}, \qquad \mathbf{E} = -\frac{\partial \mathbf{A}}{\partial t}. \tag{2.63}$$

The quantization of the radiation field in the Coulomb gauge is then based on three independent pairs of creation and annihilation operators of the type considered in the previous section and collected in the vectors $\hat{\mathbf{a}}_0^\dagger(\mathbf{k})$ and $\hat{\mathbf{a}}_0(\mathbf{k})$. These are assumed to act on a suitable domain in the tensor product $\mathcal{H} \otimes \mathcal{H} \otimes \mathcal{H}$ of three copies of the Fock space $\mathcal{H}$. If we introduce the vector $\hat{\mathbf{a}}(\mathbf{k})$ as

$$\hat{\mathbf{a}}(\mathbf{k}) = \hat{\mathbf{a}}_0(\mathbf{k}) - \left( \frac{\mathbf{k}}{|\mathbf{k}|} \cdot \hat{\mathbf{a}}_0(\mathbf{k}) \right) \frac{\mathbf{k}}{|\mathbf{k}|}, \tag{2.64}$$

we obtain the quantized vector potential $\hat{\mathbf{A}}(x)$ in the form

$$\hat{\mathbf{A}}(x) = \hat{\mathbf{A}}^{(+)}(x) + \hat{\mathbf{A}}^{(-)}(x),$$

$$\hat{\mathbf{A}}^{(+)}(x) = \frac{1}{(2\pi)^{3/2}} \int \frac{d\mathbf{k}}{2|\mathbf{k}|} \hat{\mathbf{a}}(\mathbf{k}) \exp(-ikx). \tag{2.65}$$

All commutation relations between operators $\hat{\mathbf{A}}(x)$, $\hat{\mathbf{E}}(x)$, and $\hat{\mathbf{B}}(x)$ are simple consequences of the commutation relations

$$[\hat{A}_i^{(+)}(x), \hat{A}_j^{(-)}(y)] = \int \frac{d\mathbf{k}}{2|\mathbf{k}|} \left( \delta_{ij} - \frac{k_i k_j}{|\mathbf{k}|^2} \right) \exp[-ik(x-y)]. \tag{2.66}$$

The quantized strength vectors $\hat{\mathbf{E}}$ and $\hat{\mathbf{B}}$ are given by equation (2.63) in the sense of generalized functions. The total particle number operator $\hat{N}$ and the energy operator $\hat{H}$ of the quantized free electromagnetic field are expressed with the help of the creation and annihilation operators as

$$\hat{N} = \int d\mathbf{k}\, \frac{\hat{\mathbf{a}}^\dagger(\mathbf{k})}{\sqrt{2|\mathbf{k}|}} \cdot \frac{\hat{\mathbf{a}}(\mathbf{k})}{\sqrt{2|\mathbf{k}|}} \tag{2.67}$$

and

$$\hat{H} = \int d\mathbf{k}\, |\mathbf{k}|\, \frac{\hat{\mathbf{a}}^\dagger(\mathbf{k})}{\sqrt{2|\mathbf{k}|}} \cdot \frac{\hat{\mathbf{a}}(\mathbf{k})}{\sqrt{2|\mathbf{k}|}}\,, \tag{2.68}$$

respectively. The last equations suggest to introduce the operator $\hat{\mathbf{A}}(\mathbf{k}) = \hat{\mathbf{a}}(\mathbf{k})/\sqrt{2|\mathbf{k}|}$, its adjoint $\hat{\mathbf{A}}^\dagger(\mathbf{k})$ as well as the so-called *detection* operator $\hat{\mathcal{A}}(x)$

$$\hat{\mathcal{A}}(x) = \frac{1}{(2\pi)^{2/3}} \int d\mathbf{k}\, \hat{\mathbf{A}}(\mathbf{k}) \exp(-ikx) \tag{2.69}$$

and its adjoint $\hat{\mathcal{A}}^\dagger(x)$. The total particle number operator $\hat{N}$ can then be written as

$$\hat{N} = \int d\mathbf{k}\, \hat{\mathbf{A}}^\dagger(\mathbf{k}) \cdot \hat{\mathbf{A}}(\mathbf{k}) = \int_{R^3} d\mathbf{x}\, \hat{\mathcal{A}}^\dagger(x) \cdot \hat{\mathcal{A}}(x) \tag{2.70}$$

and the commutation relations between the operators $\hat{\mathcal{A}}_i(x)$ and $\hat{\mathcal{A}}_j^\dagger(y)$ are

$$[\hat{\mathcal{A}}_i(x), \hat{\mathcal{A}}_j^\dagger(y)] = \int d\mathbf{k} \left(\delta_{ij} - \frac{k_i k_j}{|\mathbf{k}|^2}\right) \exp[-ik(x-y)]\,. \tag{2.71}$$

## 2.9 Quantum correlation functions and their properties

We finish this chapter by introducing a system of quantum correlation functions representing the basic tool for the description of quantum statistical and coherence properties of radiation. It should be noted that there is strong analogy between the correlation functions of quantum optics and the Wightman functions in the axiomatic field theory [Streater and Wightman (1964)] and their properties. The quantum correlation functions have been introduced and discussed by R. J. Glauber (1963a,b).

The quantum correlation functions are related to the corresponding quantum measurements on the light field and we can begin our considerations assuming the detection of the field by the process of absorbtion of a photon. We can consider an atom in the ground state as an ideal point detector whose sensitivity is independent of the radiating frequency. If the detected field is prepared in the initial quantum state $|i\rangle$, then after the absorbtion of a photon by the atom, the transition amplitude of the process is equal to $\langle f|\hat{A}^{(+)}(x)|i\rangle$, where $\hat{A}^{(+)}(x)$ is the annihilation operator of a photon of a linearly polarized field at a space-time point $x = (\mathbf{x}, t)$ (localized with the uncertainty of the wavelength). The transition probability per unit time from the initial state $|i\rangle$ to the final state $|f\rangle$ at the space-time point $x$ is then proportional to $|\langle f|\hat{A}^{(+)}(x)|i\rangle|^2$. But in the optical region the final state $|f\rangle$ of the field usually remains unobserved so that we must sum over all possible states $|f\rangle$ using their completeness $\sum_f |f\rangle\langle f| = \hat{1}$, which gives

$$\sum_f \langle i|\hat{A}^{(-)}(x)|f\rangle\langle f|\hat{A}^{(+)}(x)|i\rangle = \langle i|\hat{A}^{(-)}(x)\hat{A}^{(+)}(x)|i\rangle\,, \tag{2.72}$$

where $\hat{A}^{(-)}(x)$ represents the creation operator of a photon at $x$. Mostly the initial state is not a pure state but a mixed state so that we have to average over its random and unknown parameters, i.e. over all realizations of the state. Such an information about the quantum statistics of the field considered as a statistical dynamical system is contained in the statistical operator $\hat{\varrho} = \{|i\rangle\langle i|\}_{\mathbf{average}}$. Consequently we arrive at the quantum mean intensity at $x$

$$\left\{\langle i|\hat{A}^{(-)}(x)\hat{A}^{(+)}(x)|i\rangle\right\}_{\mathbf{average}} = \mathrm{Tr}\left\{\hat{\varrho}\hat{A}^{(-)}(x)\hat{A}^{(+)}(x)\right\}\,, \tag{2.73}$$

which is a particular value of the normal quantum correlation function

$$\Gamma_{\mathcal{N}}^{(1,1)}(x,x') = \mathrm{Tr}\left\{\hat{\varrho}\hat{A}^{(-)}(x)\hat{A}^{(+)}(x')\right\} \tag{2.74}$$

at $x = x'$. Also the vectorial properties of the radiation field can be taken into account and we obtain the second-rank tensor $\Gamma_{\mathcal{N},jk}^{(1,1)}$. Sometimes one considers fields obtained using polarizing filters which select photons of a polarization $\mathbf{e}$; then it is sufficient to use the scalar operator $\hat{A} = \mathbf{e}\cdot\hat{\mathbf{A}}$.

More generally, we can consider the transition probability per unit (time)$^n$ that a photon of a polarization $j_1$ is absorbed at $(\mathbf{x}_1, t_1) \equiv x_1$, a photon of a polarization $j_2$ at $x_2$ etc., and a photon of a polarization $j_n$ at $x_n$. By the same argument the probability of such a process is proportional to the particular value for $m = n$ and $x_j = x_{n+j}$, $j = 1, 2, \ldots, n$ of the $(m+n)$-th order normal correlation function (tensor)

$$\Gamma_{\mathcal{N}}^{(m,n)}(x_1, \ldots, x_{m+n})$$
$$= \mathrm{Tr}\left\{\hat{\varrho}\hat{A}^{(-)}(x_1)\ldots\hat{A}^{(-)}(x_m)\hat{A}^{(+)}(x_{m+1})\ldots\hat{A}^{(+)}(x_{m+n})\right\}\,, \tag{2.75}$$

where $x$ may also include the polarization index $j$. The operators $\hat{A}^{(-)}$ or $\hat{A}^{(+)}$ at different points can be interchanged, however, $\hat{A}^{(-)}$ and $\hat{A}^{(+)}$ are not commuting in general. The normal ordering of the field operators (all annihilation operators stand to the right of the all creation operators) in (2.75) is related to the process of detection of the field by means of absorption. The coherent states $|\{\alpha_\lambda\}\rangle$, which are eigenstates of the annihilation operator $\hat{A}^{(+)}(x)$, $\hat{A}^{(+)}(x)|\{\alpha_\lambda\}\rangle = V(x)|\{\alpha_\lambda\}\rangle$, $\langle\{\alpha_\lambda\}|\hat{A}^{(-)}(x) = \langle\{\alpha_\lambda\}|V^*(x)$, $V(x)$ being the corresponding classical wave field (Chapter 4), make it possible to express the quantum correlation functions in a "classical" form

$$\Gamma^{(m,n)}(x_1,\ldots,x_{m+n}) = \langle V^*(x_1)\ldots V^*(x_m)V(x_{m+1})\ldots V(x_{m+n})\rangle\,, \tag{2.76}$$

where the brackets $\langle\ \rangle$ mean generally the average with some quasidistribution $P(V_1,\ldots,V_{m+n})$. Thus the classical-quantum correspondence can be expressed as $\hat{A}^{(+)} \leftrightarrow V$, $\hat{A}^{(-)} \leftrightarrow V^*$, $\hat{\varrho} \leftrightarrow P$. As a consequence of the vacuum stability condition $\hat{A}^{(+)}|0\rangle = 0$, the vacuum expectation values of the normally ordered products of annihilation and creation operators are zero,

$$\Gamma^{(n,n)}_{\mathcal{N},\mathrm{vac}}(x_1,\ldots,x_n,x_n,\ldots,x_1) = \langle 0|\prod_{j=1}^{n}\hat{A}^{(-)}(x_j)\prod_{k=1}^{n}\hat{A}^{(+)}(x_k)|0\rangle = 0\,, \tag{2.77}$$

so that the physical vacuum gives no contribution to measuremets performed with the help of quadratic detectors such as photodetectors, photographic plates, eye, etc. The vacuum fluctuations do not violate the complete coherence of the field defined by the factorization condition

$$\Gamma^{(m,n)}_{\mathcal{N}}(x_1,\ldots,x_{m+n}) = \prod_{j=1}^{m}V^*(x_j)\prod_{k=m+1}^{m+n}V(x_k)\,, \tag{2.78}$$

which means that classical coherent fields are deterministic while quantum coherent fields are close to classical ones containing vacuum fluctuations (of the energy $\hbar\omega/2$ in every mode) only. These do not affect full coherence. In terms of the Hanbury Brown-Twiss correlations (Sec. 6.2) the factorization (2.78) represents the independence of counting rates of single photodetectors.

We can also consider the detection of the field by means of an atom in an excited state adopting stimulated emission as a basic measuring process. Such quantum counters were introduced by Mandel (1966) and their action is related to the antinormally ordered products (the role of the annihilation and creation operators is interchanged). The probability of such a process is proportional to antinormal correlation function

$$\Gamma^{(n,n)}_{\mathcal{A}}(x_1,\ldots,x_n,x_n,\ldots,x_1)$$
$$= \mathrm{Tr}\left\{\hat{\varrho}\hat{A}^{(+)}(x_1)\ldots\hat{A}^{(+)}(x_n)\hat{A}^{(-)}(x_n)\ldots\hat{A}^{(-)}(x_1)\right\}. \tag{2.79}$$

Of course such a detection process is sensitive to vacuum fluctuations since e.g. $\langle 0|\hat{A}^{(+)}(x)\hat{A}^{(-)}(x')|0\rangle = [\hat{A}^{(+)}(x), \hat{A}^{(-)}(x')]$, i.e. such a vacuum expectation value equals the commutator representing quantum spontaneous phenomena related to vacuum fluctuations. These and the above arguments lead to the conclusion that the probability distributions of emitted photodetectors and absorbed photons are of similar forms (excluding some pathological cases) and hence the statistics of emitted photoelectrons reflect the statistics of absorbed photons by the photodetectors, whereas the statistics of counts from a quantum counter completely differ from the photon statistics as a consequence of the physical vacuum contribution.

Also some alternating ordering of the field operators can occur if a scattering process is adopted to register the optical field [Loudon (1973)]. In such a process a photon of the frequency $\omega_1(\mathbf{k}_1)$ is absorbed, while the transition probability is proportional to the correlation function involving the alternating order,

$$\mathrm{Tr}\left\{\hat{\varrho}\hat{A}_1^{(-)}\hat{A}_2^{(+)}\hat{A}_2^{(-)}\hat{A}_1^{(+)}\right\} \approx \langle n_1\rangle\left(\langle n_2\rangle + 1\right), \tag{2.80}$$

$\langle n_1\rangle$ and $\langle n_2\rangle$ being the mean numbers of absorbed and emitted photons, respectively. In the spontaneous scattering $\langle n_2\rangle = 0$, in the stimulated scattering $\langle n_2\rangle \neq 0$. Symmetrized products of field operators are described by the Wigner quasidistributions, which are closest to classical distributions, however, they can take on negative values.

Real detectors processes require to take into account frequency sensitivity of the detector involving its response function and their spatial areas. Such detectors measure values of the field averaged over space-time regions and the corresponding integrations over the resolving time of the detector and its active area have to be performed (e.g. Knight and Allen (1983), Peřina (1991), Sec. 3.1).

### 2.9.1 Spectral properties of quantum correlation functions

We can decompose the annihilation operator $\hat{A}^{(+)}(\mathbf{x},t)$ in the form of the Fourier integral as follows

$$\hat{A}^{(+)}(\mathbf{x},t) = \int_0^\infty \hat{A}^{(+)}(\mathbf{x},\nu)\exp(-i2\pi\nu t)\,d\nu\,, \tag{2.81}$$

where only positive frequencies in the spectral components

$$\hat{A}^{(+)}(\mathbf{x},\nu) = \int_{-\infty}^{+\infty} \hat{A}^{(+)}(\mathbf{x},t)\exp(i2\pi\nu t)\,dt \tag{2.82}$$

are defined [$\hat{A}(\mathbf{x},-\nu) = \hat{A}^\dagger(\mathbf{x},\nu)$ since $\hat{A}(\mathbf{x},t)$ is Hermitian]. The spectral decomposition of the creation operator $\hat{A}^{(-)}(\mathbf{x},t)$ is obtained by means of the Hermitian conjugation. Substituting into the definition (2.75) of the correlation function we obtain

$$\Gamma_{\mathcal{N}}^{(m,n)}(x_1,\ldots,x_{m+n}) = \int_0^\infty \cdots \int_0^\infty G_{\mathcal{N}}^{(m,n)}(\mathbf{x}_1,\ldots,\mathbf{x}_{m+n},v_1,\ldots,v_{m+n}) \times \exp\left(i2\pi \sum_{j=1}^{m+n} \varepsilon_j v_j t_j\right) \prod_k dv_k\,, \tag{2.83}$$

where $\varepsilon_j = 1$ for $j = 1,\ldots,m$ and $\varepsilon_j = -1$ for $j = m+1,\ldots,m+n$, the spectral correlation function is equal to

$$\begin{aligned} G_{\mathcal{N}}^{(m,n)}&(\mathbf{x}_1,\ldots,\mathbf{x}_{m+n},v_1,\ldots,v_{m+n}) \\ &= \mathrm{Tr}\left\{\hat{\varrho}\prod_{j=1}^{m}\hat{A}^{(-)}(\mathbf{x}_j,v_j)\prod_{k=m+1}^{m+n}\hat{A}^{(+)}(\mathbf{x}_k,v_k)\right\} \\ &= \int_{-\infty}^{+\infty}\cdots\int_{-\infty}^{+\infty}\Gamma_{\mathcal{N}}^{(m,n)}(\mathbf{x}_1,\ldots,\mathbf{x}_{m+n},t_1,\ldots,t_{m+n}) \\ &\qquad \times \exp\left(-i2\pi\sum_{j=1}^{m+n}\varepsilon_j v_j t_j\right)\prod_k dt_k\,, \quad v_j \geq 0\,; \end{aligned} \tag{2.84}$$

if at least one $v_j < 0$, the spectral correlation function $G_{\mathcal{N}}^{(m,n)}$ equals zero. The formula (2.83) expresses the generalized Wiener-Khintchin theorem.

If the optical field is stationary and ergodic, the time averaging $\lim_{T\to\infty} \int_{-T}^{+T}\ldots dt$ can be adopted. The correlation functions $\Gamma_{\mathcal{N}}^{(m,n)}$ of a stationary field are invariant with respect to the translation of the time origin, i.e.

$$\Gamma_{\mathcal{N}}^{(m,n)}(t_1+\tau,\ldots,t_{m+n}+\tau) = \Gamma_{\mathcal{N}}^{(m,n)}(t_1,\ldots,t_{m+n})\,, \tag{2.85}$$

which means, with respect to (2.83), that

$$\sum_{j=1}^{m+n}\varepsilon_j v_j = 0\,. \tag{2.86}$$

This is the necessary and sufficient condition for the spectral components $\hat{A}^{(+)}(v_1),\ldots,\hat{A}^{(+)}(v_{m+n})$ of a stationary field to be correlated. Thus the correlation function $\Gamma_{\mathcal{N}}^{(m,n)}$ of a stationary field, whose statistical operator obeys the equation $i\hbar\partial\hat{\varrho}/\partial t = [\hat{H},\hat{\varrho}] = \hat{0}$, is dependent on $(m+n-1)$ time variables $\tau_2 = t_2 - t_1,\ldots,\tau_{m+n} = t_{m+n} - t_1$.

For the second order correlation function of a stationary field we have the Wiener-Khintchin theorem

$$\Gamma_{\mathcal{N}}(\mathbf{x}_1,\mathbf{x}_2,\tau) \equiv \Gamma_{\mathcal{N}}^{(1,1)}(\mathbf{x}_1,\mathbf{x}_2,t,t+\tau) = \int_0^\infty G_{\mathcal{N}}(\mathbf{x}_1,\mathbf{x}_2,v)\exp(-i2\pi v\tau)\,dv\,, \tag{2.87}$$

where the spectral correlation function is given as

$$\begin{aligned} G_{\mathcal{N}}(\mathbf{x}_1, \mathbf{x}_2, \upsilon) &\equiv G_{\mathcal{N}}^{(1,1)}(\mathbf{x}_1, \mathbf{x}_2, \upsilon, \upsilon) = \mathrm{Tr}\left\{\hat{\varrho}\hat{A}^{(-)}(\mathbf{x}_1, \upsilon)\hat{A}^{(+)}(\mathbf{x}_2, \upsilon)\right\} \\ &= \begin{cases} \int_{-\infty}^{+\infty} \Gamma_{\mathcal{N}}(\mathbf{x}_1, \mathbf{x}_2, \tau)\exp(i2\pi\upsilon\tau)\, d\tau\,, & \upsilon \geq 0\,, \\ 0\,, & \upsilon < 0\,. \end{cases} \end{aligned} \tag{2.88}$$

The Wiener-Khintchin theorem generally allows the spectral properties of radiation to be determined from the correlation measurements. Role of time ordering has to be taken into account in some cases [Knöll, Vogel and Welsch (1986), Knöll and Welsch (1992)].

### 2.9.2 Analytic properties of quantum correlation functions

It follows from the spectral decomposition (2.81) that the annihilation operator $\hat{A}^{(+)}$ is an analytic function in the lower half of the complex $t$-plane, whereas the creation operator $\hat{A}^{(-)}$ is an analytic function in the upper half of the complex $t$-plane. Therefore the quantum correlation function $\Gamma_{\mathcal{N}}^{(m,n)}$ is an analytic function in the upper half of the complex $t$-plane in variables $t_1, \ldots, t_m$ and in the lower half-plane in variables $t_{m+1}, \ldots, t_{m+n}$. Consequently the following dispersion relations hold

$$\begin{aligned} \mathrm{Im}\,\Gamma_{\mathcal{N}}^{(m,n)}(t_1, \ldots, t_{m+n}) &= -\varepsilon_k \frac{\mathcal{P}}{\pi} \int_{-\infty}^{+\infty} \frac{\mathrm{Re}\,\Gamma_{\mathcal{N}}^{(m,n)}(t_1, \ldots, t'_k, \ldots, t_{m+n})}{t'_k - t_k}\, dt'_k\,, \\ \mathrm{Re}\,\Gamma_{\mathcal{N}}^{(m,n)}(t_1, \ldots, t_{m+n}) &= \varepsilon_k \frac{\mathcal{P}}{\pi} \int_{-\infty}^{+\infty} \frac{\mathrm{Im}\,\Gamma_{\mathcal{N}}^{(m,n)}(t_1, \ldots, t'_k, \ldots, t_{m+n})}{t'_k - t_k}\, dt'_k\,, \end{aligned} \tag{2.89}$$

$\mathcal{P}$ denoting the Cauchy principal value. Here and in some cases above we have suppressed spatial variables for simplicity.

### 2.9.3 Wave equations for quantum correlation functions

In vacuum the operators $\hat{A}^{(+)}(\mathbf{x}, t)$ and $\hat{A}^{(-)}(\mathbf{x}, t)$ obey the wave equations and consequently the correlation functions and the spectral correlation functions have to obey the following equations

$$\begin{aligned} \left(\Delta_j - \frac{1}{c^2}\frac{\partial^2}{\partial t_j^2}\right)\Gamma_{\mathcal{N}}^{(m,n)}(\mathbf{x}_1, \ldots, \mathbf{x}_{m+n}, t_1, \ldots, t_{m+n}) &= 0\,, \\ \left(\Delta_j + k_j^2\right)\Gamma_{\mathcal{N}}^{(m,n)}(\mathbf{x}_1, \ldots, \mathbf{x}_{m+n}, \upsilon_1, \ldots, \upsilon_{m+n}) &= 0\,, \\ j = 1, \ldots, m+n, & \end{aligned} \tag{2.90}$$

where $\Delta_j$ is the Laplace operator with respect to the $j$-th spatial variable and $k_j = 2\pi v_j/c$ is the wave number.

For the stationary optical fields it holds that

$$\left[\Delta_1 - \frac{1}{c^2}\left(\sum_{j=2}^{m+n}\frac{\partial}{\partial\tau_j}\right)^2\right]\Gamma_{\mathcal{N}}^{(m,n)}(\mathbf{x}_1,\ldots,\mathbf{x}_{m+n},\tau_2,\tau_3,\ldots,\tau_{m+n}) = 0\,,$$
$$\left[\Delta_j - \frac{1}{c^2}\frac{\partial^2}{\partial\tau_j^2}\right]\Gamma_{\mathcal{N}}^{(m,n)}(\mathbf{x}_1,\ldots,\mathbf{x}_{m+n},\tau_2,\tau_3,\ldots,\tau_{m+n}) = 0\,,$$
$$j = 2,3,\ldots,m+n, \qquad (2.91)$$

and

$$\left[\Delta_j + k_j^2\right]G_{\mathcal{N}}^{(m,n)}(\mathbf{x}_1,\ldots,\mathbf{x}_{m+n},v_1,\ldots,v_{m+n}) = 0\,,$$
$$j = 1,\ldots,m+n, \qquad (2.92)$$

where $v_j$ are related by (2.86).

### 2.9.4 Symmetries, positive definiteness and other inequalities

From the definition of the quantum correlation function we directly have the cross-symmetry (hermitian) condition

$$\left[\Gamma_{\mathcal{N}}^{(m,n)}(\mathbf{x}_1,\ldots,\mathbf{x}_{m+n})\right]^* = \Gamma_{\mathcal{N}}^{(n,m)}(\mathbf{x}_{m+n},\ldots,\mathbf{x}_1)\,. \qquad (2.93)$$

As mentioned above the space-time variable $x$ may include the polarization index $j$. In particular for the second-order correlation in a stationary field

$$\Gamma_{\mathcal{N}}^*(\mathbf{x}_1,\mathbf{x}_2,\tau) = \Gamma_{\mathcal{N}}(\mathbf{x}_2,\mathbf{x}_1,-\tau) \qquad (2.94)$$

and also

$$G_{\mathcal{N}}^*(\mathbf{x}_1,\mathbf{x}_2,v) = G_{\mathcal{N}}(\mathbf{x}_2,\mathbf{x}_1,v)\,. \qquad (2.95)$$

It also follows from the definition that $\Gamma_{\mathcal{N}}^{(n,n)} = 0$ for $n > N$, $N$ being the number of photons in the field.

Assume any operator $\hat{B}$ with the finite Hilbert-Schmidt norm $\|\hat{B}\| = \left[\mathrm{Tr}(\hat{B}^\dagger\hat{B})\right]^{1/2}$ so that

$$\mathrm{Tr}\left\{\hat{\varrho}\hat{B}^\dagger\hat{B}\right\} \geq 0\,. \qquad (2.96)$$

Now we can choose the $\hat{B}$-operator in various forms. If we choose $\hat{B} = \hat{A}^{(+)}(x_1)\ldots\hat{A}^{(+)}(x_n)$, we obtain

$$\Gamma_{\mathcal{N}}^{(n,n)}(x_1,\ldots,x_n,x_n,\ldots,x_1) \geq 0\,. \qquad (2.97)$$

If we choose $\hat{B} = \sum_{j=1}^{n} c_j \hat{A}^{(+)}(x_j)$ with some coefficients $c_j$, we obtain the non-negative definiteness condition

$$\sum_{j,k=1}^{n} c_j^* c_k \Gamma_{\mathcal{N}}^{(1,1)}(x_j, x_k) \geq 0 . \tag{2.98}$$

Such a condition means that

$$\mathrm{Det}\left\{\Gamma_{\mathcal{N}}^{(1,1)}(x_j, x_k)\right\} \geq 0 . \tag{2.99}$$

For $n = 1$ we have for the mean intensity $\Gamma_{\mathcal{N}}^{(1,1)}(x, x) \geq 0$, for $n = 2$

$$|\gamma_{\mathcal{N}}^{(1,1)}(x_1, x_2)|^2 \leq \Gamma_{\mathcal{N}}^{(1,1)}(x_1, x_1)\, \Gamma_{\mathcal{N}}^{(1,1)}(x_2, x_2) , \tag{2.100}$$

which means that for the degree of coherence

$$\gamma_{\mathcal{N}}^{(1,1)}(x_1, x_2) = \frac{\Gamma_{\mathcal{N}}^{(1,1)}(x_1, x_2)}{\sqrt{\Gamma_{\mathcal{N}}^{(1,1)}(x_1, x_1)\, \Gamma_{\mathcal{N}}^{(1,1)}(x_2, x_2)}} \tag{2.101}$$

it holds that

$$0 \leq |\gamma_{\mathcal{N}}^{(1,1)}(x_1, x_2)| \leq 1 . \tag{2.102}$$

The optical field is the second-order coherent if the visibility of the interference fringes $|\gamma_{\mathcal{N}}^{(1,1)}(x_1, x_2)|$ is maximum.

It is obvious that the similar considerations can be performed for the spectral correlation functions and on this basis the spectral theory of coherence can be constructed [Wolf (1983, 1986)]. Such a theory shows that the spectrum of light can change during propagation (including vacuum) and may offer new interpretations of spectral changes of light from stars. Such a mechanism can simulate Doppler effect and can lead to red or blue shifts, splitting of spectral lines, etc. [Wolf (1987, 1991)].

If we choose $\hat{B} = c_1 \hat{A}^{(+)}(x_1) \ldots \hat{A}^{(+)}(x_n) + c_2 \hat{A}^{(+)}(x_{n+1}) \ldots \hat{A}^{(+)}(x_{2n})$, we arrive at

$$|\Gamma_{\mathcal{N}}^{(n,n)}(x_1, \ldots, x_{2n})|^2 \leq \Gamma_{\mathcal{N}}^{(n,n)}(x_1, \ldots, x_n, x_n, \ldots, x_1) \times \Gamma_{\mathcal{N}}^{(n,n)}(x_{n+1}, \ldots, x_{2n}, x_{2n}, \ldots, x_{n+1}) , \tag{2.103}$$

which can be used to define the higher-order degree of coherence [Klauder and Sudarshan (1968)]

$$\gamma_{\mathcal{N}}^{(n,n)}(x_1, \ldots, x_{2n}) = \Gamma_{\mathcal{N}}^{(n,n)}(x_1, \ldots, x_{2n}) \times \left[\Gamma_{\mathcal{N}}^{(n,n)}(x_1, \ldots, x_n, x_n, \ldots, x_1) \Gamma_{\mathcal{N}}^{(n,n)}(x_{n+1}, \ldots, x_{2n}, x_{2n}, \ldots, x_{n+1})\right]^{-1/2} \tag{2.104}$$

satisfying the inequality

$$0 \leq |\gamma_{\mathcal{N}}^{(n,n)}(x_1, \ldots, x_{2n})| \leq 1 . \tag{2.105}$$

The even-order quantum correlation functions $\Gamma_{\mathcal{N}}^{(n,n)}$ are determined by a system of photodetectors whose outputs are correlated (Hanbury Brown-Twiss experiment). Also the odd-order correlation functions $\Gamma_{\mathcal{N}}^{(n,m)}$, $n \neq m$ can in principle be measured using nonlinear crystals [see e.g. Peřina (1991)].

### 2.9.5 Quantum characteristic functionals

Defining the normal quantum characteristic functional

$$C_{\mathcal{N}}\{y(x)\} = \mathrm{Tr}\left\{\hat{\varrho} \exp\left[\int y(x)\hat{A}^{(-)}(x)\, d^4x\right] \exp\left[-\int y^*(x)\hat{A}^{(+)}(x)\, d^4x\right]\right\} , \tag{2.106}$$

$y(x)$ being a parametric function, the complete set of normal correlation functions can be derived with the help of functional differentiation

$$\Gamma_{\mathcal{N}}^{(m,n)}(x_1, \ldots, x_{m+n}) = \prod_{j=1}^{m} \frac{\delta}{\delta y(x_j)} \prod_{k=m+1}^{m+n} \frac{\delta}{\delta\left(-y^*(x_k)\right)} C_{\mathcal{N}}\{y(x)\}\bigg|_{y(x)\equiv 0} ; \tag{2.107}$$

here again $x \equiv (\mathbf{x}, t, j)$, so that the integrals over $x$ include also the summation over the polarization indices $j$.

Decomposing $\hat{A}^{(+)}(x)$ in terms of orthonormal functions $\{\varphi_k(x)\}$, for which

$$\int \varphi_k^*(x)\varphi_l(x)\, d^4x = \delta_{kl} , \tag{2.108}$$

we can write

$$\hat{A}^{(+)}(x) = \sum_k \hat{a}_k\varphi_k(x) , \quad \hat{a}_k = \int \hat{A}^{(+)}(x)\varphi_k^*(x)\, d^4x . \tag{2.109}$$

Introducing the parameters

$$\beta_k = \int y(x)\varphi_k^*(x)\, d^4x , \tag{2.110}$$

we can rewrite the characteristic functional in the form

$$C_{\mathcal{N}}(\{\beta_k\}) = \mathrm{Tr}\left\{\hat{\varrho} \prod_k \exp(\beta_k\hat{a}_k^\dagger) \prod_l \exp(-\beta^*\hat{a}_l)\right\} , \tag{2.111}$$

and the multimode correlation function is obtained by differentiation

$$\mathrm{Tr}\left\{\hat{\varrho}\hat{a}_1^\dagger \ldots \hat{a}_m^\dagger \hat{a}_{m+1} \ldots \hat{a}_{m+n}\right\} = \prod_{k=1}^{m} \frac{\partial}{\partial \beta_k} \prod_{l=m+1}^{m+n} \frac{\partial}{\partial(-\beta_l^*)} C_{\mathcal{N}}(\{\beta_j\})\bigg|_{\{\beta_k\}\equiv\{\beta_l^*\}\equiv 0} . \tag{2.112}$$

A similar approach can be adopted to other orderings of field operators.

### 2.9.6 Introduction to photon statistics

The correlation functions introduced above can be measured with the use of photodetectors with correlated outputs. As to the even-order correlation functions compound measurements involving nonlinear media have to be used.

Alternatively we can use a photodetector to measure a photocount distribution in a time interval $(t, t+T)$, where $T$ is a fixed resolving time of the detector. Such multifold photocount distributions may then be used to calculate moments and correlation functions. In the one-fold case the photocount distribution of detecting $n$ photons in the time interval $(t, t+T)$ equals [see e.g. Peřina (1991); more detailed discussion is presented in Sec. 3.3.1 from the point of view of quantum measurement]

$$\begin{aligned} p(n,T,t) &= \operatorname{Tr}\left\{\hat{\varrho} : \frac{(\eta \hat{W})^k}{n!} \exp(-\eta \hat{W}) :\right\} \\ &= \int_0^\infty \frac{(\eta W)^n}{n!} \exp(-\eta W) P_{\mathcal{N}}(W)\, dW\,, \end{aligned} \tag{2.113}$$

where $\eta$ is photodetection efficiency, : : represents the normal order of field operators,

$$P_{\mathcal{N}}(W) = \int \Phi_{\mathcal{N}}(\{\alpha_\lambda\})\, \delta\left(\sum_\lambda |\alpha_\lambda|^2 - W\right) d^2\{\alpha_\lambda\} \tag{2.114}$$

is the quasiprobability distribution of the integrated intensity

$$\hat{W} = \int_t^{t+T} \hat{A}^{(-)}(t')\hat{A}^{(+)}(t')\, dt'\,, \qquad W = \int_t^{t+T} |V(t)|^2\, dt \tag{2.115}$$

expressed in terms of the Glauber-Sudarshan quasidistribution $\Phi_{\mathcal{N}}$ of complex amplitudes $\{\alpha_\lambda\}$ related to normal ordering of field operators (Sec. 4.3.3). If a detector has a volume $V$ and $\mathcal{S}(x)$ represents its response function, then more generally

$$\eta W = \int_0^T \int_0^T \int_V \int_V \mathcal{S}(x' - x'') V^*(x') V(x'')\, d^4x' d^4x''\,. \tag{2.116}$$

From the photodetection equation (2.113) we find the following relation between the characteristic functions of $P_{\mathcal{N}}(W)$ and $p(n,T,t)$

$$\begin{aligned} \langle : \exp(is\hat{n}) : \rangle &= \langle \exp(is\eta W) \rangle_{\mathcal{N}} = \langle (1+is)^{\hat{n}} \rangle = \sum_{n=0}^\infty p(n,T,t)(1+is)^n\,, \\ \langle \exp(is\hat{n}) \rangle &= \langle \exp\left[\eta\left(\exp(is) - 1\right) W\right] \rangle_{\mathcal{N}}\,, \end{aligned} \tag{2.117}$$

$is$ being a parameter, which are also generating functions for the photocount distribution $p(n,T,t)$ and its factorial moments $\langle W^k\rangle_{\mathcal{N}}$ as follows

$$\begin{aligned} p(n,T,t) &= \frac{1}{2\pi}\int_0^{2\pi} \exp(-isn)\langle\exp(is\hat{n})\rangle\, ds \\ &= \frac{1}{n!}\frac{d^n}{d(is)^n}\langle\exp(is\eta W)\rangle_{\mathcal{N}}\Big|_{is=-1}, \\ \eta^k\langle W^k\rangle_{\mathcal{N}} &= \frac{d^k}{d(is)^k}\langle\exp(is\eta W)\rangle_{\mathcal{N}}\Big|_{is=0}. \end{aligned} \tag{2.118}$$

The cumulants can be defined in terms of derivatives of the natural logarithm of the generating function as

$$\kappa_j^{(n)} = \frac{d^j}{d(is)^j}\ln\langle\exp(is\hat{n})\rangle\Big|_{is=0}, \quad j=1,2,\ldots, \tag{2.119}$$

giving $\kappa_0^{(n)}=0$, $\kappa_1^{(n)}=\langle n\rangle$, $\kappa_2^{(n)}=\langle(\Delta n)^2\rangle$, etc. This means that

$$\langle\exp(is\hat{n})\rangle = \exp\left[\sum_{j=1}^{\infty}\kappa_j^{(n)}\frac{(is)^j}{j!}\right], \tag{2.120}$$

and similarly for $\kappa_j^{(\eta W)}$. It holds that $\kappa_1^{(\eta W)}=\eta\langle W\rangle_{\mathcal{N}}$, $\kappa_j^{(\eta W)}=0$, $j>1$ for coherent light and $\kappa_j^{(\eta W)}=(j-1)!\,\eta^j\langle W\rangle_{\mathcal{N}}^j$ for chaotic light of natural sources.

The photocount distribution $p(n,T,t)$ and the photon number distribution $p(n,t)=\langle n|\hat{\varrho}(t)|n\rangle$ are related by the Bernoulli distribution

$$p(n,T,t) = \sum_{m=n}^{\infty}\binom{m}{n}(\eta T)^n(1-\eta T)^{m-n}p(m,t). \tag{2.121}$$

For closed systems and a single mode $\eta W(t)=\eta|\alpha|^2T$ is to be replaced by $[1-\exp(-\eta T)]|\alpha|^2$.

For the moments we have

$$\begin{aligned} \langle\hat{n}\rangle &= \eta\langle W\rangle_{\mathcal{N}}, \\ \langle\hat{n}^2\rangle &= \eta\langle W\rangle_{\mathcal{N}}+\eta^2\langle W^2\rangle_{\mathcal{N}}, \\ \ldots &\ \ldots\ \ldots\ldots\ldots\ldots \end{aligned} \tag{2.122}$$

in correspondence with the commutation rules, e.g. $\langle\hat{n}^2\rangle=\langle(\hat{a}^\dagger\hat{a})^2\rangle=\langle\hat{a}^\dagger\hat{a}\rangle+\langle\hat{a}^{\dagger 2}\hat{a}^2\rangle$.

The variance of the number of emitted photoelectrons (absorbed photons, $\eta=1$) is expressed by the Einstein formula

$$\langle(\Delta\hat{n})^2\rangle = \langle\hat{n}\rangle+\eta^2\langle(\Delta W)^2\rangle_{\mathcal{N}}, \tag{2.123}$$

describing the wave-particle duality, since the first term $\langle\hat{n}\rangle$ represents the variance of Poissonian particles and the second term $\eta^2\langle(\Delta W)^2\rangle_{\mathcal{N}}$ represents the generalized wave fluctuations. For coherent field $\langle(\Delta W)^2\rangle_{\mathcal{N}} = 0$ and $\langle(\Delta\hat{n})^2\rangle = \langle\hat{n}\rangle$, i.e. such a field is Poissonian. Classical fields have $\langle(\Delta W)^2\rangle_{\mathcal{N}} > 0$ and they are super-Poissonian, e.g. for the chaotic field of a natural source $\langle(\Delta W)^2\rangle_{\mathcal{N}} = \langle I\rangle^2 T^2$ and $\langle(\Delta n)^2\rangle = \langle n\rangle(1 + \langle n\rangle)$, corresponding to the Bose-Einstein distibution $\langle n\rangle^n/(1 + \langle n\rangle)^{1+n}$. For quantum fields having no clasical analogues, the quasidistribution $\Phi_{\mathcal{N}}$ can be singular and can take on negative values, which may lead to $\langle(\Delta W)^2\rangle_{\mathcal{N}} < 0$ and $\langle(\Delta\hat{n})^2\rangle < \langle\hat{n}\rangle$ and such light is sub-Poissonian. The relation between the variance $\langle(\Delta\hat{n})^2\rangle$ of photoelectrons and the variance $\langle(\Delta\hat{m})^2\rangle$ of photons can be expressed in the form of the Burgess variance theorem

$$\langle(\Delta\hat{n})^2\rangle = \eta(1-\eta)\langle\hat{m}\rangle + \eta^2\langle(\Delta\hat{m})^2\rangle \,, \tag{2.124}$$

or introducing the Fano factor $F_n = \langle(\Delta\hat{n})^2\rangle/\langle\hat{n}\rangle$,

$$(F_n - 1) = \eta(F_m - 1) \,. \tag{2.125}$$

This equation expresses more generally the conservation of quality of statistics of exciting $m$-particles and emitted $n$-particles, i.e. Poissonian statistics ($F_m = 1$) remain Poissonian ($F_n = 1$), super-Poissonian statistics ($F_m > 1$) remain super-Poissonian ($F_n > 1$) and sub-Poissonian statistics ($F_m < 1$) remain sub-Poissonian ($F_n < 1$). This is the principle for the modified Franck-Hertz experiment for producing sub-Poissonian photons exciting Hg atoms by sub-Poissonian electrons [Teich and Saleh (1988)].

For experimental determination of the statistical properties of optical fields short-time measurements are often used assuming that the time $T$ of detection is much smaller than the coherence time $\tau_c = 1/\Delta v$, $\Delta v$ being the half-width of radiation spectrum. For natural sources we obtain from the central limit theorem of probability theory that the probability distribution of the complex amplitude of the field is Gaussian and therefore the probability distribution of the field intensity $I$ is negatively exponential

$$P_{\mathcal{N}}(I) = \frac{1}{\langle I\rangle_{\mathcal{N}}} \exp\left(-\frac{I}{\langle I\rangle_{\mathcal{N}}}\right) \,, \tag{2.126}$$

with the moments

$$\langle I^k\rangle_{\mathcal{N}} = k!\,\langle I\rangle_{\mathcal{N}}^k \,. \tag{2.127}$$

The substitution to the detection equation (2.113) provides the Bose-Einstein distribution of undistinguishable Bose-Einstein particles with the mean number of photoelectrons $\langle n\rangle = \eta\langle I\rangle_{\mathcal{N}} T$. In this case the distribution (2.126) determines also the distribution $P_{\mathcal{N}}(W)$ since $W \approx IT$. In the case of large detection intervals $T \gg \tau_c$ a superposition of events can be applied stating that the resulting distribution is the convolution of partial distributions and the resulting generating function equals the product of partial generating functions obtaining

$$\langle \exp(is\eta W)\rangle_{\mathcal{N}} = \frac{1}{(1 - is\langle n\rangle/M)^M}, \tag{2.128}$$

provided that the mean numbers of photoelectrons are $\langle n\rangle/M$ in every degree of freedom, $\langle n\rangle$ being the total number of emitted photoelectrons and $M \approx T/\tau_c$ ($T \gg \tau_c$) is the number of degrees of freedom (it can be multiplied by the number of spatial degrees of freedom $S/A_c$, $S$ being the detection area, $A_c$ the coherence area, and by the polarization degrees of freedom $2/(1+P^2)$, $P$ being the degree of polarization). From the generating function (2.128) we obtain the Rayleigh distribution

$$P_{\mathcal{N}}(W) = \left(\frac{\eta M}{\langle n\rangle}\right)^M \frac{W^{M-1}}{\Gamma(M)} \exp\left(-\frac{\eta W M}{\langle n\rangle}\right) \tag{2.129}$$

and from the photodetection equation the Mandel-Rice formula

$$p(n,T) = \frac{\Gamma(n+M)}{n!\,\Gamma(M)}\left(1 + \frac{M}{\langle n\rangle}\right)^{-n}\left(1 + \frac{\langle n\rangle}{M}\right)^{-M}, \tag{2.130}$$

giving the Bose-Einstein distribution for $M = 1$. Here $\Gamma$ is the gamma function. The factorial moments equal

$$\eta^k \langle W^k\rangle_{\mathcal{N}} = \langle n\rangle^k \frac{\Gamma(k+M)}{\Gamma(M)M^k} \tag{2.131}$$

and for the variance of $n$ we have

$$\langle(\Delta n)^2\rangle = \langle n\rangle\left(1 + \frac{\langle n\rangle}{M}\right). \tag{2.132}$$

The number $\delta = \langle n\rangle/M$ can be called the degeneracy parameter. For partially polarized fields $M = 2/(1+P^2)$, where the degree of polarization $P$ is given by [Born and Wolf (1965)]

$$P = \left[1 - \frac{4\mathrm{Det}\hat{J}}{(\mathrm{Tr}\hat{J})^2}\right]^{1/2} = \left[1 - \frac{4\mathrm{Det}\hat{\sigma}}{(\mathrm{Tr}\hat{\sigma})^2}\right]^{1/2}, \tag{2.133}$$

where $\hat{J} = (J_{jk})$, $J_{jk} = \langle \hat{E}_j^{(-)}(t)\hat{E}_k^{(+)}(t)\rangle$ is the coherence matrix and $\hat{\sigma} = (\sigma_{jk})$, $\sigma_{jk} = [\langle \Delta I_j(t)\Delta I_k(t)\rangle_{\mathcal{N}}]^{1/2}$, $\Delta I_j = I_j - \langle I_j\rangle_{\mathcal{N}}$ is the correlation matrix.

For the optical field of an ideal laser

$$P_{\mathcal{N}}(I) = \delta(I - \langle I \rangle_{\mathcal{N}}) \tag{2.134}$$

and the photodetection equation provides the Poissonian distribution of photocounts

$$p(n,T) = \frac{\langle n \rangle^n}{n!} \exp(-\langle n \rangle), \tag{2.135}$$

since $\langle (\Delta W)^2 \rangle_{\mathcal{N}} = 0$ and $\langle (\Delta n)^2 \rangle = \langle n \rangle$. Therefore for various kinds of fields we have the following values of the Fano factor

$$F_n = \begin{cases} 1 + \langle n \rangle, & \text{for chaotic field,} \\ 1, & \text{for coherent field,} \\ 0, & \text{for the Fock state.} \end{cases}$$

$F_n < 1$ for sub-Poisson field having no classical analogue and $F_n > 1$ for super-Poisson field. Also super-chaotic light with $F_n > 1 + \langle n \rangle$ ($\langle I^k \rangle_{\mathcal{N}} > k! \langle I \rangle_{\mathcal{N}}^k$) can violate some classical inequalities. Sub-Poisson light may also be called light with squeezed number fluctuations. With respect to (2.125) statistics of emitted photoelectrons reflect the statistics of absorbed photons (excluding some pathological cases). For $\eta = 1$ they are identical. More details about the photocount statistics can be found in monographs by Saleh (1978) and Peřina (1985).

The photocount theory from the point of view of operator ordering was discussed by Marshall and Santos (1991). A correct definition of quantum probabilities is important for decision about violation of Bell's inequalities [Santos (1991)].

## 2.10 Appendix: mathematical tools

Here we summarize basic notions and basic ideas from functional analysis used in chapter 2. We are doing it in a very condensed but we hope fully sufficient form for understanding the axiomatic formulation of quantum mechanics. We suppose that the reader is familiar with the notion of a Hilbert space. The Hilbert space $\mathcal{H}$ is always assumed to be separable and the operators are linear. The scalar product will be denoted by $\langle \ | \ \rangle$ and the corresponding norm by $\| \ \|$.

- The Hilbert spaces used in quantum mechanical models are usually infinite-dimensional and the operators representing observables are in general unbounded. In such cases we must distinguish among symmetric, self-adjoint and Hermitian operators. For technical reasons we consider in the following a linear operator $\hat{A}$ defined in a domain $D_{\hat{A}}$

which is dense in an infinite-dimensional Hilbert space $\mathcal{H}$. Then the *adjoint operator* $\hat{A}^\dagger$ is defined on the domain $D_{\hat{A}^\dagger}$ consisting of all vectors $\psi \in \mathcal{H}$ for which vectors $\psi_1 \in \mathcal{H}$ exist such that the relations $\langle\psi|\hat{A}\varphi\rangle = \langle\psi_1|\varphi\rangle$ hold for any $\varphi \in D_{\hat{A}}$. The operator $\hat{A}^\dagger$ defined by $\hat{A}^\dagger\psi = \psi_1$ is then uniquely determined because $D_{\hat{A}}$ is dense in $\mathcal{H}$. So we have in particular the equality

$$\langle\psi|\hat{A}\varphi\rangle = \langle\hat{A}^\dagger\psi|\varphi\rangle\,, \qquad \varphi \in D_{\hat{A}}\,,\ \psi \in D_{\hat{A}^\dagger}. \tag{2.136}$$

Of course it always holds that $D_{\hat{A}} \subset D_{\hat{A}^\dagger}$.

The operator $\hat{A}$ is called *symmetric* if $\hat{A} = \hat{A}^\dagger$ on $D_{\hat{A}}$. If, moreover, $D_{\hat{A}} = D_{\hat{A}^\dagger}$ then $\hat{A}$ is called *self-adjoint.*

If the self-adjoint operator is also bounded (so it is defined on the entire $\mathcal{H}$), then it is called *Hermitian.*

The operator $\hat{U}$ defined on the entire $\mathcal{H}$ is called *unitary* if $\hat{U}\hat{U}^\dagger = \hat{1}$ ($\hat{1}$ is the identity operator). In this case its range is also the entire Hilbert space $\mathcal{H}$ and $\hat{U}$ preserves the scalar product $\langle\hat{U}\varphi|\hat{U}\psi\rangle = \langle\varphi|\psi\rangle$.

A self-adjoint operator $\hat{E}$ defined on the entire Hilbert space $\mathcal{H}$ and characterized by the property $\hat{E}^2 = \hat{E}$ is called *projector*. Every projector is bounded and defines by its range $\hat{E}\mathcal{H}$ a closed subspace of $\mathcal{H}$ and vice versa every closed subspace $M$ of $\mathcal{H}$ uniquely determines a projector with the range $M$.

The *positivity* of a Hermitian operator $\hat{A}$ means that $\langle\psi|\hat{A}\psi\rangle \geq 0$ for all $\psi \in \mathcal{H}$.

The linear operator $\hat{A}$ is said to be *closed* if $\psi_n \in D_{\hat{A}}$, for all $n = 1, 2, \ldots$, $\psi_n \to 0$ and $A\psi_n \to \varphi$ for $n \to \infty$ implies that $\varphi \to 0$. The linear operator $\hat{B}$ is said to be an *extension* of the operator $\hat{A}$ if $D_{\hat{A}} \subset D_{\hat{B}}$ and if $\hat{A} = \hat{B}$ holds on $D_{\hat{A}}$. The closed linear operator $\bar{\hat{A}}$ is the *closure* of $\hat{A}$ if it is the minimal closed extension of $\hat{A}$. For any linear operator $\hat{A}$ the operator $\hat{A}^\dagger$ is closed, and if $\hat{A}$ is symmetric its closure exists and is identical to $(\hat{A}^\dagger)^\dagger$.

- *The spectral family of projectors* (*resolution of the identity*) is a function $\hat{E}_\lambda$ of a real variable $\lambda \in R$ with values being projectors which satisfy the following conditions:

  i)
$$\hat{E}_\lambda\hat{E}_\mu = \hat{E}_\mu\hat{E}_\lambda = \hat{E}_\lambda \qquad \text{if } \lambda \leq \mu\,, \tag{2.137}$$

  ii)
$$\hat{E}_{\lambda+} = \hat{E}_\lambda \tag{2.138}$$

in the strong operator topology, i.e. $\varepsilon \to 0+$ implies $\|(\hat{E}_{\lambda+\varepsilon} - \hat{E}_\lambda)\psi\| \to 0$ for all $\psi \in \mathcal{H}$,

iii)

$$\hat{E}_{-\infty} = \hat{0}\,, \qquad \hat{E}_{+\infty} = \hat{1} \tag{2.139}$$

in the strong operator topology.

Now we can formulate a very important property of self-adjoint operators which is called the *spectral decomposition.* Namely, to each self-adjoint operator $\hat{A}$ in the Hilbert space $\mathcal{H}$ a unique spectral family of projectors $\hat{E}_\lambda$ corresponds such that the domain $D_{\hat{A}}$ of $\hat{A}$ consists of all vectors $|\psi\rangle$ for which

$$\int_{\mathbf{R}} \lambda^2 \, d\langle\psi|\hat{E}_\lambda\psi\rangle < +\infty$$

and for any $\psi \in D_{\hat{A}}$ and $\varphi \in \mathcal{H}$

$$\langle\varphi|\hat{A}\psi\rangle = \int_{\mathbf{R}} \lambda \, d\langle\varphi|\hat{E}_\lambda\psi\rangle \tag{2.140}$$

holds. This is often written in shorthand notation as

$$\hat{A} = \int_{\mathbf{R}} \lambda \, d\hat{E}_\lambda \,. \tag{2.141}$$

- With the help of the resolution of the identity we can introduce the notion of a function $f(\hat{A})$ of the self-adjoint operator $\hat{A}$. Let $f$ be a measurable function of a real variable with finite values possibly with exception of a set of zero measure. Here it is understood that these properties are fulfilled with respect to the Lebesgue-Stieltjes measure $d\langle\psi|\hat{E}_\lambda\psi\rangle$ for each $\psi \in \mathcal{H}$. Then the operator $f(\hat{A})$ is defined in the domain

$$D_{f(\hat{A})} = \{\psi \in \mathcal{H} : \int_{\mathbf{R}} |f(\lambda)|^2 \, d\langle\psi|\hat{E}_\lambda\psi\rangle < +\infty\}$$

as

$$f(\hat{A}) = \int_{\mathbf{R}} f(\lambda) \, d\hat{E}_\lambda \,. \tag{2.142}$$

If the function $f$ is bounded, then the operator $f(\hat{A})$ is also bounded. If it is real, then the operator $f(\hat{A})$ is self-adjoint and if $|f(\lambda)| = 1$ for all $\lambda$, then the operator $f(\hat{A})$ is unitary.

- The characteristic function $\chi_\Delta$ of any Borel subset $\Delta$ of the real line fulfils the conditions of the previous section and we can introduce the family of projectors $\hat{E}(\Delta)$ by

$$\hat{E}(\Delta) = \int_{\mathbf{R}} \chi_\Delta \, d\hat{E}_\lambda \,. \tag{2.143}$$

These projectors are orthogonal in the sense that their composition $\hat{E}(\Delta_1)\hat{E}(\Delta_2)$ is zero for the sets $\Delta_1$, $\Delta_2$ having an empty intersection. Particularly if we take $\Delta = R$ we have for the identity operator $\hat{1}$

$$\hat{1} = \int_{\mathbf{R}} d\hat{E}_\lambda = \hat{E}(R) . \tag{2.144}$$

- The *spectrum* $\sigma(\hat{A})$ of a self-adjoint operator $\hat{A}$ is then defined as the set of points $\lambda \in R$ such that for each open interval $\Delta$ containing $\lambda$ the corresponding $\hat{E}(\Delta)$ is non-zero. So only the points of the spectrum contribute effectively to the integral in the spectral decomposition (2.141).

  The point $\lambda$ of the spectrum belongs to the so-called *point spectrum* $\sigma_p(\hat{A})$ if the projector $\hat{E}(\{\lambda\})$ is non-zero.

  The remaining part of the spectrum $\sigma_c(\hat{A})$ is called the *continuous spectrum.*

  The spectrum is always closed.

  The Hilbert space $\mathcal{H}$ can be decomposed into a direct sum of Hilbert space $\hat{E}(\sigma_p(\hat{A}))\mathcal{H}$ and its orthogonal complement $\hat{E}(\sigma_c(\hat{A}))\mathcal{H}$. The action of $\hat{A}$ leaves these subspaces invariant, so the restrictions $\hat{A}_p$ and $\hat{A}_c$ of $\hat{A}$ to these subspaces are self-adjoint and have *pure point spectrum* (the spectrum of $\hat{A}_p$ is the closure of the set of all eigenvalues of $\hat{A}$) and *pure continuous spectrum* ($\sigma_p(\hat{A}_c)$ is empty), respectively. The definition of an eigenvalue is given in the next section.

- The complex number $\lambda$ is said to be an *eigenvalue* of the linear operator $\hat{A}$, if there is at least one non-zero vector $\psi \in D_{\hat{A}}$ which satisfies the *eigenvalue equation* (*characteristic equation*)

$$\hat{A}\psi = \lambda\psi . \tag{2.145}$$

  Each vector $\psi \in D_{\hat{A}}$ satisfying the above equation is called an *eigenvector* of $\hat{A}$ with eigenvalue $\lambda$. All eigenvectors with the same eigenvalue $\lambda$ form a linear subspace $M_\lambda$ (not necessarily closed if it is infinite-dimensional), which is usually referred to as the *characteristic subspace* corresponding to the $\lambda$. If $\hat{A}$ is self-adjoint, then the characteristic subspaces are closed. All eigenvalues of a symmetric operator are real and the characteristic subspaces corresponding to different eigenvalues are mutually orthogonal. For a self-adjoint operator the set of its eigenvalues coincides with its point spectrum and $M_\lambda = \hat{E}(\{\lambda\})\mathcal{H}$ for each eigenvalue $\lambda$. If a self-adjoint operator has a pure point spectrum only, then we have a decomposition of the Hilbert space $\mathcal{H}$ into a direct sum of mutually orthogonal subspaces $M_\lambda$. If the eigenvalues are moreover simple (i.e. the characteristic spaces $M_\lambda$ are one-dimensional), then the corresponding normed eigenvectors $|e_1\rangle, |e_2\rangle, \ldots$ form a basis of the Hilbert space $\mathcal{H}$ and the spectral decomposition of the self-adjoint operator $\hat{A}$ is simply

$$\hat{A} = \sum_i \lambda_i |e_i\rangle\langle e_i| \,. \tag{2.146}$$

- We should also mention that there is a more general notion of the spectrum of a general linear operator $\hat{A}$ in the Hilbert space $\mathcal{H}$. The complex number $z \in C$ belongs to the so-called *resolvent set* of the operator $\hat{A}$ if the operator $\hat{R}_z = (\hat{A} - z\hat{1})^{-1}$ exists and if it is bounded and densely defined in $\mathcal{H}$. The operator-valued function $\hat{R}_z$ is said to be the *resolvent* of the operator $\hat{A}$. The set of points in $C$ not belonging to the resolvent set is then the *spectrum* of the operator $\hat{A}$. The spectrum of a general linear operator is disjointly divided into the *point spectrum* containing those points $z$ for which the resolvent $\hat{R}_z$ of $\hat{A}$ does not exist, the *continuous spectrum* containing the points for which the resolvent exists and is densely defined but is not bounded, and the *residual spectrum* containing the points for which the resolvent exists but is not densely defined. If $\hat{A}$ is self-adjoint, the spectrum of $\hat{A}$ is always real and the residual spectrum is empty. The definitions of the spectrum and its point and continuous parts as they were given here are then equivalent to those already used.

- For bounded operators the statement that the operators *commute* is clear. Namely, for two bounded operators $\hat{A}$ and $\hat{B}$ on $\mathcal{H}$ this means that

$$[\hat{A}, \hat{B}] = \hat{A}\hat{B} - \hat{B}\hat{A} = \hat{0} \,. \tag{2.147}$$

The generalization for the unbounded self-adjoint operators is that the projectors from their spectral families of projectors commute.

It is well known that for any set $\{\hat{A}_\alpha\}$ of mutually commuting self-adjoint operators there exist a self-adjoint operator $\hat{A}$ and real functions $f_\alpha(\lambda)$ such that $\hat{A}_\alpha = f_\alpha(\hat{A})$.

For a set of commuting operators $\{\hat{A}_\alpha\}$ there is also a good notion of the function $f(\hat{A}_1, \hat{A}_2, \ldots, \hat{A}_n)$ if the function $f$ fulfils the same conditions as given in the definition of the function of one self-adjoint operator. The operator $f(\hat{A}_1, \hat{A}_2, \ldots, \hat{A}_n)$ is defined on the domain

$$D_{f(\hat{A}_1,\hat{A}_2,\ldots,\hat{A}_n)} = \{\psi \in \mathcal{H} : \int_{\mathbf{R^n}} |\psi(\lambda_1,\lambda_2,\ldots,\lambda_n)|^2 \times d\langle\psi|\hat{E}^{(1)}_{\lambda_1}\hat{E}^{(2)}_{\lambda_2}\ldots\hat{E}^{(n)}_{\lambda_n}\psi\rangle < +\infty\}$$

as

$$f(\hat{A}_1,\hat{A}_2,\ldots,\hat{A}_n) = \int_{\mathbf{R^n}} f(\lambda_1,\lambda_2,\ldots,\lambda_n)\, d\hat{E}^{(1)}_{\lambda_1} d\hat{E}^{(2)}_{\lambda_2}\ldots d\hat{E}^{(n)}_{\lambda_n}\,. \tag{2.148}$$

The spectrum $\sigma(f)$ of the operator $f(\hat{A}_1,\hat{A}_2,\ldots,\hat{A}_n)$ can then be expressed in terms of the spectra of operators $\hat{A}_1,\hat{A}_2,\ldots,\hat{A}_n$ as

$$\sigma(f) = \overline{f(\sigma(\hat{A}_1)\times\sigma(\hat{A}_2)\times\ldots\times\sigma(\hat{A}_n))}\,, \tag{2.149}$$

where $\overline{\Delta}$ means the closure of the set $\Delta$. There is analogous correspondence between the properties of the function $f(\lambda_1,\lambda_2,\ldots,\lambda_n)$ and the operator $f(\hat{A}_1,\hat{A}_2,\ldots,\hat{A}_n)$ as in the case of one-variable.

- The vector $\psi$ in $\mathcal{H}$ is said to be a *cyclic vector* for the set of self-adjoint mutually commuting operators $\hat{A}_1,\hat{A}_2,\ldots,\hat{A}_n$, if the smallest closed subspace of $\mathcal{H}$ containing $\psi$ and invariant with respect to all resolvents $\hat{R}^{(1)}_{z_1},\hat{R}^{(2)}_{z_2},\ldots,\hat{R}^{(n)}_{z_n}$ ($z_i \in C-R$, $i=1,2,\ldots,n$) coincides with the entire space $\mathcal{H}$.

  The set of commuting self-adjoint operators is said to be a *complete set of operators* if they have a cyclic vector.

  The complete sets of operators have nice properties:

  i) If a self-adjoint or bounded operator $\hat{A}$ commutes with all operators of a complete set, then it is a function of these operators.

  ii) If each operator which commutes with a set of self-adjoint mutually commuting operators is a function of these operators, then the set of commuting operators is complete.

  iii) The completeness of a set consisting of one self-adjoint operator $\hat{A}$ with a pure point spectrum only implies that the eigenvalues of the operator are simple. In this case the eigenvectors $|e_i\rangle$ corresponding to the eigenvalues $\lambda_i$ constitute a basis in the Hilbert space $\mathcal{H}$ and each vector $\psi \in \mathcal{H}$ is uniquely determined by coefficients $\{c_i\}$ expressing it as a linear combination of the elements of the basis. In this way we obtain a unitary transformation of the original Hilbert space $\mathcal{H}$ onto the Hilbert space $l^2$ of square-summable sequences $c = \{c_1, c_2, \ldots\}$, $\sum_i |c_i|^2 < \infty$. The scalar product of two such sequences $c$ and $\tilde{c}$ corresponding to vectors $\psi$ and $\tilde{\psi} \in \mathcal{H}$ is then given by $\langle\psi|\tilde{\psi}\rangle = \langle c|\tilde{c}\rangle = \sum_i c_i^* \tilde{c}_i$. The operator

$\hat{A}$ acts on such sequences simply by multiplication with the eigenvalues $\hat{A}\{c_i\} = \{\lambda_i c_i\}$ and the domain of $\hat{A}$ in $l^2$ consists of sequences $\{c_i\}$ with the finite sum $\sum_i |\lambda_i c_i|^2$. For a complete set consisting of several commuting self-adjoint operators the same role is played by the set of their common eigenvectors.

- In the situation of a complete set of general self-adjoint operators $\hat{A}_1, \hat{A}_2, \ldots, \hat{A}_n$ the similar statement holds. Let us assume first a self-adjoint operator $\hat{A}$ with simple spectrum (not necessarily pure point) in $\mathcal{H}$. We know that there exists a cyclic vector $\varphi$ and that each self-adjoint or bounded operator $\hat{B}$ commuting with $\hat{A}$ is a function of $\hat{A}$, $\hat{B} = f(\hat{A})$. Let $\hat{E}_\lambda$ be the spectral decomposition of the operator $\hat{A}$ and let us consider the Hilbert space $L^2(\sigma(\hat{A}), d\mu)$ of all functions $f(\lambda)$, $\lambda \in \sigma(\hat{A})$ such that

$$\int_{\mathbf{R}} |f(\lambda)|^2 \, d\mu(\lambda) < +\infty \,,$$

where the integration measure is defined as $d\mu(\lambda) = d\langle\varphi|\hat{E}_\lambda\varphi\rangle$. Because of the definition and the properties of the resolvent operators $\hat{R}_z$, the linear subspace spanned by $\varphi$ and all vectors of the form $\hat{R}_z\varphi$, $z \in C - R$, is dense in $\mathcal{H}$ and each vector $\psi$ from this subspace is of the form $\psi = f(\hat{A})\varphi$ for some function $f \in L^2(\sigma(\hat{A}), d\mu)$. Making use of the definition of $f(\hat{A})$ we have

$$\|\psi\|^2 = \|f(\hat{A})\varphi\|^2 = \int_{\mathbf{R}} |f(\lambda)|^2 \, d\mu(\lambda) \,. \tag{2.150}$$

The correspondence $\psi \leftrightarrow f(\lambda)$ can be extended to a unitary mapping $\hat{U}$ of the entire space $\mathcal{H}$ onto $L^2(\sigma(\hat{A}), d\mu)$. The image of the operator $\hat{A}$ under this unitary isometry is the operator $\hat{U}\hat{A}\hat{U}^{-1} : u(\lambda) \to \lambda u(\lambda)$ with the domain

$$\{f(\lambda) \, : \, \int \lambda^2 |f(\lambda)|^2 \, d\mu(\lambda) < +\infty\} \,.$$

Different cyclic vectors yield unitarily equivalent representations.

If the spectrum of the self-adjoint operator $\hat{A}$ is not simple, then the Hilbert space $\mathcal{H}$ can be decomposed into a countably direct sum of its subspaces $\mathcal{H}_i$ which are invariant under the action of $\hat{A}$ and such that the spectrum of each restriction $\hat{A}_i$ of $\hat{A}$ to $\mathcal{H}_i$ is simple. It follows that there exist a set $M$ with positive measure $d\mu$ and a real measurable function $a(m)$ defined and finite on a dense subset of $M$ and a unitary isometry $\hat{U} : \mathcal{H} \to L^2(M, d\mu)$ such that

$$\psi \in D_{\hat{A}} \qquad \text{if and only if} \qquad a(m)\hat{U}\psi \in L^2(M, d\mu)$$

and the image $\hat{U}A\hat{U}^{-1}$ of the operator $\hat{A}$ under $\hat{U}$ is the multiplication with the function $a(m)$. $M$ is simply the union of the spectra of $\hat{A}_i$ in $\mathcal{H}_i$ and the measure $d\mu$ is defined by the corresponding measures $d\mu_i$. The function $a(m)$ is equal to the $\lambda$ for each subspace. In the case of pure point spectrum, the interpretation is straightforward. Let us take an orthonormal basis in each characteristic subspace $M_\lambda$ and order it in any way. The subspaces described above are then obtained by collecting together all first vectors from each basis, then the second ones, etc.

The generalization of the above statements to the sets of commuting operators and complete sets of commuting operators $\hat{A}_1, \hat{A}_2, \ldots, \hat{A}_n$ is straightforward if we consider functions of variables $\lambda_1, \lambda_2, \ldots, \lambda_n$ and the measure defined with the help of $d\hat{E}^{(1)}_{\lambda_1} d\hat{E}^{(2)}_{\lambda_2} \ldots d\hat{E}^{(n)}_{\lambda_n}$.

- As we have already seen for the self-adjoint operator $\hat{A}$ with the pure point spectrum only, an orthonormal basis can be formed in $\mathcal{H}$ with the help of eigenvectors of $\hat{A}$. In the case of continuous spectrum there are no eigenvectors (an eigenvector should, by definition, belong to $\mathcal{H}$) corresponding to the points of the continuous part of the spectrum. However, generalized eigenvectors can be defined even for the continuous part of the spectrum.

  Let $\hat{A}$ be a self-adjoint operator in $\mathcal{H}$. In this case one can always find two additional spaces, a Banach space $\mathcal{H}_+$ and a space of all continuous antilinear functionals $\mathcal{H}_-$ on this Banach space such that the spaces $\mathcal{H}_+ \subset \mathcal{H} \subset \mathcal{H}_-$ form the so-called *Gelfand triple.* $\mathcal{H}_+$ is a dense subspace of $\mathcal{H}$ defined as $\mathcal{H}_+ = \hat{K}\mathcal{H}$ for some *Hilbert-Schmidt operator* $\hat{K}$ on $\mathcal{H}$ (i.e. for a bounded linear operator such that for each basis $e_1, e_2, \ldots$ of $\mathcal{H}$ the sum $\sum_i \|\hat{K}e_i\|^2$ is finite) such that $\hat{K}$ is injective with dense range. Then $\hat{K}$ has an inverse and the norm on $\mathcal{H}_-$ is given by $\|\psi\|_+ = \|K^{-1}\psi\|$. The space $\mathcal{H}_-$ is identified with the completion of the space $\mathcal{H}$ in the norm $\|\psi\|_- = \|\hat{K}^\dagger\psi\|$ and the pairing between $\mathcal{H}_+$ and $\mathcal{H}_-$ is a prolongation of the scalar product on $\mathcal{H}$. It means that the value of the functional $\psi_- \in \mathcal{H}_-$ on the vector $\psi_+ \in \mathcal{H}_+$ is $\psi_-(\psi_+) = \langle\psi_-|\psi_+\rangle$ whenever $\psi_-$ belongs to $\mathcal{H} \subset \mathcal{H}_-$. Let $\hat{U}$ be the unitary isometry of $\mathcal{H}$ induced by $\hat{A}$, under which $\hat{U}\mathcal{H} = L^2(M, d\mu)$ and $\hat{U}\hat{A}\hat{U}^{-1} = a(m)$. We consider the restriction of the isometry $\hat{U}$ onto the space $\mathcal{H}_+$ and define elements $\Phi(m), m \in M$ of $\mathcal{H}_-$ by the relation

$$\Phi(m)(\psi_+) = (U\psi_+)(m)\,. \tag{2.151}$$

  The vector function $\Phi(m)$ is then called the *complete system of generalized eigenvectors* (*eigenfunctions* of the operator $\hat{A}$ and the function $\tilde{\psi}_+(m) = (\hat{U}\psi_+)(m)$ is called the *Fourier transform* of $\psi_+$. The Fourier transform can be extended to a unitary isometry of $\mathcal{H}$ onto $L^2(M, d\mu)$ such that the inversion formula

$$|\psi\rangle = \int \tilde{\psi}(m)\Phi^*(m)\, d\mu \tag{2.152}$$

holds for each $\psi \in \mathcal{H}$ in the weak sense. Namely, for every $\psi_+ \in \mathcal{H}_+$ we have

$$\langle\psi_+|\psi\rangle = \int \tilde{\psi}(m)\overline{\Phi(m)(\psi_+)}\, d\mu\,.$$

Thus each element of the Hilbert space $\mathcal{H}$ can be decomposed in this complete system of generalized eigenvectors. Moreover there exists a dense subset of $M$ on which

$$\Phi(m)(\hat{A}\psi_+) = a(m)\Phi(m)(\psi_+)\,,$$

$$\psi_+ \in \mathcal{H}_+ \cap D_{\hat{A}}, \hat{A}\psi \in \mathcal{H}_+\,. \tag{2.153}$$

This equation suggests that the functionals $\Phi(m)$ can be considered as (generalized) eigenvectors of $\hat{A}$. For this it is sufficient to remind that for $\hat{A}$ self-adjoint, $\psi \in \mathcal{H}$ is its eigenvector with eigenvalue $\lambda$ if and only if $\langle\psi|\hat{A}\varphi\rangle = \lambda\langle\psi|\varphi\rangle$ for every $\varphi \in D_{\hat{A}}$.

If this construction is applied to the orthogonal complement $\hat{E}(\sigma_c(\hat{A}))\mathcal{H}$ in $\mathcal{H}$ of the subspace $\hat{E}(\sigma_p(\hat{A}))\mathcal{H}$ of $\mathcal{H}$ spanned by all eigenvectors of $\hat{A}$, we get an expansion of each $\psi \in \mathcal{H}$ in terms of eigenvectors and generalized eigenvectors of $\hat{A}$.

- We say that a set of unitary operators $\{\hat{U}(t)\}_{t\in R}$ in Hilbert space $\mathcal{H}$ is *a strongly continuous one-parameter group of unitary transformations*, if it satisfies

$$\hat{U}(t)\hat{U}(s) = \hat{U}(t+s) \tag{2.154}$$

for all $s, t \in R$, $\hat{U}(0) = \hat{1}$ and the operator-valued function $\hat{U}(t)$ is continuous in the strong operator topology.

An example of the above notion is provided by the one-parameter group $\hat{U}(t) = \exp(it\hat{A})$, where $\hat{A}$ is self-adjoint. In the domain $D_{\hat{A}}$ of $\hat{A}$ the derivative $d(\hat{U}\psi)/dt$ at $t = 0$ is then defined as the limit $t \to 0$ (in the norm) of $[\hat{U}(t)\psi - \psi]/t$. We say that $\hat{A}$ is the *generator* of $\hat{U}(t)$. Conversely we have the so-called *Stone theorem:*

Any strongly continuous one-parameter group of unitary transformations is generated by some self-adjoint operator.

- In the general case of infinite-dimensional Hilbert space $\mathcal{H}$ with some basis $\{e_\alpha, \alpha = 1, 2, \ldots\}$ the *trace* of an operator $\hat{A}$ is defined by

$$\mathrm{Tr}(\hat{A}) = \sum_\alpha \langle e_\alpha | \hat{A} e_\alpha \rangle \tag{2.155}$$

under the condition that the above expression is finite and does not depend on the choice of the basis. The operator $\hat{A}$ is then called *trace class operator.* The trace class self-adjoint operator $\hat{A}$ has always pure point spectrum only. If $\{\lambda_1, \lambda_2, \ldots\}$ are its eigenvalues, then the trace is equal to the sum

$$\mathrm{Tr}(\hat{A}) = \sum_i \lambda_i \,. \tag{2.156}$$

The spectral decomposition for the trace class self-adjoint operator $\hat{A}$ has the form

$$\hat{A} = \sum_i \lambda_i M_{\lambda_i} \,, \tag{2.157}$$

with all characteristic subspaces $M_{\lambda_i}$ of finite dimension (the trace should be finite). We can then take some orthogonal basis in every characteristic subspace and join them to get an orthonormal basis $\{e_1, e_2, \ldots\}$ of the entire Hilbert space $\mathcal{H}$. Considering every eigenvalue with its multiplicity and denoting the projector on the one-dimensional subspace generated by $e_i$ as $|e_i\rangle\langle e_i|$, we obtain the spectral decomposition of the trace class self-adjoint operator $\hat{A}$ in the form

$$\hat{A} = \sum_i \lambda_i |e_i\rangle\langle e_i| \,. \tag{2.158}$$

If $\hat{A}$ is a trace class operator and $\hat{B}$ is bounded, the following important formula holds

$$\mathrm{Tr}(\hat{A}\hat{B}) = \mathrm{Tr}(\hat{B}\hat{A}) \,. \tag{2.159}$$

- In this book the notation introduced by Dirac is frequently used. To a vector $\psi$ of $\mathcal{H}$ the symbol $|\psi\rangle$ called *ket vector* is associated such that the action of an operator $\hat{A}$ is given by

$$\hat{A}|\psi\rangle = |\hat{A}\psi\rangle \,.$$

The continuous antilinear functional $f_\psi$ on $\mathcal{H}$ corresponding to each $\psi \in \mathcal{H}$ (according to the well-known Riesz lemma) by the equality

$$f_\psi(\varphi) = \langle \psi | \varphi \rangle$$

is then denoted by $\langle \psi |$ and called *bra vector.* This justifies notations already used for the scalar product and projectors. According to the definition of the adjoint operator we can write also

$$\langle \hat{A}^\dagger \psi | = \langle \psi | \hat{A} \,.$$

Also the Gelfand triple is well adapted to the Dirac notation; the elements of $\mathcal{H}_+$ are assumed to be the ket vectors and the elements of $\mathcal{H}_-$ are assumed to be the bra vectors (or vice-versa) according to the definition

$$\langle \psi_+ | \psi_- \rangle = \overline{\langle \psi_- | \psi_+ \rangle}\,.$$

# Chapter 3

# Quantum theory of measurement

Inspite of its abstract mathematical structure quantum mechanics is above all the pragmatic theory which enables to predict the results of the experiments. The weakness of classical physics to explain the behavior of microsystems initialized the revolution in physics at the beginning of this century. The experiments, both "gedanken" or feasible, are playing the primary role in development of quantum mechanics. Quantum measurement represents therefore an extraordinarily important part of quantum theory which links the fundamental phenomena in microworld to those which can be classically detected and recognized according to the rules of classical physics. Of course, this treatment implicitly assumes that such a separation of the Universe to the classical and quantum parts is possible. This point represents a contraverse issue, which is unsatisfactory mainly from the philosophical viewpoint. Moreover, the border between the classical and quantum worlds seems to be fuzzy. Quantum observation theory linking hypothetical quantum features to the classical phenomenon of everydays life is considered from this reason as uncomplete and there are live discussions concerning this topic in existing literarure. In spite of the existence of conceptual problems, any future explanation should take into account the agreement of the pragmatic Copenhagen interpretation of quantum mechanics with existing experiments. In our treatment we will therefore focus on the operational aspects of the quantum measurement taking the existence of quantum and classical worlds as granted by experience. In Sec. 3.1 the foundation of measurement within quantum mechanics will be given, adopting the famous von Neumann (1955) formulation. The fundamental principles will be demonstrated in the following sections. The mathematical consequences of the von Neumann treatment will be established in Sec. 3.2. Here we will specify the states, on which the measurement is performed, and introduce the general formalism for probabilistic description of quantum measurement. Special attention will be payed to the influence of the measurement process on the measured system. The general theory will be applied to the optical processes in Sec. 3.3. The detection of photon number, homodyne and heterodyne measurement will be detailed here explicitly. Special attention will be paid to the problems originating from optics (and from the classical theory of electromagnetic fields in general), impact of which is, nevertheless, much more broader. The concepts of quantum phase including geometric phase will be discussed in Sec. 3.4, linear amplification processes will be dealt with in Sec. 3.5 and the general theory of the back-action-evading measurement will be addressed in the subsequent Sec. 3.6. In the last Sec. 3.7, the developed theory will be applied to the analysis of noises in the Michelson interferometer — a possible high sensitivity device designed for detection of gravitational waves.

## 3.1 Foundation of quantum measurement

Quantum mechanics was built by its inventors as the pragmatic theory and experiments — thought or feasible, triggered its development. This could be demonstrated for example by the Dirac quantum interpretation [Dirac (1927)] of the famous Young experiment of the interference of light on the double-slit aperture. Dirac concludes that the existence of the interferometric pattern should be interpreted as "the interference of a photon with itself". Principle of superposition (or "quantum interference of wave function") belongs to the fundamental rules of quantum mechanics. Formal quantum mechanical description of Young experiment was given by Walls (1977) and the generalization of original Dirac statement may be found in the paper by Paul (1986). The formulation of quantum mechanical predictions and rules influenced strongly also the understanding of measurement process itself. The famous example is given by the Heisenberg inequality (2.25) interpreted in quantum measurement as fundamental uncertainty principle preventing simultaneous observation of observables represented by noncommuting operators. The Heisenberg uncertainty relation represents the mathematical condition for noises associated with noncommuting observables. As quite good measure of noise the root-mean-square of respective operator is assumed here. Let us point out that Heisenberg relations *do not determine* the minimum noise of simultaneous detection of two noncommuting observables. This was recognized later [see for example Arthurs and Kelly (1965), She and Heffner (1966), Caves (1982)] (see Sec. 3.5 in this chapter). This understanding represents the substantial progress in development of quantum measurement. Anyway, Heisenberg relations may be interpreted as fundamental impossibility to measure all the quantities with unlimited precision given only by

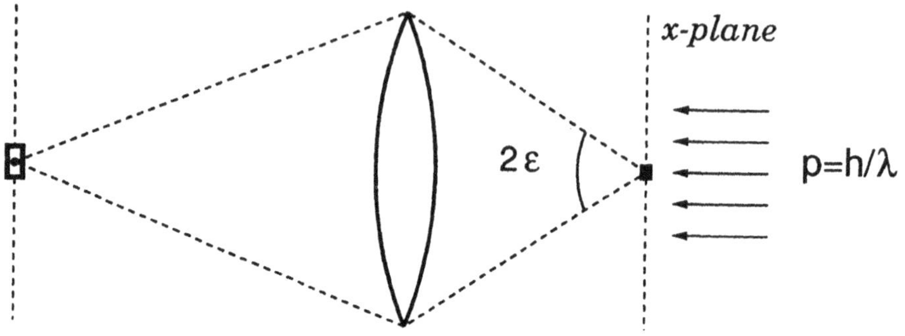


Figure 3.1: Heisenberg microscope.

our technical skill. The coordinate and momentum of free particle represent an ordinary example of noncommuting observables. The role of uncertainty relations may be intuitively demonstrated on the example of the hypothetical devices [von Neumann (1955), de Broglie (1978), Braginsky and Khalili (1992)]. Let us mention explicitly the so-called Heisenberg microscope designated for determination of the coordinate.

The setup of the thought experiment is sketched in Fig. 3.1. The measured object is located in the front of the microscope. The numerical aperture of the microscope is $\sin(\varepsilon)$. The object is illuminated by the monochromatic light with the wavelength $\lambda$. Assuming the ideal aberrationfree optical system, the position of the probe particle may be localized with the accuracy

$$\delta x \geq \frac{\lambda}{2\sin(\varepsilon)}$$

due to the diffraction of the light on the aperture tending to the Airy spot. Process of illumination may be considered as the scattering of the photons on the probe particle, which is assumed at rest at the beginning of the measurement. The conservation of the momentum in the $x$-plane then reads

$$P_x = p' \sin(\alpha)\,,$$

where $P_x$ denotes the $x$-component of the momentum of the particle after the scattering and $p'$ is the momentum of the photon after the scattering. For the purpose of the measurement, the scattered wave should have the same wavelength as the illuminating wave tending to the relations

$$p' \sin(\alpha) \approx p \sin(\alpha) = \frac{h \sin(\alpha)}{\lambda}\,,$$

where $h$ being the Planck constant. Since only the light scattered within the cone $|\alpha| \leq \varepsilon$ can be registered by the microscope, the uncertainty of the perturbation of the momentum may be estimated as

$$\delta P_x \approx \frac{2h\sin(\varepsilon)}{\lambda}.$$

Consequently, both the estimation of position and momentum uncertainties can be joined to the relation

$$\delta P_x \delta x \geq h \tag{3.1}$$

resembling the Heisenberg uncertainty principle (2.25). It is remarkable from the philosophical point of view that the Heisenberg uncertainty appears as a natural consequence of the wave character (duality) of the photon. The wave property was used for estimation of the diffraction Airy spot, whereas particle character was implicitly assumed considering the momentum conservation. Assumptions connected with this semiclassical analysis will be useful for our further explanation, especially for the discussion of simultaneous measurement of noncommuting observables and nondemolition measurements presented in Secs. 3.5 and 3.6.

### Probabilistic interpretataion of quantum mechanics

The comprehensive presentation of measurement within the framework of quantum mechanics was formulated in von Neumann famous book *Mathematical Foundations of Quantum Mechanics* [von Neumann (1955)]. Let us gather shortly his conclusion concerning this topic. Interpretation of measurement involves the probabilistic interpretation of wave function. The possible results of measurement are according to the fundamental principles of quantum mechanics associated with spectrum of self-adjoint operator. The probability of occurrence of detected data is predicted by the square of absolute value of probabilistic amplitude, which is given as the scalar product of wave function and eigenvector of observable corresponding to respective measured eigenvalue. The same is true also for simultaneous measurement of many observables. Von Neumann argues that such measurement has sense only for commuting set of operators of observables yielding common orthogonal complete basis. In the following we will see that this statement naturally includes also the case of approximate simultaneous measurement of noncommuting observables.

### Projection postulate

Von Neumann further analyzed and interpreted the sequence of two measurements of the same physical quantity, when one measurement follows im-

mediately the other. He distinguished three possibilities, which could be hypothetically expected:
i) The measured output could be entirely statistical in the sense that the second measurement, even if taken immediately after the first one, could yield also a dispersion regardless of the value found initially; for example — its dispersion might be equal to the original one.
ii) The first measurement could yield dispersion, but the second immediately subsequent measurement is constrained to reproduce the result of the first one.
iii) The measurement could be determined causally.

The experiment showed that only the statement ii) is consistent with the statistical theory. Such a measurement was performed prior to the formulation of quantum theory by Compton and Simons. Light was scattered by electrons and in their setup energy and momenta were carefully measured for both the electrons and scattered light quanta. The path before collision was known and the measurement confirmed the mechanical laws of collision. Significantly, this confirmation was achieved by two different ways — by capture of the light and of the electrons. These two measurements did not occur simultaneously since light and electrons did not arrive at the same time. In all the cases the first result was recovered by the second subsequent measurement. The statistical nature and validity of the conclusion ii) is now obvious. Even if before the first measurement the result was completely undetermined and only statistical statement could be made, the correlation between first and second measurement is perfectly sharp and causal. The mathematical formulation of this behavior is known as *reduction (collaps) of wave function* in the process of measurement. In the axiomatic formulation of quantum mechanics the term *projection postulate* is also used. Von Neumann denotes this intervention in a quantum system as the process **1**, which can be represented by transformation of statistical operator $\hat{\varrho}$, on which the measurement is performed, to the new state $\hat{\varrho}'$,

$$\hat{\varrho} \rightarrow \hat{\varrho}' = \sum_n \langle\varphi_n|\hat{\varrho}|\varphi_n\rangle \, |\varphi_n\rangle\langle\varphi_n| \, . \tag{3.2}$$

Here summation runs over complete set of orthogonal states $|\varphi_n\rangle$ corresponding to the spectrum of measured observable. As is shown from the von Neumann textbook, this process increases the entropy of the system and characterizes therefore an irreversible process. On the contrary, the second possible intervention in quantum system is called in von Neumann terminology as process **2** and describes the time evolution of the statistical operator

$$\hat{\varrho}(t) = \hat{U}(t)\hat{\varrho}\hat{U}^{-1}(t) \, , \tag{3.3}$$

$\hat{U}$ being the unitary evolution operator. This process is reversible and does not change the entropy of the system.

## Division of the world

The interpretation of quantum measurement needs to distinguish several parts in the whole world relevant to the description of measurement process. In the von Neumann notation system I is a system actually observed, system II is a measuring apparatus and system III is an actual observer. Consequently, the experience then makes only the statements of the type: an observer has made a certain subjective observation; and never any like this: physical quantity has certain value. Moreover, the pragmatic theory requires application of different physical laws in different regions to establish the relation between theory and observed reality. The existence of *classical devices* in the world of observer guarantees that typical quantum efffects, such as interference of macroscopic states, are not observed in everyday life. These interpretational difficulties following from quantum mechanics were recognized in the very beginning of quantum mechanics. The classical states corresponding to interference of two different macroscopic states are called "Schrödinger cats". It is well known that Einstein had doubts about the completeness of quantum theory and stimulated deep discussions about the foundations of quantum mechanics. He expressed his opinion concerning quantum measurement in the analogy: "It turns out that the Moon is there even when no one looks at it." The border between different subsystems is plausible and is given rather by our pragmatic reasons and impossibility to describe the world as one unit in the framework of the only theory valid everywhere, than by another fundamental reasons. Even if the existing quantum mechanics predicts excellently all the results of quantum measurements and up to this time there is no observation of any deviation from this rule, the conceptual weakness of the philosophical foundation of quantum measurement is source of dissatisfaction in the community of physicists. There are therefore attempts to describe the whole process of measurement by means of quantum mechanics within the time evolution, without any additional assumptions about collapse of wave function. An interesting overview is given for example by Zurek (1991), where the necessary application of classical mechanics is explained on the basis of decoherence of wave function. How live and contraverse this topic is the reader can learn from the subsequent discussion [Physics Today (1993)]. There are still many other plausible arguments interpreting the process of irreversible measurement from the quantum mechanical viewpoint as a consequence of the lack of information about the initial state of the whole system describing the measurement. An interesting antropometric justification is presented in the explanation of Braginsky and Khalili (1992). The observer cannot write down even in principle the wave function for himself. To do so, it would be necessary to specify the initial states of all the elementary particles, from which the observer is made. However, the amount of information the observer is able to comprehend is limited by the number of neurons in his brain to the order of about $10^{10}$. This number is smaller compared to many orders than the number of particles which should be taken into account. The averaging over inaccessible information, just like in the classical probabilistic theory, inevitably tends to the irreversible and stochastic behavior of quantum measurement. Of course, the explicit inclusion of a human observer is not the question of principle. The arrangement of the measurement may be done so that the results of registration are collected and are evaluated afterwards without any changes in general argumentation.

These are the difficulties one should face trying to describe also the system III in the notation of von Neumann in the intimately coherent fashion. We will not follow this fundamental approach in our explanation. Instead of this, we will accept the portion of intuition characteristic for pragmatic interpretation. The rules of quantum mechanics, formulated mathematically in Chap. 2, will be applied to the description of measurement process taking into account the quantum nature of measured signal (system I) and apparatus (system II).

## 3.2 Formalism of quantum measurement

Theory of quantum measurement links the events of quantum world with the classically observable effects. Evolution in quantum domain is described by the Schrödinger equation, which is reversible in time and predicts deterministically the evolution of the wave function from the initial to the final states. Measurement process itself obviously does not possess this property. Quantum measurement is characterized by the irreversible changes, such as the creation of silver grain in photoemulsion or emission of a photoelectron from the surface of photodetector. This obvious discrepancy between quantum and classical features raises the question whether or not quantum mechanics is able to give satisfactory description of macroscopic world. We will not address this fundamental problem and instead we will focus our attention on the operational aspects of quantum measurement theory. The classification of possible measurements respective to the quantum fields involved in the measurement process will be more important for further description. This question will be adressed shortly in Sec. 3.2.1. In our explanation we will follow the interpretation raised by von Neumann (1955). In the next Sec. 3.2.2 we will show how the measurement of one and more observables may be naturally extended to involve any physically feasible measurement. This formulation of the so-called generalized measurement was introduced to quantum mechanics by Davies and Lewis (1970) [see also Davies (1976)]. This method was further developed and appplied by Helstrom (1976) and by Yuen (1982). As the main goal, this formulation encompasses the problem of approximate measurement of noncommuting observables and measurements associated with the non-Hermitian operators. In Sec. 3.2.3 we address the question how the back-action of the performed measurement influences the measured quantum system. Explicitly, the von Neumann projection postulate will be reformulated to involve the case of continuous measurement, characterized by destruction of quantum state and by nonunitary evolution of the state in the signal space. The developed method will be further utilized in Sec. 3.6. Here, we will focus on the complementary aspect of the back-action of the measurement. We will clarify the conditions under which the back-action-evading may be achieved, eliminating the influence of the measurement on the measured system. The so-called *quantum nondemolition measurements* also represent, besides verification of quantum mechanical predictions, powerful tools for future high-sensitivity measurements.

### 3.2.1 Quantum systems

Adopting the classical description of the observer III as the classical part of the measurement device, the two broad classes of measurements can be distinguished:
i) The measurement in which the quantum object interacts directly with the classical part of the measurement device will be called *direct measurement*, using the terminology of Braginsky and Khalili (1992). Only the system I should be described according to the laws of quantum mechanics. Apparatus represented by the system II in von Neumann terminology could be joined to the remaining classical part III. An obligatory example of such a process is the registration of a photon (particle) on the screen with photoemulsion. Since the quantum object interacts with the huge (classical) object with many degrees of freedom, the measuring device perturbs strongly the state of the microscopic system. Moreover, the device perturbs usually not only the component which is measured, but often also the components associated with other degrees of freedom. In the case of optical photon, the interaction energy needed for exposition of photosensitive layer is comparable with the total photon energy. Process of detection tends to the destruction of the measured signal field, since the photon is completely absorbed.
ii) The second class is called *indirect measurements* [Braginsky and Khalili (1992)], and, as included in the name, information about the quantum system I is obtained indirectly after the interaction with another quantum field represented by the measurement apparatus II. This additional quantum system is also called *quantum probe* or *auxiliary field.* The measurement consists of two steps. In the first step the signal field, in which measurement we are interested, is interacting with the probe field prepared in a well defined quantum state. The interaction of these two quantum objects is governed by the

Schrödinger equation. In the second step, the classical measurement is performed on the correlated state of signal and probe fields. Knowing the initial state of the probe field and quantum interaction between both the quantum fields, the information about signal may be inferred. Obviously, such measurements are more gentle than direct measurements, since more influences should be kept under control. Explicitly, all the impurities in preparation of initial probe field and deviations of quantum interaction from the assumed model (e.g. dissipations) are affecting the inferred result. In general, to achieve the best strategy of indirect measurement, two independent steps should be well separated, i.e. the quantum interaction between the quantum probe and signal should finish before the classical measurement. A more deeper discussion of this measurement involving semiclassical electric devices yielding the indirect measurement of electric charge on a capacitor may be found in monography by Braginsky and Khalili (1992). We will detail this question in Secs. 3.3 and 3.5 on the examples relevant to quantum optics. In the following we will focus our attention on the mathematical aspects of the measurement.

### 3.2.2 Probability operator measure

The interaction of macroscopic measurement device with a microscopic system introduces irreducible noise. The randomness of the measured output is therefore inevitably attributed to any measurement performed on quantum systems and consequent experimental error cannot be eliminated even if the resolution of apparatus is further increased. In this section we formulate the theory which enables us to describe effectively the probability of occurrence of output data in dependence on conditions of measurement. This description governs both the realistic cases of realization which can happen: number of independent systems of the same kind may be prepared in the same quantum state and then the measurement with certain instrument may be performed on each system; another implementation assumes the repeated measurement on the system only, which is prepared before, and it is in the same conditions during each measurement. An example of these two experimental treatments may be the interference experiment with laser light shining on the screen through an aperture. In the first case the laser beam strong enough is assumed, which creates the diffraction pattern in the real time. Nevertheless the same result will be observed if the light intensity will be extremely low, but the detection time is appropriately longer [Taylor (1909)].

According to quantum mechanics, each state of measured system can be represented by the state vector in the case of pure state or more generally, by the statistical operator on the certain Hilbert space $\mathcal{H}$. The cases distinguished in the previous section as *direct* and *indirect* measurements reflect now in the structure of this Hilbert space. Direct measurement is characterized by choosing $\mathcal{H} = \mathcal{H}_s$, whereas indirect measurement should be described on the tensor product of two independent spaces $\mathcal{H} = \mathcal{H}_s \otimes \mathcal{H}_a$, where $\mathcal{H}_s$ and $\mathcal{H}_a$ being the Hilbert spaces of system and apparatus, respectively. The mathematical description of measurement should take into account the probabilistic interpretation of quantum mechanics, where the probability of finding the system in an arbitrary state is given by the square of absolute value of wave function in the given state. If the probability distribution $p(x)$ of measuring any variable $x$ is to be generated in the similar way, the generalized observable should be associated with a linear operator $\hat{X}(x)$ acting on $\mathcal{H}$ via the relation

$$p(x) = \mathrm{Tr}\left[\hat{\varrho}\hat{X}(x)\right] . \tag{3.4}$$

Here $\hat{\varrho}$ denotes the statistical operator acting on the Hilbert space. Operators defined on the separated Hilbert spaces $\mathcal{H}_s$ or $\mathcal{H}_a$ should be interpreted as acting on $\mathcal{H}$ via tensor product, e.g. $\hat{\varrho} = \hat{\varrho}_a \otimes \hat{1}_s$ in the case of indirect measurement. The properties of probability distribution function reflect now in the conditions putted on $\hat{X}$. Since $p(x)$ should be real non-negative normalized function of the measured variable, the linear operator $\hat{X}$ should be self-adjoint and positively defined,

$$\hat{X}(x) \geq \hat{0} , \tag{3.5}$$

and it should represent also the decomposition of identity operator

$$\int_\Delta \hat{X}(x)\, dx = \hat{1} ; \tag{3.6}$$

integration runs over the range of all possible measured values $\Delta$. When $x$ takes only discrete values, the integral should be reinterpreted as the sum. The linear operator $\hat{X}$ is called probability operator measure (sometimes also positive operator measure). Let us demonstrate that this treatment reasonably spreads the definition of measurement in ordinary quantum mechanics.

Assume the direct measurement associated with the self-adjoint operator $\hat{A}$ on the signal Hilbert space $\mathcal{H}_s$. According to the spectral theorem (2.146), there exists a complete system of orthogonal eigenvectors with real eigenvalues

$$\hat{A}\,|\lambda_i\rangle = \lambda_i|\lambda_i\rangle , \qquad i = 0, 1, 2, \ldots . \tag{3.7}$$

The respective probability operator measure is then given by the orthogonal projectors

$$\begin{aligned} \hat{\Pi}_i &= |\lambda_i\rangle\langle\lambda_i| , \\ \hat{\Pi}_i\hat{\Pi}_j &= \delta_{ij}\hat{\Pi}_i . \end{aligned} \tag{3.8}$$

The probability of finding the value $\lambda_i$ is simply given by the probability amplitude $|\langle\lambda_i|\psi\rangle|^2$, when the measurement was performed in the pure quantum state $|\psi\rangle$. The notion of probability operator measure includes therefore naturally the ordinary axiomatic theory of measurement associated with self-adjoint operators.

Nevertheless, the meaning of probability operator measure becomes non-trivial in the description of indirect measurement, where the respective Hilbert space of states includes additional degrees of freedom associated with the apparatus space $\mathcal{H}_a$. Let us apply the above mentioned formalism to this case. Suppose we measure the observable represented by self-adjoint operator $\hat{M}$ on $\mathcal{H} = \mathcal{H}_s \otimes \mathcal{H}_a$ with orthogonal eigenkets $|\mu\rangle$. The "system plus apparatus state" is now described by the statistical operator $\hat{\varrho} = \hat{\varrho}_s \otimes \hat{\varrho}_a$. The measurement statistics $\langle\mu|\hat{\varrho}|\mu\rangle$ can be written in the form

$$p(\mu) = \mathrm{Tr}_s \left(\hat{\varrho}_s \langle\mu|\hat{\varrho}_a|\mu\rangle\right) , \tag{3.9}$$

where the suffix $s$ in the operation Tr indicates that the traced operator $\hat{\varrho}_s\langle\mu|\hat{\varrho}_a|\mu\rangle$ is formally assumed as acting on the signal Hilbert space $\mathcal{H}_s$ only. We could easily conlude that $\langle\mu|\hat{\varrho}_a|\mu\rangle$ is the positive operator on $\mathcal{H}_s$ yielding the resolution of identity operator

$$\int_\Delta \langle\mu|\hat{\varrho}_a|\mu\rangle \, d\mu = \hat{1}_s \,. \tag{3.10}$$

To verify this explicitly, it is sufficient to decompose the eigenvector $|\mu\rangle$ on the full Hilbert space $\mathcal{H}$ using complete orthogonal basis of respective components of tensor product as

$$|\mu\rangle = \sum_{i,j} \zeta_{ij} |s_i\rangle |a_j\rangle \,,$$

where $|s_i\rangle$, $|a_j\rangle$ are basis vectors in $\mathcal{H}_s$, $\mathcal{H}_a$, respectively. Consequently, the operator

$$\hat{\Pi}(\mu) = \langle\mu|\hat{\varrho}_a|\mu\rangle \tag{3.11}$$

fulfils all the requirements putted on the probability operator measure (3.5) and (3.6). Nevertheless, there is an important mathematical difference between direct and indirect measurements, since the probability operator measure given by (3.11) is not an orthogonal projector satisfying the condition (3.8).

Up to now we have assumed the measurement corresponding to the detection of single variable only, when the measurement is associated with the only self-adjoint operator on $\mathcal{H}$. Nevertheless, this treatment may by simply generalized in the framework of conventional quantum mechanics in the case of simultaneous detection of $m$ *commuting* operators. The previous mathematical description does not need any deep changes. The single observable $\hat{M}$ should be replaced by the set of self-adjoint operators $\hat{M}_k$ commuting on $\mathcal{H}$

$$[\hat{M}_k, \hat{M}_l] = \hat{0} \,, \tag{3.12}$$

$k$ and $l$ being $1, 2, \ldots m$. The measured output is described by the statistics of respective eigenvalues $\mu_k$, what reflects in integrations by formal replacing $d\mu$ by $d^m\{\mu\} \equiv \prod_{i=1}^m d\mu_i$ and $|\mu\rangle$ by $|\{\mu\}\rangle \equiv \prod_{i=1}^m |\mu_i\rangle$. Nevertheless, the restriction of measurement onto signal space has deep consequences, since the operators

$$\hat{\mathcal{M}}_k = \mathrm{Tr}_a \left[\hat{\varrho}_a \hat{M}_k\right] \equiv \int \mu_k \langle\{\mu\}|\hat{\varrho}_a|\{\mu\}\rangle \, d^m\{\mu\} \tag{3.13}$$

obviously do not commute on $\mathcal{H}_s$,

$$[\hat{\mathcal{M}}_k, \hat{\mathcal{M}}_l] \neq \hat{0}\,, \tag{3.14}$$

for $k \neq l$. We see that the probability operator measure describes the feasible measurement of *commuting* operators $\hat{M}_k$ on $\mathcal{H}$ and on the other hand it determines, via spectral decomposition (3.13), the noncommuting self-adjoint operators $\hat{\mathcal{M}}_k$ on $\mathcal{H}_s$. This formalism is therefore sometimes called *generalized quantum measurement* or *approximate simultaneous measurement of noncommuting observables* [Yuen (1982)]. Particularly, for $m = 2$ one can formally construct the operator $\hat{\mathcal{M}} = \hat{\mathcal{M}}_1 + i\hat{\mathcal{M}}_2$ and the generalized measurement may be associated with the measurement of non-Hermitian operator.

Theory of measurement is closed with respect to the notion of probability operator measure, since any quantum measurement may be represented in this framework. Let us further notice that single generalized measurement on $\mathcal{H}$ determines the sequence of ordinary measurements on $\mathcal{H}_s$ only by specification of various apparatus states $\hat{\varrho}_a$.

Moreover, more general statement is valid. It is a remarkable fact that for each measurement described by the probability operator measure $\hat{X}(x)$ fulfilling requirements (3.4)–(3.6) on certain $\mathcal{H}_s$ there exists a Hilbert space $\mathcal{H}_a$, an apparatus state $\hat{\varrho}_a$ and commuting self-adjoint operators on $\mathcal{H} = \mathcal{H}_s \otimes \mathcal{H}_a$ with orthogonal eigenstates $|x\rangle$ such that

$$\mathrm{Tr}_s \left[\hat{\varrho}_s \hat{X}(x)\right] = \langle x|\hat{\varrho}_s \otimes \hat{\varrho}_a|x\rangle\,. \tag{3.15}$$

This statement, representing in fact the inversion respective to the above mentioned treatment, is known as the Naimark extension [Holevo (1982)]. More physically, any generalized measurement may be assumed as measurement of commuting self-adjoint operators acting on some enlarged Hilbert space. Nevertheless, the formal proof guarantees the existence of such operators, but does not represent any prescription how to find physically feasible measurement. It will not be therefore carried out here and may be found in the Holevo book. Let us notice that the Naimark theorem justifies the von Neumann description of quantum measurement. We can therefore conclude that any realistic model of detection has to be based on the measurement of *commuting* operators on enlarged Hilbert space, or — what is effectively equivalent — on the notion of the probability operator measure fulfilling the conditions (3.4)–(3.6) on the signal Hilbert space. And since it is often more convenient to fix the Hilbert space for description of the experiment, the probability operator measure represents the tool for description of quantum measurements.

Let us demonstrate the meaning of the Naimark extension on the detection of noncommuting operators in the following example. The two noncommuting Hermitian operators $\hat{\mathcal{A}}$ and $\hat{\mathcal{B}}$,

$$[\hat{\mathcal{A}}, \hat{\mathcal{B}}] \neq \hat{0}\,, \tag{3.16}$$

acting on the same Hilbert space may be subsequently and independently measured in the following setup: let us assume that the quantum state, in which the measurement is peformed, is well controlled during the whole thought experiment. At first, the observable $\hat{\mathcal{A}}$ can be measured in prepared quantum state with the resulting statistics $p(a)$. The results of measurement are memorized and afterwards the measurement of the observable $\hat{\mathcal{B}}$ is repeated in the same quantum state yielding the statistics $p(b)$. One may assume the statistics

$$p(ab) = p(a)p(b)$$

as a result of *approximate simultaneous measurement of noncommuting observables* $\hat{\mathcal{A}}$, $\hat{\mathcal{B}}$. This statement may be essentially correct providing that we will interpret this experiment as the measurement of independent commuting operators $\hat{A} = \hat{\mathcal{A}} \otimes \hat{1}_B$ and $\hat{B} = \hat{1}_A \otimes \hat{\mathcal{B}}$ acting on the enlarged Hilbert space $\mathcal{H} = \mathcal{H}_A \otimes \mathcal{H}_B$ — a tensor product of two identical Hilbert spaces on which both the operators are defined. We can conclude that all observables can be measured simultaneously, taking into account the portion of the noise inevitably associated with the increased degrees of freedom of the system and apparatus. Such measurement should be interpreted in accordance with von Neumann statement as simultaneous measurement of commuting operators.

### 3.2.3 Back-action of the measurement

We have demonstrated in the previous section that quantum mechanics predicts the result of measurement within the statistical theory. No doubts that such treatment may be applied repeatedly if the subsequent measurements are performed on the system. To clarify this question, it is necessary to specify in which state the system is left after observation. Two qualitatively different classes of quantum measurement can be distinguished for the purpose of our classification. An ordinary example, which could be given, is absorption of a photon by photosensitive surface. The quantum state is destroyed after the observation and is therefore lost for further investigation. Measurements of this kind are traditionally called *measurements of the second kind.* We will show in the following how to describe the time evolution of the system under this kind of measurement. The von Neumann projection postulate concerns the so-called *measurement of the first kind,* which can be characterized as instantaneous observation performed by an ideal device.

Let us assume for simplicity the observation described by a self-adjoint operator $\hat{A}$ with the corresponding resolution of unity

$$\sum_m \hat{\Pi}_m = \hat{1}\,. \tag{3.17}$$

Moreover, let the probability operator measure is given by orthogonal projectors

$$\hat{\Pi}_m \hat{\Pi}_n = \delta_{nm} \hat{\Pi}_n\,. \tag{3.18}$$

Provided that the measurement was performed and the value $a_m$ was found,

$$\hat{A}\hat{\Pi}_m = a_m \hat{\Pi}_m\,,$$

and the initial state described by the statistical operator $\hat{\varrho}$ is projected by the measurement into the final state described by $\hat{\varrho}_f$

$$\hat{\varrho} \rightarrow \hat{\varrho}_f = \frac{\hat{\Pi}_m \hat{\varrho} \hat{\Pi}_m}{\mathrm{Tr}\left(\hat{\varrho}\hat{\Pi}_m\right)}\,. \tag{3.19}$$

Obviously, this mathematical construction reproduces with certainty the measured output $a_m$ if the measurement of observable $\hat{A}$ is subsequently performed on the state described by $\hat{\varrho}_f$.

Von Neumann projection postulate can now be naturally extended in the same sense as in the case of generalized measurement. We need not restrict our description to the projector-valued probability operator measure only, but we can encompass generalized resolutions of unity as well. Let us demonstrate this treatment on the general microscopic model of continuous measurement proposed by Imoto, Ueda and Ogawa (1990). The main idea is that continuous measurement can be accomplished by the simultaneous progress of the system-apparatus coupling and readout of the measuring apparatus. The whole process may be decomposed into the sequence of subsequent infinitesimal processes, described within von Neumann projection postulate and time evolution of coupled system-apparatus. Let us consider that the system and apparatus are coupled via the interaction hamiltonian $\hat{H}_{\mathrm{int}}$. The coupled statistical operator after the interaction then becomes

$$\hat{\varrho}_{s-a}(\Delta t) = \hat{U}\hat{\varrho}_s(0) \otimes \hat{\varrho}_a(0)\hat{U}^{\dagger}\,, \tag{3.20}$$

where $\hat{\varrho}_{s-a,a,s}$ are the statistical operators of coupled system, apparatus and system in respective times; $\hat{U} = \exp\left(-i\hat{H}_{\rm int}\Delta t/\hbar\right)$ denotes the unitary evolution operator of coupled system and apparatus, $\Delta t$ being the duration of interaction. After the time $\Delta t$ the measurement on apparatus is performed. Assuming the measurement of apparatus observable $\hat{X}$ with eigenvalues $X$ and respective projector

$$\hat{\Pi}_X = |X\rangle\langle X|,$$

then the state of the system can easily be specified as

$$\hat{\varrho}_s(\Delta t) = \frac{{\rm Tr}_a\left[\hat{\varrho}_{s-a}(\Delta t)\hat{\Pi}_X\right]}{{\rm Tr}_{s-a}\left[\hat{\varrho}_{s-a}(\Delta t)\hat{\Pi}_X\right]}. \tag{3.21}$$

Symbols ${\rm Tr}_a$ and ${\rm Tr}_{s-a}$ denote traces over the system and over the system and apparatus, respectively. After the apparatus observable is measured, the state of the apparatus is reset to the initial state and the whole measurement process can be repeated for statistical operator of the system (3.21). If we write the reduced statistical operator (3.21) formally using the "superoperator $\hat{M}$" (i.e. operator acting on operators) $\hat{\varrho}_s(\Delta t) = \hat{M}\left[\hat{\varrho}_{s-a}(\Delta t)\right]$, we can characterize compactly the change of the statistical operator of the system as

$$\hat{\varrho}_s(0) \rightarrow \hat{\varrho}_s(\Delta t; X) = \hat{M}\left[\hat{U}\hat{\varrho}_s(0)\otimes\hat{\varrho}_a(0)\hat{U}^\dagger\right], \tag{3.22}$$

$\hat{\varrho}_s(\Delta t; X)$ being the statistical operator of the system in given time and with measurement readout $X$. Evidently, the whole process can be repeated yielding sequence of readout values $X_1, X_2, X_3, \ldots$ for every time interval $\Delta t$. The time evolution of the system statistical operator after the $N$-th step may be described by the recursion relation

$$\hat{\varrho}_s(N\Delta t; X_N) = \hat{M}_N\left\{\hat{U}\hat{\varrho}_s\left[(N-1)\Delta t; X_{N-1}\right]\otimes\hat{\varrho}_a(0)\hat{U}^\dagger\right\}. \tag{3.23}$$

If we make the limit $\Delta t \rightarrow 0$, $N \rightarrow \infty$ and $N\Delta t = t$, the readout becomes a function of $t$ as $X(t)$. The non-unitary time evolution of the system depends in general on the initial state of the system statistical operator $\hat{\varrho}_s(0)$ and on the readouts $X(\tau)$ for $\tau \leq t$.

We can shortly summarize our conclusions concerning formulation of quantum measurement. We have classified the measurements as direct and indirect according to the structure of the respective Hilbert space:

i) The application of the von Neumann treatment on the *direct measurement* represents the ordinary formulation of quantum measurement problem as a measurement associated with the Hermitian operator, since the respective probability operator measure is projection-valued.

ii) The application of the von Neumann treatment on the *indirect measurement*, when the system-apparatus state is represented by the correlated quantum state, tends to the notion of generalized measurement with the probability operator measure on signal space fulfilling the requirements (3.4)–(3.6).

In the following section 3.3, the theory of measurement will be demonstrated on physically relevant examples of optical measurements.

## 3.3 Quantum optical measurements

In this section the formalism of generalized measurement will be applied to problems of quantum optics. Let us remember that the fundamental ideas of quantum mechanics were tested just on the example of electromagnetic field [Dirac (1958)]. The quantum concept of wave-particle duality of light represents important milestone in understanding of nature. From the historical point of view it reflects Newton-Huygens discussions about the corpuscular-wave nature of light. Contemporary optics allows much more to be explained than at the beginning of this century — in the time of foundation of quantum mechanics. Rapid progress has been done above all in the technology of generation and detection of light. Invention of the laser opened new possibilities for fundamental experiments, triggered the development of nonlinear optics and contributed to far-reaching applications which are visible in everyday life. It is therefore important to notice that physical background of all the detection methods of light is connected with photon counting, based on the corpuscular behavior of light. This detection methods were worked up to perfection as we have already learned in Sec. 2.9. The key features of this basic quantum optical measurement will be remembered in Sec. 3.3.1. Combination of more photon counters makes it possible to construct sophisticated detection methods. In Sec. 3.3.2 the detection of quadrature operators (homodyne detection) and measurement of complex amplitude (heterodyne detection) will be addressed. These methods are examples of indirect measurements on a prepared quantum system and have far-reaching consequences in contemporary quantum optics. The measurement of fundamental optical quantities will be completed by the general theory of quantum phase detection presented in Sec. 3.4. In Sec. 3.5 the theory of linear amplifiers will be developed demonstrating explicitly the fundamental enhancing of noise, when one tries to amplify the quantum signal to clasically detectable output. In all the cases of quantum measurement we will concentrate on the principle questions only; the experimental aspects will be addressed in Chap. 8. Let us also notice that the measurement methods discussed in this section are tightly related to the subject of Chap. 5 dealing with the nonclassical phenomena. As we have learned in Sec. 3.1, the pragmatic interpretation of quantum mechanics does not distinguish between measured effects and measurement itself. Since the only purpose of our treatment is to emphasize the effects (and measurements) which have no classicaly analogy, such an interpretation is suitable for our treatment.

### 3.3.1 Photon counting

Photon counting experiments started the fast development of quantum optics triggered by invention of the laser in sixties. The setup of this simplest quantum measurement is following: light beam under investigation falls onto a photosensitive surface (e.g. phototube) which, in accordance with the Einstein theory of induced emission, produces the electrons (the so-called photoelectrons). The duration of exposition $T$ is experimentally controlled for example by a shutter. Suitable electronics in the circuit evaluates the photocurrent, which can be expressed in terms of the total number of registered electrons. This detection is repeated many times keeping the same conditions for measurement and the measured data are recorded in each shot. When the experiment is complete, the obtained data are expressed in the form of photocount probability distribution $p(n,T)$. The mathematical theory can be found in standard textbooks, see for example books by Loudon (1973) and Peřina (1991) and the main results are noticed in Sec. 2.9. For the purpose of our explanation it is enough to remember the expression (2.121) relating the measured photocount distribution to the diagonal elements of the statistical operator

$$p(n) = \langle n|\hat{\varrho}|n\rangle$$

by means of the Bernoulli distribution. Assuming the ideal process of detection with unit efficiency $\eta = 1$ (the experimental possibilities could be really very close to this value), the statistics of electrons reflect the probability distribution of $p(n)$ as

$$p(n,T) = p(n)\,.$$

We can easily interpret this result in the spirit of the language of probability operator measure introduced in equations (3.4)–(3.6). An ideal photon counting measurement represents the direct quantum measurement on signal field associated with the probability operator measure

$$\hat{\Pi}_n = |n\rangle\langle n|\,. \tag{3.24}$$

Here $|n\rangle$, $n$ being integer, is the orthonormal complete Fock basis on signal Hilbert space yielding the resolution of unity

$$\sum_{n=0}^{\infty} |n\rangle\langle n| = \hat{1}\,. \tag{3.25}$$

Obviously, this detection may be interpreted as a measurement of self-adjoint observable of the photon number represented by the operator

$$\hat{n} = \sum_{n=0}^{\infty} n \, \hat{\Pi}_n \, .$$

Photon counting plays a fundamental role in quantum optics, since all the optical quantities should be inferred only from the information obtained by the detection of photons in various setups. A straightforward generalization — the measurement of the correlation functions already mentioned in Sec. 2.9 — is an important example revealing the quantum behavior of light.

Let us follow our general scheme for description of quantum measurement and let us clarify how the performed quantum measurement influences the quantum state of the system. The quantum theory of photodetection, which takes the measurement back action into account was first treated by Mollow (1968). His treatment simplifies the description, and following the von Neumann postulates **1** and **2**, the unitary evolution of the field-detector system is assumed until the photoelectron number is measured. The quantum theory of continuous detection of photon number was initiated by Srinivas and Davies (1981). They postulated the model of state reduction for one-count and no-count processes, which was justified by an explicit quantum model of interaction of light with a two-level atom discussed by Imoto, Ueda and Ogawa (1990). In this model the change of the statistical operator $\hat{\varrho}_1$ after counting one photon is described by the transformation

$$\hat{\varrho}(t) \rightarrow \hat{\varrho}_1 \propto \lambda \hat{a} \hat{\varrho}(t) \hat{a}^{\dagger} \, , \tag{3.26}$$

$\lambda$ being the constant characterizing the magnitude of the coupling between the field and the detector (atom). Assuming that one-count and no-count events form an exclusive exhaustive set of events, the change of statistical operator after no-count process is given by the transformation

$$\hat{\varrho}(t) \rightarrow \hat{\varrho}_0(t+\tau) \propto \exp\left[(-i\omega - \lambda/2)\hat{n}\tau\right] \hat{\varrho}(t) \exp\left[(i\omega - \lambda/2)\hat{n}\tau\right] \, , \tag{3.27}$$

where $\omega$ is the optical frequency of the mode under consideration, $\hat{n} = \hat{a}^{\dagger}\hat{a}$ is the photon-number operator and $\tau$ is the time interval during which no photon was counted. Evidently, this model cannot be described by the application of von Neumann projection postulate directly on the signal space only, since the annihilation operator is not obviously any projector. Photon counting is an example of the measurement of the second kind. The detection cannot be described by instantaneous projections, but rather as a continuous measurement process taking into account the quantum description of detector. The mathematical model was presented in Sec. 3.2.3 as indirect measurement on coupled signal and apparatus fields. Explicitly, Imoto, Ueda and Ogawa (1990) used the quantum model of the detector considering a beam of two level atoms interacting with light inside the cavity. Atoms are prepared in their ground states. Each atom passes through the cavity and interacts with the light inside the cavity during the time interval $\tau$. After passing through the field, each atom is measured, what effectively reduces the information about the quantum state of light. An atom in the upper state corresponds to the one-count process, an atom in the ground state corresponds to the no-count process. The straightforward application of perturbation theory verifies the desired form of the transformation of the statistical operator by no-count and one-count processes. Consequently, any continuous photon counting measurement may be decomposed using these two fundamental events yielding a non-unitary time evolution of field under continuous photodetection process [Imoto, Ueda and Ogawa (1990)]. In other words, continuous measurement of photon number is equivalent to a sequence of infinitesimal processes composed of a unitary evolution via the electric dipole interaction, and the projection onto the atomic level. We see again that the main idea of the von Neumann projection postulate is kept when reduction of wave function is applied on the whole space of coupled system and measuring apparatus.

### 3.3.2 Homodyne and heterodyne detections

The great advantage of quantum optical measurement is that the measurement by means of independent photodetectors may easily be combined yielding information about the electromagnetic field. The fundamental role of photodetection was recognized at the very beginning stage of quantum optics [Kelley and Kleiner (1964)]. The quantum description of detection methods, developed recently in the radiofrequency range, was further considered in connection with investigation of novel sources of nonclassical light (Chap. 8). A complete theory of optical homodyne and heterodyne measurements can be found in the articles by Yuen and Shapiro (1978), Shapiro, Yuen and Machado Mata (1979) and Yuen and Shapiro (1980). Simplified self-contained quantum mechanical calculations of noises were performed by Yuen and Chan (1983) and Schumaker (1984). Further generalization of this ingenious optical method is possible as multiport homodyne detection [Walker and Carroll (1986), Walker (1987a)]. Nevertheless for the purpose of our considerations, we will focus on the principle considerations assuming ideal lossless optical devices and we will relate these measurements to the theory developed in the previous sections. We will demonstrate how the quadrature operator and the complex amplitude may be measured in the framework of such a scheme.

## Homodyne detection

As the first example of more sophisticated measurement setup let us consider model of homodyne detection of the electromagnetic field. The scheme of the measurement is skeched in Fig. 3.2. The signal electromagnetic field is mixed on the beam splitter with the strong coherent monochromatic laser beam. This so-called local oscillator field may be described as a classical wave with stabilized frequency and well controlled phase dependence. Each of the two beam splitter output ports can serve for detection of mean intensity of the quadrature operator of input signal field, since this information is carried by the dominant part of the output intensities.

In the one-port homodyne experiment, the only detector placed in one output port is used to monitor the output intensity. The two-port homodyne scheme monitors the intensities from both output ports of a 50/50 beam splitter and then subtracts the two photodetector outputs. Both the schemes — one-port and two-port homodyning — permit direct observation of input signal quadrature, but the two-port homodyning does not require such rigorous conditions on the local oscillator performance as does the conventional one-port homodyning [ Schumaker (1984)]. Hence, the two port scheme provides a more practical means for observation of squeezing, as is shown in Sec. 5.1 and Chap. 8. Let us describe this quantum measurement following the treatment of Schumaker (1984). Assuming the single-mode electromagnetic field with frequency $\Omega$, the simplified strength operator can be written as

$$\begin{aligned} \hat{E}(x,t) \;\propto\; & \frac{1}{2}\left\{\hat{a}\exp\left[-i\Omega(t-x)\right] + \hat{a}^{\dagger}\exp\left[i\Omega(t-x)\right]\right\} \\ & = \hat{a}_1\cos\left[\Omega(t-x)\right] + \hat{a}_2\sin\left[\Omega(t-x)\right] . \end{aligned} \tag{3.28}$$

Here $\hat{a}_1$, $\hat{a}_2$, $\hat{a} = \hat{a}_1 + i\hat{a}_2$ denote the quadrature operators relevant to the respective annihilation operator of the signal field. Analogous relation for annihilation operator $\hat{b}$ describes the field of the local oscillator. The beam splitter transforms the input modes with the annihilation operators $\hat{a}$, $\hat{b}$ into the two output modes with the annihilation operators $\hat{c}$, $\hat{d}$ as

$$\begin{bmatrix} \hat{c} \\ \hat{d} \end{bmatrix} = \hat{U} \begin{bmatrix} \hat{a} \\ \hat{b} \end{bmatrix} . \tag{3.29}$$

Here $\hat{U}$ denotes $2 \times 2$ matrix

$$\hat{U} = \left(\hat{U}^{\dagger}\right)^{-1} = \begin{bmatrix} \sqrt{1-\mathcal{T}}\exp(i\mu) & \sqrt{\mathcal{T}} \\ \sqrt{\mathcal{T}} & -\sqrt{1-\mathcal{T}}\exp(-i\mu) \end{bmatrix} ,$$

$\mathcal{T}$ being the transmittivity of the beam splitter and $\mu$ being the phase shift between transmitted and reflected waves. Let us introduce some useful abbreviation. The mean values of the annihilation and quadrature operators of

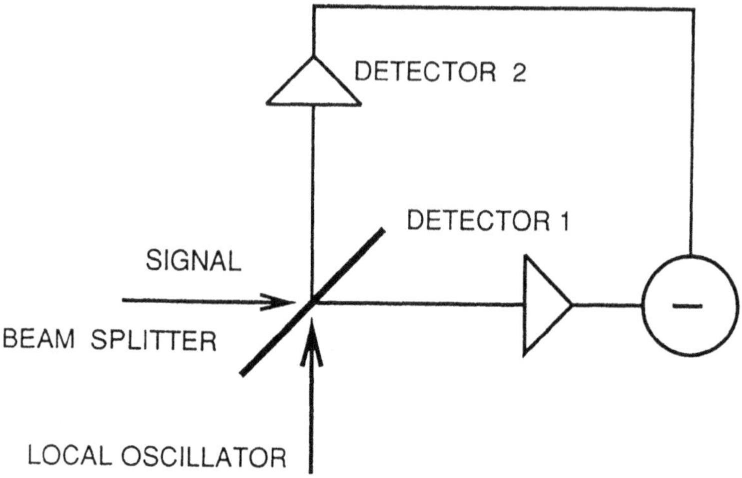


Figure 3.2: Homodyne detection.

the signal field will be denoted as

$$A = A_1 + iA_2 = \langle \hat{a} \rangle = \langle \hat{a}_1 \rangle + i\langle \hat{a}_2 \rangle$$

and analogous notation will be used for the field of the local oscillator $B$, with the real and imaginary parts $B_1$ and $B_2$, respectively. Part of the annihilation operator carrying the noise will be formally denoted as $\Delta\hat{a} \equiv \hat{a} - A$, etc. and in the following we will also use the abbreviation for root-mean square of the operator as $\Delta A \equiv \sqrt{\langle (\Delta\hat{A})^2 \rangle}$, etc. Choosing advantageously the phase of the local oscillator so that the quantity

$$B = \langle \hat{b} \rangle \exp(-i\mu)$$

is real, we can easily specify the detected difference of intensities on both the outputs as

$$\langle \hat{P} \rangle = \langle \hat{c}^\dagger \hat{c} \rangle - \langle \hat{d}^\dagger \hat{d} \rangle = 2BA_1 . \tag{3.30}$$

The noise associated with this measurement is given as

$$(\Delta P)^2 = 4B^2(\Delta a_1)^2 + 4A_1^2(\Delta b_1)^2 + \langle X \rangle , \tag{3.31}$$

where

$$\langle X \rangle = (\Delta a_1)^2(\Delta b_1)^2 + (\Delta a_2)^2(\Delta b_2)^2 - \frac{1}{8}$$
$$+ 2\langle \Delta\hat{a}_1\Delta\hat{a}_2 + \Delta\hat{a}_2\Delta\hat{a}_1 \rangle \langle \Delta\hat{b}_1\Delta\hat{b}_2 + \Delta\hat{b}_2\Delta\hat{b}_1 \rangle .$$

Assuming the strong coherent field of the local oscillator, the dominant terms in (3.30) and (3.31) yield directly information about the quadrature component of the signal field $\hat{a}_1$. Such a measurement may be associated with the resolution of identity operator on the signal Hilbert space

$$\int_{-\infty}^{\infty} dx\, |x\rangle_{1\,1}\langle x| = \hat{1}\,, \tag{3.32}$$

where formally $|x\rangle_1$ denotes the eigenvector of the quadrature operator,

$$\hat{a}_1|x\rangle_1 = x|x\rangle_1\,.$$

The respective probability operator measure

$$\hat{X}(x) = |x\rangle_{1\,1}\langle x| \tag{3.33}$$

represents an orthogonal projector, since the eigenvectors are normalized as $\langle x|x'\rangle = \delta(x - x')$. This measurement of quadrature operator $\hat{a}_1$ determines the shift $\nu$ induced by the canonically conjugated quadrature operator $\hat{a}_2$

$$|\psi(\nu)\rangle = \exp\left(-2i\nu\hat{a}_2\right)|\psi\rangle\,, \tag{3.34}$$

$|\psi\rangle$ being an arbitrary quantum state in which the measurement was performed. This physical interpretation follows from the following considerations: the transformation (3.34) maps the eigenstate of observable $\hat{a}_1$ into another eigenstate

$$|x + \nu\rangle_1 = \exp\left(2i\nu\hat{a}_2\right)|x\rangle_1\,, \tag{3.35}$$

as can easily be verified by application of the commutation rule between quadrature operators $[\hat{a}_1, \hat{a}_2] = i/2$. Consequently, the measurement (3.33) performed on the shifted state (3.34) yields the conditional probability density

$$p(x - \nu) = |_1\langle x - \nu|\psi\rangle|^2 \tag{3.36}$$

of the detection value $x$ provided that the true value of induced shift was $\nu$. Performing for example such a measurement in the coherent state with the complex amplitude $\alpha$, the respective distribution is determined by Gaussian function

$$|\langle\alpha|x - \nu\rangle|^2 = \sqrt{\frac{2}{\pi}}\exp\left(-2[x - \nu - \mathrm{Re}(\alpha)]^2\right)$$

with the root-mean-square deviation $\Delta x = \sqrt{\langle x^2\rangle - \langle x\rangle^2} = 1/4$. Particularly, this is also the noise associated with the special case of vacuum field for $\alpha = 0$. Experimental realization corresponds to the closed signal port. Surprisingly from the classical point of view, such a measurement does not yield minimum noise. Nevertheless this observation is natural in quantum mechanics if one realizes that the Heisenberg relation (2.25) for quadrature operators restricts the product of variances but not variances itself. Reduction of one canonically conjugated observable is then compensated by the increase of the noise of remaining component. Quantum states exhibiting this behavior are known as squeezed states of light (Sec. 4.4). The substantial success of quantum optics is that such purely quantum effect has been already observed, as will be discussed in Chapter 8. Quantum nondemolition aspects of homodyne detection of squeezed light will be dealt with in Sec. 3.6.

The statistics of one quadrature operator represent the maximum amount of information which could be obtained on the basis of single homodyne detection. Of course, such knowledge is not sufficient for recovering the complete information included in the quantum state (or statistical operator). This is obvious since the homodyne detection ignores the canonically conjugated component. Nevertheless, it is a remarkable fact that statistical operator may be characterized by means of homodyne detection of rotated quadrature operator [Vogel and Risken (1989)]. This ingenious method enabled to determine the quasidistribution function of the field from the measured data using optical homodyne tomography. This topic will be addressed in Sec. 5.6.

### Heterodyne detection

For gathering more complete information about signal field, more sophisticated method should be applied and the noncommuting quadrature components should be measured simultaneously. General quantum theory developed in Sec. 3.2, especially the Naimark extension, makes it possible to perform such measurements, provided that the quantum description of apparatus is involved, too. Let us demonstrate this treatment on the heterodyne detection. The setup is very similar to that of homodyne detection (Fig. 3.3) [Yuen and Shapiro (1980), Shapiro (1985), Shapiro and Wagner (1984)]. The signal and auxiliary (sometimes called also as image) fields feeding the same port of a beam splitter are mixed with a field of the local oscillator. On the contrary to the homodyne detection, the frequency of the single mode signal, auxiliary and local oscillator fields, respectively, are different and may be formally represented as

$$\begin{aligned}
\hat{E}_s(x,t) &\propto \frac{1}{2}\left\{\hat{a}\exp\left[-i\Omega_s(t-x)\right] + \hat{a}^\dagger \exp\left[i\Omega_s(t-x)\right]\right\} , \\
\hat{E}_a(x,t) &\propto \frac{1}{2}\left\{\hat{b}\exp\left[-i\Omega_a(t-x)\right] + \hat{b}^\dagger \exp\left[i\Omega_a(t-x)\right]\right\} , \\
\hat{E}_L(x,t) &\propto \frac{1}{2}\left\{\hat{a}_L\exp\left[-i\Omega_L(t-x)\right] + \hat{a}_L^\dagger \exp\left[i\Omega_L(t-x)\right]\right\} , \quad (3.37)
\end{aligned}$$

where the frequencies are $\Omega_s = \Omega_L + \Omega$, $\Omega_a = \Omega_L - \Omega$, $\Omega_L \gg \Omega$ and $\Omega$ is the frequency difference. To simplify notation, the annihilation operators of the

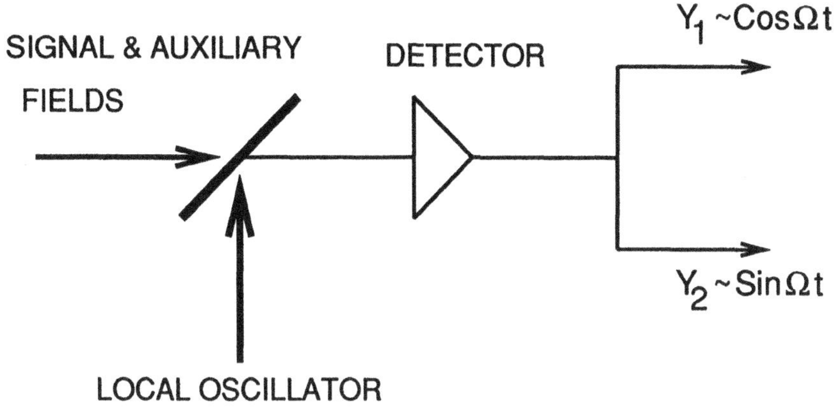


Figure 3.3: Heterodyne detection.

signal, auxiliary and local oscillator fields were denoted $\hat{a}$, $\hat{b}$, $\hat{a}_L$, respectively. A broad band detector placed in one output port of the beam splitter detects beats on the frequency $\Omega$. After demodulation the time-dependent components proportional to $\cos(\Omega t)$ and $\sin(\Omega t)$ can be detected simultaneously yielding the measured variables $Y_1$ and $Y_2$, respectively. Quantum description performed in the analogous way as in the case of homodyne detection yields the quantum counterpart of the classical complex amplitude $Y = Y_1 + iY_2$ as

$$\hat{Y} = \hat{a} \otimes \hat{1}_b + \hat{1}_a \otimes \hat{b}^\dagger . \tag{3.38}$$

Formally, the same quantum measurement may be represented as double homodyne detection [Yuen and Shapiro (1980), Hradil (1992)]. In this setup both the signal and auxiliary fields are not distinguished in the time-domain (by means of different frequencies), but in the spatial domain as fields feeding different input ports. All fields are assumed to have the same frequency. Signal field is mixed with the auxiliary field on the beam splitter and different quadrature components are registered on both the output ports using homodyne detection. This scheme needs obviously four detectors. According to classification of Sec. 3.2, the quantum measurement (3.38) represents an indirect measurement, since the registered quantum state depends on both the signal and apparatus quantum states. Consequently, choice of the apparatus state tends to the particular realization of quantum measurement on the signal Hilbert space. Let us specify the resolution of identity operator generated by the measurement (3.38). For this the method of characteristic function may be advantageously used [Helstrom (1976), Hradil (1993)].

The symmetrically ordered characteristic function of the measurement (3.38) is given as

$$C(y) = \mathrm{Tr}\left[\hat{\varrho}\exp\left(y\hat{Y}^\dagger - y^*\hat{Y}\right)\right] . \tag{3.39}$$

Since the quantum state is specified as the tensor product of two independent states $\hat{\varrho} = \hat{\varrho}_s \otimes \hat{\varrho}_a$, the characteristic function may be factorized as

$$C(y) = C_s(y)\, C_a(-y^*) = \mathrm{Tr}\left[\hat{\varrho}_s \exp\left(y\hat{a}^\dagger - y^*\hat{a}\right)\right] \mathrm{Tr}\left[\hat{\varrho}_a \exp\left(y\hat{b} - y^*\hat{b}^\dagger\right)\right] , \tag{3.40}$$

$C_s$, $C_a$ being the symmetrically ordered characteristic functions of the signal and apparatus fields. We are looking for the probability operator measure acting on the signal Hilbert space only. Therefore we can project the statistical operators $\hat{\varrho}_a$ into the formally same statistical operator $\hat{\varrho}_a^s$ acting on the signal Hilbert space. The equality (3.40) may then be interpreted as the product of characteristic functions defined on the same (signal) space. The resulting field is therefore a superposition of independent fields with Glauber-Sudarshan quasidistribution $\Phi_{\mathcal{N}}$

$$\Phi_{\mathcal{N}}(\xi) = \Phi_{\mathcal{N}_s} * \Phi_{\mathcal{N}_a}^{(*)} = \int d^2\eta\, \Phi_{\mathcal{N}_s}(\eta)\Phi_{\mathcal{N}_a}(\xi^* - \eta^*) , \tag{3.41}$$

where $\Phi_{\mathcal{N}_s}$, $\Phi_{\mathcal{N}_a}$ are quasidistributions of statistical operators $\hat{\varrho}_s$, $\hat{\varrho}_a$, existing in general as ultradistributions [Peřina (1985), Sec. 13.3]. The quasidistribution function $\Phi_{\mathcal{A}}$,

$$\Phi_{\mathcal{A}}(\alpha) = \frac{1}{\pi}\int d^2\xi\, \Phi_{\mathcal{N}}(\xi)\exp\left(-|\xi - \alpha|^2\right) , \tag{3.42}$$

related to antinormal ordering may be, after little algebra, rewritten to the form

$$\Phi_{\mathcal{N}}(\alpha) = \frac{1}{\pi}\mathrm{Tr}\left[\hat{\varrho}_s\hat{D}(\alpha)\hat{\varrho}\hat{D}(-\alpha)\right] . \tag{3.43}$$

The statistical operator $\hat{\varrho} = \int d^2\xi\, \Phi_{\mathcal{N}a}(-\xi^*)\, |\xi\rangle\langle\xi|$ is generated by the statistical operator of the apparatus, but acts on the signal Hilbert space. Since the relation (3.43) holds for each signal input fields, the operator determining the probability of detection of the complex amplitude $\alpha$ is given by the probability operator measure

$$\hat{\Pi}(\alpha) = \frac{1}{\pi}\hat{D}(\alpha)\hat{\varrho}\hat{D}(-\alpha) . \tag{3.44}$$

This is the desired general resolution of unity generated by the measurement process (3.38). The special example is given as the famous resolution of identity in coherent states, eq. (4.29),

$$\hat{1} = \frac{1}{\pi}\int d^2\alpha\,|\alpha\rangle\langle\alpha|, \tag{3.45}$$

if the apparatus state is represented by the vacuum described by $\hat{\varrho}_a = |0\rangle\langle 0|$. Such measurement is called (vacuum) heterodyne detection [Yuen and Shapiro (1980)]. The resolution of unity for the squeezed state may be derived analogously specifying the apparatus state as the squeezed state.

The simultaneous measurement of complex amplitude represents a sophisticated method which is widely used nowadays. Since this quantum measurement is in principle able to determine the quasidistribution $\Phi_{\mathcal{A}}$ of a given quantum state, the statistical operator may be reconstructed yielding the maximum possible information about the quantum state.

## 3.4 Quantum phase

Phase detection represents one of the most accurate schemes of measurement currently available and has played a major role through the history of physics. The measurements realized by Michelson were crucial for the creation of the special theory of relativity and triggered further development. On the other hand, the interference played also an important role in quantum mechanics. Laser interferometry of gravitational waves is expected to change our view on the structure of the Universe. A proper characterization of the quantum phase is therefore essential. This problem is as old as quantum mechanics itself [Dirac (1927)] and there are therefore various concepts that differ substantially. These treatments may be grouped into two broad classes: those based on the definition of phase operator with certain properties [Susskind and Glogower (1964), Carruthers and Nieto (1968), Levy-Leblond (1976), Newton (1980), Pegg and Barnett (1989), Shapiro and Shepard (1991), Hall (1991)] and theories based on the experimentally feasible schemes [Bandilla and Paul (1969), Paul (1974), Caves (1981), Bondurant and Shapiro (1984), Shapiro and Wagner (1984), Walker and Carroll (1986), Yurke, McCall and Klauder (1986), Walker (1987), Braunstein and Caves (1990), Noh, Fougères and Mandel (1992), Hradil (1992), Ban (1992), Freyberger and Schleich (1993), Holland and Burnett (1993), Beck, Smithey and Raymer (1993), Smithey, Beck, Cooper and Raymer (1993)]. An overview with more representative references may be found in a special issue of Physica Scripta, edited by Schleich and Barnett (1993). In the subsequent Sec. 3.4.1 we will generally formulate the problem of quantum phase measurement within the framework of pragmatic interpretation of quantum mechanics developed in the previous sections. Quantum phase detection will be associated with the measurement of two commuting Hermitian operators. Traditional "quantum phase difficulties", ascribed sometimes to the lack of uniqueness of the Hermitian phase operator, are related to the classical attempt to distinguish between quantum effect and measurement itself. Different ideal quantum phase concepts are determined by the reducible representations of an Euclidean algebra on Hermitian operators. Possible interferometric measurements of phase are special realizations of such a treatment. This formulation emphasizes the ideal aspects of quantum phase problem. The realistic aspects of quantum phase experiments, associated with discrete nature of measurement — counting of photons — will be addressed in Sec. 3.4.1. Here the estimation of the counted phase shift will be given. The developed theory will be illustrated by the example of Mach-Zehnder interferometer in Sec. 3.4.2, which represents the simplest possibility how to determine the phase shift. Quantum phase discussed in Secs. 3.4.1 and 3.4.2 should be interpreted as the non-random c-number phase shift of the respective operators in the Heisenberg picture, induced by certain phase shift transformation. Berry (1984) addressed another aspect of phase shift transformation, the so-called problem of geometric phase in quantum processes. We will focus on this problem in Sec. 3.4.3 showing that the geometric phase may be interpreted, analogously to the quantum phase problem, as non-random c-number phase shift of the wave function in the Schrödinger picture. This transformation appears as a consequence of the anholonomy of the system, which is not well isolated from the influences of surrounding environment. Quantum phase concepts will be briefly discussed in Chap. 5 devoted to nonclassical phenomena (Sec. 5.5). Here we treat fundamental aspects of quantum phase measurements.

### 3.4.1 Ideal phase measurement

Let us describe the operator (Heisenberg) picture of the phase detection. Suppose that the general measurement of a pair of observables represented by commuting Hermitian operators $\hat{Y}_1$, $\hat{Y}_2$,

$$[\hat{Y}_1, \hat{Y}_2] = \hat{0}\,, \tag{3.46}$$

is available. This measurement will be called a *feasible measurement* and will be represented in the framework of ordinary quantum mechanics by a probability operator measure [Shapiro and Shepard (1990), Helstrom (1976), Yuen (1982)]. Operators are acting on the full Hilbert space

$$\mathcal{H} = \mathcal{H}_s \otimes \mathcal{H}_a \tag{3.47}$$

created by the direct product of the signal ($\mathcal{H}_s$) and apparatus ($\mathcal{H}_a$) Hilbert spaces. For the sake of brevity, any quantum state is assumed to be defined on the full Hilbert space $\mathcal{H}$. Operators acting on $\mathcal{H}_{s,a}$ spaces are interpreted as acting on $\mathcal{H}$ via tensor product with the identity operator (e.g. $\hat{a}_s \equiv \hat{a}_s \otimes \hat{1}_a$). Assuming that the operator $\hat{Y} \equiv \hat{Y}_1 + i\hat{Y}_2$ has continuous spectrum, the relations of completeness and orthogonality in $\mathcal{H}$ read

$$\int d^2y\,|y\rangle\langle y| = \hat{1}\,,$$

$$\langle y|y'\rangle = \delta^2(y-y')\,. \tag{3.48}$$

Here again for the sake of brevity the notation does not show explicitly the possible degenerations of the eigenvectors $|y\rangle$ and in the following degeneration will be assumed only for the value $y = 0$. The probability distribution function of finding the complex amplitude $y$ by performing the measurement on a general quantum state $|\psi\rangle$ is given by

$$p(y) = |\langle\psi|y\rangle|^2\,. \tag{3.49}$$

The purpose of an arbitrary phase detection is to determine the non-random c-number displacement parameter $\theta \in (-\pi, \pi]$ entering the displacement transformation [Shapiro and Shepard (1991), Helstrom (1976)] of the operators for complex amplitude $\hat{Y}$ ,

$$\hat{Y}(\theta) = \exp\left(-i\theta\hat{N}\right)\hat{Y}\exp\left(i\theta\hat{N}\right)\,, \tag{3.50}$$

$\hat{N}$ being a Hermitian displacement operator and $\theta$ is the true value of the phase shift.

Let us specify the physically required properties of the Hermitian operators $\hat{Y}_1$, $\hat{Y}_2$ and $\hat{N}$. The expectation value of $\hat{Y}$ in the limit of classical field should provide information about $\theta$. For concreteness only models for which classically $\theta = \arg(Y)$ will be addressed. In any case the known classical dependence of $\arg(Y)$ on $\theta$ makes it possible to obtain statistical information about the phase shift via the measurement of the phase-like variable $\phi = \arg(Y)$. In principle, a number of non-equivalent quantum phase models corresponding to the same classical behavior could appear, among which the model with the best resolution should be selected. To assess the accuracy of the available phase data according to mathematical statistics [Bandilla and Paul (1969), Rao (1973), Lukš and Peřinová (1991), Hradil (1992)], the dispersion

$$D^2 = 1 - |\langle\exp(i\phi)\rangle|^2 \tag{3.51}$$

may be used. The phase information is then included in the normalized exponential phase operator

$$\hat{R}(\theta) = \begin{cases} \hat{Y}(\theta)\left[\hat{Y}^\dagger(\theta)\hat{Y}(\theta)\right]^{-1/2} & \text{for } \mathcal{H}_0^\perp \\ \hat{0} & \text{for all states in } \mathcal{H}_0\,. \end{cases} \tag{3.52}$$

Here $\mathcal{H}_0$ is the subspace spanned by the eigenstates with zero complex amplitude — $\hat{Y}(\theta)|y = 0, k\rangle = 0$, $k$ being the quantum number specifying the degeneration — and $\mathcal{H}_0^\perp$ is the orthogonal complement of $\mathcal{H}_0$,

$$\mathcal{H} = \mathcal{H}_0 \oplus \mathcal{H}_0^\perp . \tag{3.53}$$

The prescription (3.52) represents the quantum counterpart of the polar transformation of Cartesian coordinates. There is no specific phase associated with the origin, or alternatively, all values are possible. The detection of zero complex amplitude corresponds to the registration of uniformly distributed noise for the phase variable. This is the only way how to obtain zero expectation value for the exponential phase operator arbitrarily and without respect to a particular model of the detection. The values of cosine and sine, which are equal to zero, cannot be discarded even if they do not bring any information about the phase shift. Such a treatment would tend to renormalization of the quantum state on which the phase measurement was performed [Hradil (1993a)].

The phase detection on $\mathcal{H}$ should be described separately on the orthogonal subspaces in (3.53). The respective probability operator measures are given by marginal phase distributions of the identity operator (3.48). The "phase measurement" on $\mathcal{H}_0$ is described by the positive operator measure

$$\hat{\Pi}_0(\phi) = \frac{1}{2\pi}\hat{1}_0 , \tag{3.54}$$

where $\hat{1}_0 = \sum_k |y = 0, k\rangle\langle k, y = 0|$ denotes the identity operator on $\mathcal{H}_0$. Obviously such measurement does not yield any information about phase shift. The measurement on $\mathcal{H}_0^\perp$ is given by the probability operator measure

$$\hat{\Pi}_\perp(\phi) = \int d^2y\, \delta\left[\phi - \arg(y)\right] |y\rangle\langle y| = \int_{0+}^{\infty} d|y|\, |y|\, ||y|\exp(i\phi)\rangle\langle |y|\exp(i\phi)| , \tag{3.55}$$

which defines an orthogonal projector

$$\hat{\Pi}_\perp(\phi)\,\hat{\Pi}_\perp(\phi') = \delta(\phi - \phi')\hat{\Pi}_\perp(\phi) . \tag{3.56}$$

Let us emphasize that the relation analogous to (3.56) is not correct on the full Hilbert space $\mathcal{H}$ for

$$\hat{\Pi}(\phi) = \hat{\Pi}_0(\phi) + \hat{\Pi}_\perp(\phi) .$$

Significantly, the phase measurement may be described using the spectral decomposition on the Hilbert space $\mathcal{H}_0^\perp$ of states with a meaningful phase.

Here the Hermitian phase operator is defined by

$$\hat{\phi} = \int_{-\pi}^{\pi} d\phi\, \phi\, \hat{\Pi}_\perp(\phi) , \tag{3.57}$$

and an arbitrary function F of the phase operator fulfils the functional identity

$$F(\hat{\phi}) \equiv \hat{F}, \tag{3.58}$$

where

$$\hat{F} = \int_{-\pi}^{\pi} d\phi \, F(\phi)\hat{\Pi}_{\perp}(\phi).$$

To understand the "traditional phase difficulties" we need to formulate the adequate theory on the *signal Hilbert space $\mathcal{H}_s$ only.* Once the quantum state of the apparatus is specified, all the operators may be reduced by performing the partial trace over the apparatus state $\hat{\varrho}_a$. Consequently, the reduced probability operator measure acting only on the signal space $\mathcal{H}_s$ is

$$\hat{\mathcal{P}}(\phi) = \mathrm{Tr}_a \left[\hat{1}_s \otimes \hat{\varrho}_a \hat{\Pi}(\phi)\right]. \tag{3.59}$$

Obviously, this reduced probability operator measure is no longer an orthogonal projector since

$$\hat{\mathcal{P}}(\phi)\hat{\mathcal{P}}(\phi') \neq \delta(\phi - \phi')\hat{\mathcal{P}}(\phi).$$

The Hermitian phase operator on the reduced Hilbert space may be formally introduced as

$$\hat{\phi}_s = \int_{-\pi}^{\pi} d\phi \, \phi \hat{\mathcal{P}}(\phi), \tag{3.60}$$

but the analogue to the identity (3.58) is not valid. This operator cannot predict the "measurable functions" of the phase operator

$$\hat{\mathcal{F}}(\phi) = \int_{-\pi}^{\pi} d\phi \, F(\phi)\hat{\mathcal{P}}(\phi). \tag{3.61}$$

Consequently, the notion of Hermitian phase operator is useful only on the *Hilbert space $\mathcal{H}_0^{\perp}$ of states with meaningful phases.* Here the phase operator can describe well the phase experiments since the respective probability operator measure is the orthogonal projector (3.56). On the *full Hilbert space $\mathcal{H}$* or on the *signal Hilbert space $\mathcal{H}_s$* the measurement should be represented by the respective probability operator measure, whereas the notion of "Hermitian phase operator" is redundant.

Let us develop the quantum phase theory further. The commutator $[\hat{Y}, \hat{N}]$ predicts the sensitivity of the measurement $\hat{Y}$ with respect to the phase shift, $\hat{N}$ being the self-adjoint operator inducing the phase shift. In general, as follows from eqs. (3.51) and (3.52), the dispersion depends on the detected phase shift $\theta$. This dependence may be eliminated by imposing the condition

$$[\hat{Y}, \hat{N}] = \hat{Y}, \tag{3.62}$$

in which case obviously $\hat{Y}(\theta) = \exp(i\theta)\hat{Y}$ and the dispersion (3.51) is independent of the true value of phase shift. The Hermitian operators $\hat{Y}_1$, $\hat{Y}_2$ and $\hat{N}$ fulfilling the commutation relations (3.46) and (3.62) are generators of the Euclidean group E(2) in two dimensional space [Vilenkin (1965), Kim and Noz (1991)]. Measurements satisfying these relations will be called *ideal feasible measurements.* To classify them, let us repeat briefly the theory of unitary irreducible representations of the group of rotations and translations in two dimensions. Every group element $g(\mathbf{a}, \alpha)$ is determined by the displacement vector $\mathbf{a} = (r\cos\phi, r\sin\phi)$ and by the angle of rotation $\alpha$. The representation of this group may be given on the Hilbert space $\mathcal{D}$ of infinitely differentiable functions $f(\mathbf{x})$ defined on the unit circle $\mathbf{x} = (\cos\varphi, \sin\varphi)$. Introducing the notation $f(\mathbf{x}) \equiv f(\varphi)$, a representation of the group E(2) is given by

$$T_R(g)f(\varphi) = \exp\left[iRr\cos(\varphi - \phi)\right] f(\varphi - \alpha)\,, \tag{3.63}$$

R being an arbitrary real number $R \geq 0$. The infinitesimal generators are simply the operators

$$\begin{aligned} \hat{A}_1 &= R\cos(\varphi)\,, \\ \hat{A}_2 &= R\sin(\varphi)\,, \\ \hat{A}_3 &= -\frac{d}{d\varphi}\,, \end{aligned} \tag{3.64}$$

fulfilling the commutation relations

$$[\hat{A}_1, \hat{A}_2] = \hat{0}\,, \qquad [\hat{A}_1, \hat{A}_3] = \hat{A}_2\,, \qquad [\hat{A}_2, \hat{A}_3] = -\hat{A}_1\,.$$

Relation (3.63) gives for $R \neq 0$ the unitary irreducible representation of the group E(2). If $R = 0$, the representation (3.63) has the form of a reducible representation of the rotational group, which is a subgroup of E(2),

$$T_0(g)f(\varphi) = f(\varphi - \alpha) \tag{3.65}$$

and may be further decomposed as the direct sum of one–dimensional unitary irreducible representations

$$T_0 = \sum_{k=-\infty}^{\infty} \oplus T_{0k}\,,$$

$$T_{0k} = \exp(k\alpha)\,, \tag{3.66}$$

where $k$ is an integer $-\infty < k < \infty$. The representations $T_R$ and $T_{0k}$ exhaust all the unitary irreducible representations of E(2).

Let us discuss the physical consequences. Any ideal phase measurement represents the algebra of the group E(2), since the operators $\hat{Y}_1$, $\hat{Y}_2$ and $\hat{N}$

fulfil the commutation relations of the generators $\hat{A}_1$, $\hat{A}_2$ and $\hat{A}_3$, respectively. The operator $\hat{Z} = \hat{Y}^\dagger \hat{Y}$ commutes obviously with all the generators since

$$[\hat{Z}, \hat{N}] = \hat{0} \tag{3.67}$$

and is therefore a Casimir operator. The ideal phase measurement is in general a reducible representation. Let us suppose, without loss of generality, that the operators $\hat{N}$ and $\hat{Z}$ determine a complete system of commuting operators with a common set of eigenvectors $|n, z\rangle$. Every ideal measurement may be represented in this basis on the orthogonal subspaces $\mathcal{H}_z$. Irreducible representations indexed by the parameter $z$ are isomorphic to representations (3.64) with $z = R^2$. Explicitly, choosing the phase of the eigenstates $|n, z\rangle$ through the relation $|n-1, z\rangle = \hat{R}\,|n, z\rangle$, the exponential phase operator $\hat{R}$ on $\mathcal{H}_z$ reads

$$\hat{R} = \begin{cases} \sum_n |n-1, z\rangle\langle z, n| & \text{for } z \neq 0\,, \\ \hat{0} & \text{for } z = 0\,. \end{cases} \tag{3.68}$$

The representation of the rotational subgroup does not contribute to $\hat{R}$ and corresponds to a measurement with indefinite phase. Consequently, the exponential phase operator on $\mathcal{H}$ is expressed as

$$\hat{R} = \int dz \sum_n |n-1, z\rangle\langle z, n|\,. \tag{3.69}$$

This operator form has been proposed by Ban (1991, 1992) for both the continuous and discrete spectrum of the Casimir operator $\hat{Z}$. Explicit calculations establishing the relation between phase concepts with continuous and discrete spectrum of the Casimir operator were performed by Hradil (1993).

Let us discuss shortly the practical aspects of ideal quantum measurements. Realistic schemes for interferometric measurement of a phase shift are intimately connected with the SU(1,1) and SU(2) symmetries [Yurke, McCall and Klauder (1986), Kim and Noz (1991)]. The fact that the group E(2) may be obtained by contractions of these Lie groups [Vilenkin (1965), Kim and Noz (1991)] establishes the connection between the experimentally measurable and the ideal phase concepts. In particular, the multimode homodyne phase shift detection proposed and performed by Noh, Fougères and Mandel (1992) (Sec. 5.5) approximates well an ideal phase measurement even in ordinary experimental conditions. The limiting phase concept may be characterized by the complex amplitude

$$\hat{Y}_{SW} = \hat{a}_s + \hat{a}_a^\dagger \tag{3.70}$$

with the displacement operator $\hat{N} = \hat{a}_s^\dagger \hat{a}_s - \hat{a}_a^\dagger \hat{a}_a$, $\hat{a}_{s,a}$ being the annihilation operators of the signal and auxiliary modes. This phase model represents the simplest realization of the ideal phase measurement and is known as the Shapiro and Wagner (1984) phase concept. As there are obviously no eigenstates of $\hat{Y}_{SW}$ with zero eigenvalue this reducible representation does not include any representation of the rotational subgroup. The exponential phase operator $\hat{R}$ is unitary on $\mathcal{H}$. Let us emphasize that the classification of measurements into ideal and nonideal ones has nothing to do with the ultimate resolution and both classes of models can yield, in principle, the same accuracy. Any phase-sensitive experiment may be considered as a basis for a quantum phase concept. The notion of quantum phase is no longer ambiguous if we impose the requirement of canonical conjugation between the exponential phase operator and the operator of phase shift in the sense of the commutation relation $[\hat{R}, \hat{N}] = \hat{R}$. The only physically relevant meaning of this condition is that the accuracy of the phase measurement is independent of measured phase shift. The simplest model of this kind is the Shapiro-Wagner phase detection which represents the algebra of the group E(2) on the two-mode Hilbert space. The simple theory of feasible phase measurements shows that all the problems traditionally associated with the phase operator appear when one considers the quantum phase by itself without reference to a particular measurement. The present interpretation of the "quantum phase problem" supports the pragmatic quantum mechanical point of view of not distinguishing between the measured effect and the measurement itself. In a forthcoming section we shall analyse the deviation of realistic measurements, based on the discrete nature of measured phase data.

### 3.4.2 Realistic phase measurement

Provided that the phase measurement cannot be described as an ideal phase concept, the phase shift should always be inferred from the performed measurement. Particularly, we will establish general method how to estimate the phase shift from the experimentally measurable quantities. As an explicit example the accuracy of phase measurement using the Mach-Zehnder interferometer will be evaluated. In the realistic quantum phase measurement the information about phase shift is obtained via counting of photons and measured phase parameter is therefore in principle *discrete*. On the other hand, the induced phase shift is a *continuous* parameter. The measured output cannot therefore be straightforwardly interpreted in terms of possible values of the phase shift. The realistic quantum measurement should be therefore completed by the *quantum estimation of phase shift*. The quantum estimation procedure is crucial for phase concepts based on the detection of different-than-phase-variable, where the phase shift is rather inferred than measured. On the other hand, only the ideal quantum phase concepts yield information about the phase shift directly without any quantum estimation.

The treatment is formally analogous to the continuous case mentioned in previous section. Let us point out the main differences only. Suppose that the general measurement of a pair of observables represented by commuting Hermitian operators $\hat{Y}_1$, $\hat{Y}_2$, $[\hat{Y}_1, \hat{Y}_2] = 0$ is available. Assuming the discrete spectrum of the complex-valued operator $\hat{Y} \equiv \hat{Y}_1 + i\hat{Y}_2$, the relations of completeness and orthogonality for eigenstates $|y_m\rangle$ can be formulated analogously to the continuous case. The measured phase-like variable $\varphi_k = \arg(Y)$ can achieve only the discrete values $\varphi_k$ consistent with the exponential operator (3.52). The probability for this to occur can be easily expressed as the sum of all the contributions $p_m$, for which $\arg(y_m) = \varphi_k$. The conditional probability distribution of measuring the value $\varphi_k$ when $\eta$ is true is then given as

$$p_k(\eta) = \sum_{\arg(y_m)=\varphi_k} |\langle y_m| \exp\left(-i\eta\hat{N}\right) \psi\rangle|^2 . \tag{3.71}$$

This distribution is normalized with respect to the parameter $k$ for each phase shift $\eta$, $\sum_k p_k(\eta) = 1$. The Bayes rule may be now used asking what actual value of phase shift $\phi$ could be inferred if the phase $\varphi_k$ was detected. Desired probability distribution is

$$p(\phi|\varphi_k) = \frac{1}{C_k} p_k(\phi) , \tag{3.72}$$

with the normalization $C_k = \int_{-\pi}^{\pi} d\theta\, p_k(\theta)$. Consequently, the conditional probability density of inferring the phase shift $\phi$ when the true value was $\eta$ is given as

$$P_1(\phi|\eta) = \sum_k \frac{1}{C_k} p_k(\phi) p_k(\eta) , \tag{3.73}$$

representing possible estimation of the phase shift. The known dependence of the measured statistics on the phase shift $p_k(\eta)$ fully determines the statistics of the inferred phase shift $P_1(\phi|\eta)$. The inferred probability density is symmetric for the true and inferred phase shifts, but it *is not shift-invariant*, it is dependent only on the difference $\eta - \phi$. The estimation tending to shift-invariant probability distribution was suggested by Noh, Fougères and Mandel (1993). Let us rewrite formally the registered probability as conditional probability in the form

$$p_r(\varphi|\eta) = \sum_k p_k(\eta)\, \delta(\varphi_k - \varphi) . \tag{3.74}$$

The estimation is based on the following interpretation of the registered probabilities (3.74): the inferred value of phase shift is $\phi$ whenever for the true phase shift $\eta + \theta$ the output $\varphi = \phi + \theta$ is registered, $\theta$ being an arbitrary phase within the interval $[0, 2\pi)$. Consequently, the resulting probability distribution of inferred phase shift $\phi$ is given as the sum of all independent contributions

$$P_2(\phi|\eta) = \frac{1}{2\pi}\int_0^{2\pi} d\theta\, p_r(\phi+\theta|\eta+\theta)\,. \tag{3.75}$$

Specifying the explicit form of discrete registered distribution (3.74), we obtain the shift-invariant conditional probability distribution of inferred phase shift as

$$P_2(\phi|\eta) = \frac{1}{2\pi}\sum_k p_k(\eta - \phi + \varphi_k)\,. \tag{3.76}$$

But the phase shift may be estimated also after each trial, provided that quantum description of the performed experiment is known. Let us suppose that the measurement of $\varphi_{k_i}$, $i = 1, 2, \ldots n$, is performed repeatedly $n$ times in the given quantum state under the identical conditions for the same true value of phase shift $\eta$. The measured output $\varphi_{k_i}$ should be interpreted as the estimation of the phase shift with the probability distribution $p(\phi|\varphi_{k_i})$ in accordance with the relation (3.72). This information may be used as the prior probability distribution for the subsequent $(i+1)$-th measurement. Then assuming the uniform prior distribution before the first trial, the probability distribution after the $n$-th measurement is given as the normalized product (likelihood function)

$$P_3(\phi|\eta) \propto \prod_{i=1}^{n} p(\phi|\varphi_{k_i})\,. \tag{3.77}$$

The dependence on the true value of phase shift $\eta$ is hidden in the statistics of measured outputs $\varphi_{k_i}$ and may be shown explicitly for sufficiently large number of trials $n$. Then each value $\varphi_{k_i}$ appears about $np_{k_i}(\eta)$ times in the likelihood function and the distribution (3.77) reads

$$P_3(\phi|\eta) = \frac{1}{C_n(\eta)}\left\{\prod_k [p_k(\phi)]^{p_k(\eta)}\right\}^n, \tag{3.78}$$

where the normalization is

$$C_n(\eta) = \int_{-\pi}^{\pi} d\phi \left\{\prod_k [p_k(\phi)]^{p_k(\eta)}\right\}^n.$$

Suffix $k$ exhausts all the possible values $\varphi_k$ appearing with non-zero probability. The last proposal how to infer the statistics of phase shift from the performed measurement represents the application of the von Neumann reduction postulate [von Neumann (1955)]. The total expenses needed for this measurement are $n$ times larger than in the case of single trial. The resulting conditional probability distribution $P_3(\varphi|\eta)$ is neither shift-invariant nor symmetric. One could therefore ask what is the relation between the distributions

(3.73) and (3.78). This is in general a nontrivial problem. The estimation (3.73) evaluates the content of the phase information respective to the particular measurement and particular quantum state. We therefore intuitively expect that $P_3 \to P_1$ for sharp phase measurement, large $n$ and provided that the same average energy is needed for the total measurement. This behavior will be illustrated on the example of Mach-Zehnder interferometer.

Nevertheless, we conclude that the discrete variable cannot be directly used as a good estimator of continuous phase shift. However, a more general statement is also valid. Detection of any phase sensitive (continuous or discrete) variable may in principle be used for evaluation of the phase shift. The argumentation and the derived probability distributions (3.73) and (3.78) will not change, if we disregard the phase-like character of measured output assuming that $k$ represents a general phase sensitive variable. The phase shift is always partially measured and partially inferred [Holland and Burnett (1993), Beck, Smithey and Raymer (1993), Smithey, Beck, Cooper and Raymer (1993)] even in the case of classical interferometry. A phase measurement represents one of the most accurate detection techniques currently available and the investigation of involved noises deserves therefore respective attention.

### 3.4.3 Phase detection in Mach-Zehnder interferometer

Let us apply our theory to the description of SU(2) interferometer [Yurke, McCall and Klauder (1986)]. Similar quantum detection of the phase shift was also presented by Holland and Burnett (1993). The ordinary measurement set-up is depicted in Fig. 3.4. Let us restrict ourselves to the case of single mode quasimonochromatic fields on each port for simplicity.

Within the domain of classical optics, the input fields with complex amplitudes $V_1$, $V_2$ are combined on the 50/50 ideal beam splitter and the phase shift $\eta$ is induced in arms of the interferometer. Assuming the closed auxiliary input port 2, then $V_2 = 0$, the integrated intensities on output ports $W_3$, $W_4$ may be easily specified as $W_3 = W_1 \cos^2(\eta/2)$, $W_4 = W_1 \sin^2(\eta/2)$, where $W_1$ is the integrated intensity of the input field introduced in Sec. 2.9.6. In the classical interferometry the average value of energy is registered on each output ports 3 and 4 in dependence on the phase shift. The phase difference $\eta$ is then inferred from this interferometric pattern. The accuracy is given in semiclassical approximation by fluctuations of the photon number [Caves (1981), Bondurant and Shapiro (1984), Yurke, McCall and Klauder (1986)] (compare with the photon-counting error discussed in Sec. 3.7). Quantum

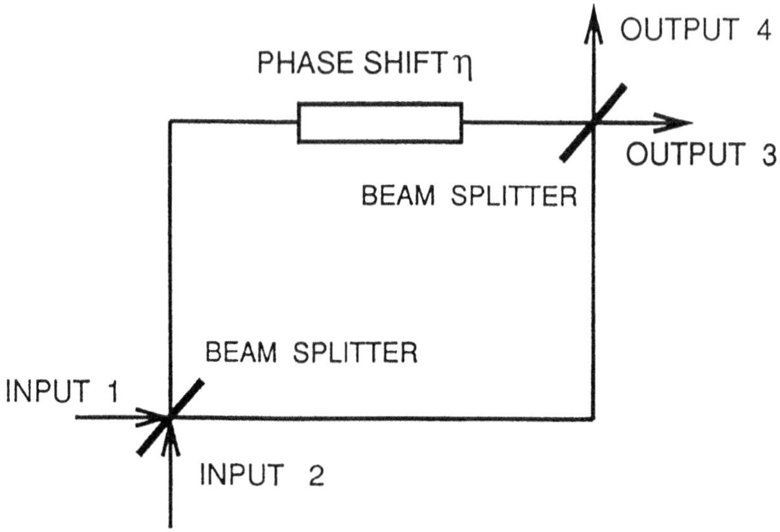


**Figure 3.4: Setup of Mach-Zehnder interferometer.**

counterpart may easily be established assuming the classical relation

$$\tan\left(\frac{\eta}{2}\right) = \sqrt{\frac{W_4}{W_3}}. \tag{3.79}$$

Desired phase shift can be deduced from the phase of the complex amplitude $Y_{MZ} = W_3 + iW_4$. Assuming canonical quantization and ideal process of photodetection, quantum model may be obtained by formal replacing complex amplitudes and integrated intensities by annihilation operators $\hat{a}_i$ and photon number operators $\hat{N}_i$, respectively, $i = 1, 2, 3, 4$, provided that normal operator ordering is adopted. Quantum analogy of the closed auxiliary port 2 is then expressed by the condition $\langle \hat{a}_2 \rangle = 0$. The complex amplitude $Y_{MZ}$ may be quantized introducing advantageously the generators of the SU(2) algebra [Yurke, McCall and Klauder (1986)]. Let us define $\hat{J}_1 = (1/2)(\hat{a}_1^\dagger \hat{a}_2 + \hat{a}_1 \hat{a}_2^\dagger)$, $\hat{J}_2 = (1/2i)(\hat{a}_1^\dagger \hat{a}_2 - \hat{a}_1 \hat{a}_2^\dagger)$, $\hat{J}_3 = (1/2)(\hat{n}_1 - \hat{n}_2)$, and the Casimir operator is $\hat{J}^2 = \hat{J}_1^2 + \hat{J}_2^2 + \hat{J}_3^2 = \hat{C}(\hat{C}+1)$, where $\hat{C} = (1/2)(\hat{n}_1 + \hat{n}_2)$. Then the measured complex amplitude is

$$\hat{Y}_{MZ} = \hat{N}_3 + i\hat{N}_4 = (1+i)\hat{C} + (1-i)\exp\left(-i\eta\hat{J}_2\right)\hat{J}_3\exp\left(i\eta\hat{J}_2\right) \tag{3.80}$$

and $\hat{J}_2$ acts then as the displacement operator $\hat{N}$. The measured quantum phase-like variable $\varphi$ tending to the expression (3.79) in the classical limit may be defined as

$$\tan^2\left(\frac{\varphi}{2}\right) = \tan\left[\arg(Y_{MZ})\right] . \tag{3.81}$$

The sine and cosine of the phase are given using the measured photoelectron counts $N_3$, $N_4$ as

$$\begin{aligned} \sin(\varphi) &= \pm\frac{2\sqrt{N_3 N_4}}{N_3+N_4}, \\ \cos(\varphi) &= \frac{N_3-N_4}{N_3+N_4}. \end{aligned} \tag{3.82}$$

Obviously, this detection cannot distinguish between the signs of $\varphi$.

## Phase resolution of classical interferometry

Let us specify for concreteness the quantum measurement in the state with closed auxiliary input port 2 and $N$ photons on the input port 1, $|\psi\rangle_{\rm in} = |N\rangle_1 \otimes |0\rangle_2$. The joint probability distribution of counting $N_3 = N - k$ and $N_4 = k$ photons is given by binomial distribution

$$\mathcal{P}(N_3 = N-k, N_4 = k) = \binom{N}{k}\left[\cos\left(\frac{\eta}{2}\right)\right]^{2(N-k)}\left[\sin\left(\frac{\eta}{2}\right)\right]^{2k} . \tag{3.83}$$

Detection of photons $k = 1, 2, \ldots, N$ on the port 4 should be interpreted as registration of phases $\varphi_k$ and $\varphi_{-k} = -\varphi_k$,

$$\varphi_k = 2\arctan\sqrt{\frac{k}{N-k}} . \tag{3.84}$$

Phases with different signs are detected with equal probabilities $p_k(\eta) = p_{-k}(\eta) = (1/2)\mathcal{P}(N-k, k)$. Zero phase variable $\varphi_0 = 0$ for $k = 0$ is detected with the conditional probability $p_k(\eta) = \mathcal{P}(N, 0)$. Consequently, the sharpest distribution appears when the true phase shift is $\eta = 0$, since only the value $\varphi_k = 0$ is registered, $p_k(\eta = 0) = \delta_{k0}$. On the contrary the broadest distribution appears at $\eta = \pi/2$.

Let us specify explicitly the case of the sharpest phase measurement. The inferred conditional probability distribution (3.73) reduces to the one non-zero term yielding

$$P_1(\phi|\eta = 0) = \frac{(2N)!!}{2\pi\,(2N-1)!!}\left[\cos\left(\frac{\phi}{2}\right)\right]^{2N} . \tag{3.85}$$

Dispersion of the inferred phase shift is simply

$$D^2_{MZ} = 1 - \left|\int_{-\pi}^{\pi} d\phi \, \exp(i\phi) P(\phi|\eta = 0)\right|^2 = \frac{2N+1}{(N+1)^2} \approx \frac{2}{N}, \tag{3.86}$$

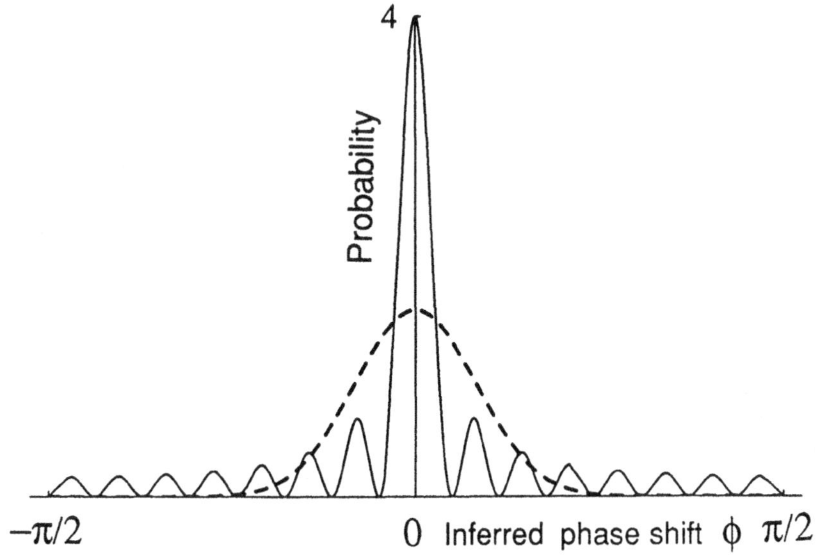


Figure 3.5: The inferred probability distributions $P_1(\phi|\eta = 0) \equiv P_3(\phi|\eta = 0)$ normalized on the interval $[-\pi, \pi)$ (dashed line) for classical interferometry; $N = 30$; and $P_1(\phi|\eta = 0)$ normalized on the interval $[-\pi/2, \pi/2)$ suggested by Holland and Burnett for enhancing resolution (full line); $N_1 = N_2 = 15$.

in accordance with the accuracy of the classical interferometry [Yurke, McCall, Klauder (1986)]. One can also easily verify that the estimation (3.78) tends to the same probability distribution in this special case. Of course, the number of particles in each trial should be renormalized to $N/n$ to compare the performance with the same total energy $N$. The probability distribution of inferred phase shift is depicted in Fig. 3.5. The estimation (3.76) is not suitable in this measurement since the detected phase data are not phase-shift invariant. This is obvious from comparison of detected distributions for various true values of phase shift; for example the distribution $p_k(\eta = 0)$ yields one peak, whereas $p_k(\eta = \pi/2)$ exhibits two peaks.

In our example the calculations were performed for the Fock state as the input state. Nevertheless one can easily conclude that there will be apparently no significant differences for the coherent state on the open input port. Obviously, the coherent states in the input ports of Mach-Zehnder interferometer will be transformed into some coherent output states on the output ports. As is well known from mathematical statistics, the Poissonian distribution as the photocount distribution of coherent states represents the asymptotical limit of the binomial distribution (3.83) for $N \to \infty$ and one can expect comparable results.

### Enhanced phase resolution

A potential improvement of the sensitivity of the Mach-Zehnder interferometer was suggested recently by Holland and Burnett (1993). In their setup both the input ports of the inteferometer in Fig. 3.4 are fed by the field in the Fock states with the same photon number N, $|\psi_{\rm in}\rangle = |N\rangle_1 \otimes |N\rangle_2$. One can easily express the probability of detection of $N-q$ photons on the port 3 and $N+q$ photons on the port 4 as

$$\mathcal{P}(N_3 = N-q, N_4 = N+q|\eta) = |\langle\psi_{\rm in}|N-q\rangle_3\,|N+q\rangle_4|^2 \tag{3.87}$$

$$= \frac{1}{N!}\sum_{l=0}^{N-q}\binom{N}{l+q}\binom{N}{l}(-1)^{l+q}\sqrt{(N-q)!(N+q)!}$$
$$\times\left[\cos\left(\frac{\eta}{2}\right)\right]^{2N-(2l+q)}\left[\sin\left(\frac{\eta}{2}\right)\right]^{2l+q} \tag{3.88}$$

$$= \frac{(N-q)!}{(N+q)!}\left[P_N^q(\cos\eta)\right]^2 . \tag{3.89}$$

Here $P_N^q$ denotes the associated Legendre polynomials [Vilenkin (1965)] and $q$ being an integer. It can be shown using the symmetry of associated polynomials that the photocount distribution (3.89) yields the same result for phase shifts $\pm\eta$ and $\pi\pm\eta$. The measurement is therefore not able to distinguish among these four possibilities on the interval $[0,2\pi)$. To achieve meaningful resolution, the prior knowledge about the initial distribution of phase shift should be restricted to the interval $[0,\pi)$. Provided that this condition is fulfilled, the sharpest measurement occurs just like in the case of classical interferometry for $\eta = 0$, when only $q = 0$ is detected. Let us estimate the phase information which one can learn from such a detection. If $q \ll N$ the associated polynomials may be asymptotically replaced by Bessel functions [Gradshteyn and Ryzhik (1965)] as

$$P_N^q(\cos\phi) \approx J_q(N\phi) .$$

Consequently, the inferred probability distribution in the most favourable case reads, in accordance with the relation (3.73),

$$P_1(\phi|\eta = 0) \propto J_0^2(N\phi) . \tag{3.90}$$

This distribution, shown in Fig. 3.5 in comparison with the result of the classical interferometry, seems to be concentrated in the region restricted by the first zero point. Consequently, one could expect the uncertainty of phase shift of the order $D \approx 1/N$. Nevertheless, such a conclusion is incorrect. Bessel functions are not power–square integrable, $\int_0^\infty dx\, J_0^2(x) = \infty$ and the probability included in the central peak cannot be considered as dominant contribution. More careful treatment of dipersion is therefore necessary. Denoting

$$Z_1 \equiv \int_{-\pi/2}^{\pi/2} d\phi\, J_0^2(N\phi) = \frac{2}{N}\int_0^{N\pi/2} dx\, J_0^2(x)\,,$$
$$Z_2 \equiv \int_{-\pi/2}^{\pi/2} d\phi\, J_0^2(N\phi)\cos(\phi) \approx \frac{2}{N}\int_0^{N\pi/2} dx\, J_0^2(x)\left[1-\frac{x^2}{2N^2}\right],$$

the disperison may be specified as

$$D_{HB}^2 = 1-\left(\frac{Z_2}{Z_1}\right)^2 \approx \frac{1}{N^2}\frac{\int_0^{N\pi/2} dx\, J_0^2(x)x^2}{\int_0^{N\pi/2} dx\, J_0^2(x)}.$$

Assuming now the decomposition of involved integrals as

$$\int_0^{N\pi/2} dx\, J_0^2(x) = \int_0^K dx\, J_0^2(x) + \int_K^{N\pi/2} dx\, J_{0,as}^2(x)$$

and the asymptotical behavior of Bessel function as

$$J_{0,as}^2(x) \approx \frac{1+\cos(2x)}{\pi x},$$

the dispersion may be roughly estimated as

$$D_{HB}^2 \propto \frac{1}{\ln(N)}.$$

Consequently, the resolution is slightly worse than in case of ordinary classical measurement discussed in the previous section.

Nevertheless there is a way how to beat the limit of classical interferometry $D \propto 1/\sqrt{N}$. This famous technique, discovered by Caves (1981), was suggested as a method how to increase the sensitivity of the Michelson interferometer suited for detection of gravitational waves. We will discuss this model separately in Sec. 3.7. The method is known as *squeezed state interferometry* and the resolution

$$D^2 \propto \frac{1}{N^2} \tag{3.91}$$

is achieved by the interference of squeezed states. The quantum phase aspects of this nonclassical measurement were investigated for example by Shapiro and Wagner (1984), Bondurant and Shapiro (1984) and Yurke, McCall and Klauder (1986).

It is a remarkable fact that the distribution predicted by Holland and Burnett (1993) yields very narrow central peak, which is embedded in the noisy background. One can therefore ask whether it is possible to beat the limit of squeezed state interferometry by some more sophisticated detection method, which would be sensitive to such a peak only. Unfortunately, the answer is not known presently, but there are doubts about this possibility. Existing suggestions how to do it [Shapiro, Shepard and Wong (1989), Shapiro and Shepard (1991), Yuen (1992)] have not yet been confirmed by current investigations [Hradil and Shapiro (1992), Braunstein, Lane and Caves (1992), Lane, Braunstein and Caves (1993), Braunstein (1992)].

### 3.4.4 Geometric phase

In the previous section we dealt with the phase shift transformation (3.50) which, in the ideal case, performs the c-number phase shift transformation of the quantum complex amplitude as

$$\hat{Y} \to \exp(i\theta)\hat{Y}\,.$$

In the corresponding Schrödinger picture the same mapping may be described as the transformation of the input quantum state in which the measurement was performed in the following way

$$|\psi\rangle_{\rm in} \to \exp\left(-i\theta\hat{N}\right)|\psi\rangle_{\rm in}\,, \tag{3.92}$$

$\theta$ being the induced phase shift and $\hat{N}$ the respective displacement operator. As we have demonstrated such phase shift may be induced and subsequently measured (estimated) by means of various interferometric methods and detection techniques.

Nevertheless, the phase shift transformation (3.92) is not the only transformation one can imagine. Another one has already been noticed in the formulation of fundamental principles of quantum mechanics in Chap. 2. The orthonormal basis vectors in the Hilbert space are determined up to an arbitrary phase factor and consequently, the quantum states $|\psi\rangle$ and $\exp(-i\theta)|\psi\rangle$ are therefore in standard textbooks of quantum mechanics considered as equivalent. Peculiarity of the phase transformation

$$|\psi\rangle \to \exp(-i\theta)|\psi\rangle \tag{3.93}$$

was formulated by Berry (1984). He discovered that in the quantum adiabatic theorem there exists an extra phase, in addition to the usual dynamical phase, which appears after the system has gone through closed circuit in the parameter space of the hamiltonian. Consequently, the quantum state is transformed according to the transformation (3.93), where the phase $\theta$ depends on the trajectory in the parameter space. Since the explanation is given using topological arguments, the induced phase (3.93) is called as *geometric phase* (also Berry or Berry-Pancharatnam phase), similarly to the phase change in the Bohm-Aharonov effect.

The essential assumptions for appearing the geometric phase is that the evolution of the hamiltonian should be cyclic and adiabatic. The feature of this effect is very general and appears in all fields of physics [Berry (1990)] — its precursors may be found in fields of polarized light, radio waves, molecules and curved surfaces. Nevertheless, the most famous example represents the anholonomy resulting from parallel transport in classical mechanics. The constraint is called holonomic if it can be integrated yielding the conservation law, which reduces number of degrees of freedom. Otherwise it is anholonomic as a consequence of its non-integrability. A physical example of the mechanical effect equivalent to the geometric phase is the Foucault pendulum. The direction of swing is slaved to parallel transport, it means that the direction of the swing has no component of angular velocity along the vertical axes connecting the observer with the center of Earth. Hence after a day the swing direction does not return to the original one. The anholonomy is the angle between the initial and final directions and is equal to $2\pi\cos(\lambda)$, where $\lambda$ is the latitude of the location of the Foucalt pendulum. In the language of geometry, the anholonomy is equal to the solid angle corresponding to the closed trajectory $C$ of the pendulum — to the diurnal circuit with the latitude $\lambda$.

Berry formulated the change of the geometric phase based on the example of particles with the spin in adiabatically changing magnetic field $\mathbf{B}(t)$. As the magnetic field adiabatically changes its direction keeping the magnitude constant, the energy $E$ of the system remains constant. The vector $\mathbf{B}(t)$ traces out the closed circuit C in the configurational space $(B_x, B_y, B_z)$. After one round trip, the system must come back to its original state. Apart from the ordinary dynamical phase shift $\exp(-iEt/\hbar)$ also the geometrical phase factor $\exp[i\gamma(C)]$ appears, $\gamma(C)$ being the Berry's (geometric) phase. Berry showed that

$$\gamma(C) = -m_s\Omega(C)\,, \tag{3.94}$$

where $m_s$ is the projection of the spin to the direction of $\mathbf{B}(t)$ and $\Omega(C)$ represents the solid angle corresponding to surface closed by the trajectory $C$ in the parameter space $(B_x, B_y, B_z)$.

The optical proposal was formulated by Chiao and Wu (1986), the measurement was performed by Tomita nad Chiao (1986). They extended the Berry's treatment to the photon, which is massless spin-1 boson. The helicity is defined as the projection of the spin $\mathbf{s}$ onto the direction of propagation,

$\mathbf{s} \cdot \mathbf{k}$ and it can only be $\pm 1$. The adiabatical switching of the magnetic field, to which the photon is not coupled, is replaced by the constraint on the direction of propagation of the photon with wave vector $(k_x, k_y, k_z)$. Since the helicity remains $+1$ or $-1$, the spin of the photon should follow the direction of propagation. Slow changes in the environment affecting the propagation of the photon will effectively yield the same effect of geometric phase as in the Berry's proposal. Moreover, there is another advantage in this optical realization. The state with helicity $\pm 1$ corresponds in the classical electrodynamics to two orthogonal circularly polarized orientations of the electromagnetic wave. Polarization may be therefore interpreted as a "spin effect" which originates at the quantum level, but survives the correspondence principle $\hbar \to 0$ up to the classical level described by the Maxwell equations. As the photon propagates smoothly down a helical waveguide, $\mathbf{k}$ is constrained to remain parallel to the local axis of the waveguide and also the direction $\mathbf{s}$ is constrained to follow this direction. Hence tracing out a loop $C$ on the surface of a sphere in the parameter space $(k_x, k_y, k_z)$, the Berry's argument leads to the appearing of geometric phase

$$\gamma(C) = -\sigma\Omega(C)\,, \tag{3.95}$$

where $\Omega(C)$ is the solid angle substended by the loop $C$ with respect to the origin $\mathbf{k} = \mathbf{0}$ and $\sigma = \pm 1$ is the helicity. Especially for uniform helix it holds that

$$\gamma(C) = 2\pi N[1 - \cos(\Theta)]\,, \tag{3.96}$$

where $N$ is the winding number of the helix and $\Theta$ is the angle between local waveguide and the axis of the helix.

A possible way to observe this geometric phase of photon is to split the beam of photons with the given circular polarization into two coherent beams and send them along two oppositely coiled fibres. The induced geometric phase shift may then be observed by means of interferometric pattern. In the experiment [Chiao and Wu (1986), Tomita and Chiao (1986)] another setup was used. The linearly polarized light

$$|x\rangle = \frac{1}{\sqrt{2}}[|\sigma = +1\rangle + |\sigma = -1\rangle] \tag{3.97}$$

was sent through the single helically wound optical fiber. The effect of the geometrical phase (3.95) rotates the polarization plane of linearly polarized light as

$$|x\rangle \to |x'\rangle = \frac{1}{\sqrt{2}}[\exp(i\gamma_+)|\sigma = +1\rangle + \exp(-i\gamma_+)|\sigma = -1\rangle]\,. \tag{3.98}$$

The geometric phase shift was observed in many different optical realizations using laser interferometry [Bhandari and Samuel (1988)], non-planar

Mach-Zehnder interferometer [Chiao, Antaramian, Ganga, Jiao, Wilkinson and Nathel (1988)] or its dynamical manifestation as a frequency shift [Simon, Kimble and Sudarshan (1988)]. The effect was significantly demonstrated also on the quantum level as nonclassical Berry's phase of one photon [Kwiat and Chiao (1991)]. Quantum "non-local" character of the Berry-Pancharatnam phase was experimentally shown by Grayson, Torgerson and Barbosa (1994).

## 3.5 Linear amplifiers

In this section we will describe the linear amplifier as a general device which increases the signal without significant degrading the signal-to-noise ratio. The noise after amplification is much larger than minimum noise permitted by quantum mechanics. The amplified signal may be therefore considered as the classical field and further measurement performed on it does not change its state, on the contrary to the conclusion of quantum mechanics. Amplifiers are able to relate the quantum effects to the classical phenomena. In this treatment we will restrict ourselves to the linear theory only where the output fields are linear combinations of input fields. The crucial question is how much the quantum signal is influenced by the process of amplification. The development of maser permitted to use amplifiers that were much quieter than other contemporary devices. The analysis of noises showed that amplification decreases the noise of any signal. This is valid for all linear amplifiers and the theory found various applications in detection of gravitational waves, theory of nondemolition measurement and in design of amplifiers based on the Josephson junction (the so-called SQUID) with sensitivity closed to quantum mechanical limit. Detection of nonclassical light and especially of nearly one-photon states (see Chap. 8) is another demonstration of successful controlling quantum noises in amplification of weak signals.

Motivation for more detailed investigation of quantum mechanical noises may be found also among the old problems of quantum mechanics. In Sec. 3.1 the Heisenberg microscope was characterized as a device, which determines the position of free particle and consequently disturbs its impulse. The intuitive treatment tends to qualitative estimation that the uncertainty product for both the position and impulse should be in accordance with the Heisenberg uncertainty principle. One can therefore ask, what is the limitation of total noises of canonically conjugated observables, which can be actually achieved. The important conclusion of general theory of linear amplifiers is the prediction of the total noise of simultaneous measurement of noncommuting observables. The first quantum mechanical analysis of this problem was given by Arthurs and Kelly (1965). Let us start therefore our considerations with a brief review of their main results using an explicit model of linear interaction between the measured signal field having the operators of position $\hat{q}$ and momentum $\hat{p}$ and two single mode systems $\hat{x}$, $\hat{y}$ representing the apparatus. The meter variables bringing information about the measured system are the momentum operators $\hat{P}_x$, $\hat{P}_y$ since the interaction between the meter and observed field has the form

$$\hat{H}_{\text{int}} = K(\hat{q}\hat{P}_x + \hat{p}\hat{P}_y), \tag{3.99}$$

the interaction constatnt $K$ being assumed large. The meter wave functions (in $x$- and $y$-representations, respectively) are prepared in the initial time as Gaussian functions

$$\langle\varphi_x|x\rangle = \sqrt{\frac{2}{\pi b}}\exp\left(-\frac{x^2}{b}\right), \qquad \langle\varphi_y|y\rangle = \sqrt{\frac{2b}{\pi}}\exp\left(-bx^2\right), \tag{3.100}$$

$b$ being a parameter determining the width of distribution. The measured signal field is assumed to be in a general quantum state. The simultaneous measurement of noncommuting signal operators $\hat{q}$, $\hat{p}$ runs as follows: the interaction (3.99) is switched on in the time $t = 0$ and after time interval $t = 1/K$ the ideal simultaneous measurement of commuting position operators $\hat{x}$, $\hat{y}$ is made. It is straightforward matter to calculate the time evolution of respective fields. An expected value of $\hat{x}$ is equal to an expected value of $\hat{q}$ before the interaction and similarly, an expected value of $\hat{y}$ is equal to that of $\hat{p}$. Significantly the variances of the measured meter variables after the interaction may be related to those of signal field before the interaction as

$$\begin{aligned} (\Delta x)^2 &= (\Delta q)^2 + \frac{b}{2}, \\ (\Delta y)^2 &= (\Delta p)^2 + \frac{1}{2b}. \end{aligned} \tag{3.101}$$

Measured variances are therefore larger as a consequence of enhanced noise caused by joint detection. We may recover exactly the non-perturbed noise only for each component $\hat{q}$, $\hat{p}$ separately for $b \to 0$ or $b \to \infty$, disturbing the noise of the remaining quadrature component. An important conclusion may be derived for the uncertainty products of measured components. From the uncertainty relation for the signal field

$$(\Delta q)^2(\Delta p)^2 \geq \frac{1}{4} \tag{3.102}$$

and the relations (3.100) one can easily deduce that

$$(\Delta x)^2(\Delta y)^2 \geq 1. \tag{3.103}$$

Consequently, the measurement process added the noise equivalent at least to doubling of vacuum fluctuations of each component. The Heisenberg inequality (3.102) cannot be therefore interpreted as uncertainty relation for simultaneous detection of noncommuting variables, but rather as a mathematical property of wave function and the respective non-commuting operators. The same conclusion as expressed in (3.103) may be derived using different methods [She and Heffner (1966), Arthurs and Goodman (1988)]. The problem of simultaneous detection belongs to current problems investigated in quantum optics in connection with the measurement of distribution functions and extracting optimum information [Yamamoto and Haus (1986), Lai and Haus (1989), Stenholm (1992)].

The above discussed model of Arthurs and Kelly can be treated as the phase insensitive linear amplifier. Such devices were discussed in connection with laser operation by Haus and Mullen (1962) and Heffner (1962). The comprehensive analysis and classification of noises in the process of amplification was given by Caves (1982). Following his considerations let us present the main results of the linear amplification theory. For the sake of simplicity we restrict ourselves to the case of narrow-band linear amplifiers, where input and output are nearly sinusoidal oscillations (their frequencies $\omega_I$ and $\omega_O$ need not be necessarily equal). In this model one can specialize to single input and output modes of electromagnetic fields. In the following we will denote as $\hat{a}$ and $\hat{b}$ the annihilation operators of input and output fields, respectively. The linear amplifier may then be characterized as a device which relates the input field to the output one as follows

$$\hat{b} = M\hat{a} + L\hat{a}^\dagger + \hat{\mathcal{F}}, \tag{3.104}$$

where $\hat{\mathcal{F}}$ is an operator responsible for an additive noise. Fortunately, for estimation of quantum noise involved in the process of amplification its explicit form is not important. The only physically relevant condition follows from the boson nature of the operators $\hat{a}$, $\hat{b}$ in (3.104) as the relation

$$1 = |M|^2 - |L|^2 + [\hat{\mathcal{F}}, \hat{\mathcal{F}}^\dagger]. \tag{3.105}$$

Let us specify now the Heisenberg uncertainty relation for the pair of quadrature components of noise operator $\hat{\mathcal{F}}$ defined as

$$\hat{\mathcal{F}} = \hat{\mathcal{F}}_1 + i\hat{\mathcal{F}}_2 .$$

Denoting further

$$|\Delta\mathcal{F}|^2 \equiv (\mathcal{F}_1)^2 + (\mathcal{F}_2)^2, \tag{3.106}$$

the Heisenberg uncertainty relation (2.25) reads

$$(\mathcal{F}_1)^2(\mathcal{F}_2)^2 \geq \frac{1}{4}|\langle[\hat{\mathcal{F}}_1, \hat{\mathcal{F}}_2]\rangle|^2,$$

which may easily be rewritten to the form

$$|\Delta\mathcal{F}|^2 \geq \frac{1}{2}|\langle[\hat{\mathcal{F}}, \hat{\mathcal{F}}^\dagger]\rangle|^2 . \tag{3.107}$$

Let us return to the linear amplifiers. Combining relation (3.105) with (3.107), the lower limit of noise fluctuations may be estimated as

$$|\Delta\mathcal{F}|^2 \geq \frac{1}{2}\left|1 - |M|^2 + |L|^2\right| . \tag{3.108}$$

This inequality represents the central result of the linear amplifier theory and the rest of this section pursues the consequences of this observation. For further considerations it is convenient to introduce quadrature components of the input and output fields as

$$\begin{aligned} \hat{a} &= \hat{X}_1 + i\hat{X}_2 , \\ \hat{b} &= \hat{Y}_1 + i\hat{Y}_2 . \end{aligned} \tag{3.109}$$

The behavior of the output field with respect to the phase transformation of the input field

$$\hat{a} \rightarrow \exp(-i\varphi)\hat{a} \tag{3.110}$$

is important for the classification of possible amplification processes. The phase shift $\varphi$ may be induced in the process of measurement (e.g. in the arms of interferometer) or can be interpreted as the result of free evolution during the time interval $t$ ($\varphi = \omega_I t$). The input quadrature components rotate in the configuration space $(X_1, X_2)$ as

$$\begin{aligned} \hat{X}_1(\varphi) &= \hat{X}_1 \cos(\varphi) + \hat{X}_2 \sin(\varphi) , \\ \hat{X}_2(\varphi) &= -\hat{X}_1 \sin(\varphi) + \hat{X}_2 \cos(\varphi) . \end{aligned} \tag{3.111}$$

More detailed investigation of the phase shift transformation and description of the phase sensitive noise using the so-called "noise ellipse" may be found in the paper by Caves (1982) and this will be treated in Secs. 5.1 and 8.4.

According to the response of the amplified field to the phase shift transformation, the phase-insensitive and phase-sensitive linear amplifiers may be distinguished. The fundamental property of a phase-insensitive amplifier is that when the input signal has phase-insensitive noise, the output shows no phase preference. More mathematically, the properties may be parametrized by the following two conditions:

i) The output field transforms according to the same rule (3.111) as input field with the same phase shift $\varphi_O = \varphi_I$ (phase-preserving amplifiers) or $\varphi_O = -\varphi_I$ (phase-conjugating amplifiers).

ii) If the input signal is phase-insensitive, the output is also phase-insensitive and $\langle \hat{a}^2 \rangle = \langle \hat{a} \rangle^2$ implies the equality $\langle \hat{b}^2 \rangle = \langle \hat{b} \rangle^2$ (coherence is conserved). All the amplifiers which do not fulfil these conditions are phase-sensitive.
The consequences of the conditions i) and ii) are easy to work out. The former one implies that $L = 0$ for phase preserving amplifiers or $M = 0$ for phase conjugating ones. The condition ii) requires

$$\langle \hat{\mathcal{F}}^2 \rangle = 0 \,. \tag{3.112}$$

The equation (3.104) for amplified field may be formally rewritten as

$$\begin{aligned} \hat{Y}_1 &= G_1 \hat{X}_1 + \hat{\mathcal{F}}_1 \,, \\ \hat{Y}_2 &= G_2 \hat{X}_2 + \hat{\mathcal{F}}_2 \,, \end{aligned} \tag{3.113}$$

where gains of respective quadrature variables are defined as $G_1 = |M + L|^2$, $G_2 = |M - L|^2$; the mean gain is defined as $G = (1/2)(G_1 + G_2) = |M|^2 + |L|^2$.

The consequencies for phase-insensitive amplifiers are obvious. Since the gain is independent of phase, $G_1 = G_2$ as a consequence of the condition i), $\Delta\mathcal{F}_1 = \Delta\mathcal{F}_2$ as a consequence of (3.112) and both the relations in (3.113) are identical. The total noise of amplified field reads

$$|\Delta b|^2 = G|\Delta a|^2 + |\Delta\mathcal{F}|^2 \,. \tag{3.114}$$

The added noise may be characterized by the added noise number

$$A = \frac{(\Delta\mathcal{F})^2}{G} \tag{3.115}$$

and the fundamental theorem for phase-insensitive linear amplifiers follows from inequality (3.108) in the form

$$A \geq \frac{1}{2}\left|1 \mp \frac{1}{G}\right| . \tag{3.116}$$

The upper (lower) sign holds for phase-preserving (phase-conjugating) amplifiers. Consequently, only passive ($G = 1$) phase-preserving device does not add any noise.

Analogous considerations for added noise numbers of each quadrature component for phase-sensitive amplification,

$$A_j = \frac{(\Delta\mathcal{F}_j)^2}{G_j} \,, \qquad j = 1, 2 \,, \tag{3.117}$$

can be done leading to the amplifier uncertainty principle

$$\sqrt{A_1 A_2} \geq \frac{1}{4}\left|1 + \frac{\sigma}{\sqrt{G_1 G_2}}\right| , \tag{3.118}$$

where $\sigma = \text{sgn}(|L| - |M|)$.

Let us notice that phase-insensitive amplifier is described by the special case of this relation, just like the explicit model of Arthurs and Kelly (1965) characterized by $G_1 = G_2 = 1$. Let us point out another interesting possibility allowed by the amplifier uncertainty principle. One can consider an amplifier with $G_1 G_2 \gg 1$. Preparing the input state with well localized component $\Delta X_1 \ll 1/2$, the amplification with $G_1 \gg 1$ could reduce the delivered noise as $A_1 \ll 1/4$. Consequently, using the phase-sensitive detection, the information coded in the quadrature component could be recovered with the accuracy much better than the phase-insensitive technique provides. Of course, the price paid for this improvement is the increased noise delivered to the canonically conjugated variable $\Delta X_2 \gg 1/2$, $A_2 \gg 1/4$. This idea is referred to as the "back-action-evading" or "nondemolition" measurement and we will pursue this investigation in the forthcoming Sec. 3.6.

The amplification of quantum field serves as a general method for investigation of quantum fields. We touched only the main points of this topic. More detailed information, including multimode description, characterization of noise temperature and comparison of earlier treatments may be found in the literature [Caves (1982), Yamamoto and Haus (1986)].

## 3.6 Quantum nondemolition measurement

The main motivation for the theory of measurement is to increase the performance of the apparatus up to the ultimate limit allowed by the respective theory. Let us illustrate this general statement on the explicit example of position measurement in optics. The accuracy predicted by geometrical optics is restricted by the appearing of the wave effects, as mentioned in Sec. 3.1. This famous limitation, the so-called Airy spot, predicts the localization of the particle within the accuracy of the order of light wavelenght $\lambda$. Nevertheless, using the interferometric measurement setup, this resolution may be further improved and the next limitation follows from quantum mechanics as uncertainty principle, similarly to the case of Heisenberg microscope discussed in Sec. 3.1. The determination of the position reflects itself in the enhanced uncertainty of the canonically conjugated quantity and momentum, which appears in the enhanced noise of the subsequent position measurement. Such measurements are self-affected by the back-action of the performed detection and are therefore called as *demolition* ones. Let us analyse the ultimate limit of such a measurement of the position of a free particle.

### Demolition measurements

The mathematical consequence of the demolition nature of the measurement are following. Suppose that in the time $t = 0$ the position of the particle was measured and the value $X(0) = \langle \hat{X}(0) \rangle$ was found with the uncertainty $\Delta X^2(0) = \langle \hat{X}^2(0) \rangle - \langle X(0) \rangle^2$; analogous notation will be used also for canonically conjugated momentum $\hat{P}$. The evolution of a free mass is given by

$$\hat{X}(t) = \hat{X}(0) + \frac{t}{m}\hat{P}(0), \tag{3.119}$$

$m$ being the mass of tested free particle. Repeating the position measurement after the time $\tau$ in a state with uncorrelated momentum and positions, the respective variance of the position may be expressed as

$$\Delta X(\tau) = \left[\Delta X^2(0) + \frac{\tau^2}{m^2}\Delta P^2(0)\right]^{1/2}, \tag{3.120}$$

$$\Delta X(\tau) \geq \left[\Delta X^2(0) + \frac{\hbar^2 \tau^2}{4m^2 \Delta X^2(0)}\right]^{1/2} \geq \sqrt{\frac{\hbar\tau}{m}}. \tag{3.121}$$

In the derivation of inequality (3.121) the Heisenberg uncertainty

$$\Delta P \geq \frac{\hbar}{2\Delta X}$$

has been used. Expression on the right-hand side of (3.121) is known as the *standard quantum limit* of free particle position. The natural question is: does this quantity represent any ultimate limit or can be overcome within quantum mechanics? Yuen (1983) pointed out that the standard quantum limit is consequence of the assumption that the moment and position states of the free particle are uncorrelated. Assuming the full form of the position variation of the free evolution (3.119), there is a term

$$\frac{\tau}{m}\left[\langle \hat{X}(0)\hat{P}(0) + \hat{P}(0)\hat{X}(0) \rangle - 2X(0)P(0)\right]$$

omitted in relation (3.120). Yuen's point of view is that some measurements of $X$ leave the free mass in a state for which this part is negative. He calls such states as *contractive states* because the uncertainty of $X$ component decreases for certain time. Contractive states may be specified, for example, as the squeezed states associated with the operators $\hat{X}$ and $\hat{P}$. Consequently, there are measurements of $X$ that beat the standard quantum limit.

Nevertheless, Caves (1985) improved the formulation of the standard quantum limit and found another reason for which it is relevant to the position measurement. His heuristic argument is based on including the effect of imperfect resolution $\sigma$ of the measuring device, which is linearly coupled to measured variable $\hat{X}$. This assumption is important since then the total *measured* variance $\Delta$ is given as the sum of apparatus resolution and variation of $X$-coordinate,

$$\Delta^2 = \sigma^2 + \Delta X^2 . \tag{3.122}$$

Consider now two subsequent measurements of $\hat{X}$ in $t = 0$ and $t = \tau$, made with identical measuring apparatus. Let the first measurement give some value $X$ with the best possible accuracy $\sigma$. The key point for understanding the precision limitation is the question what does it mean that "the resolution of $X$ is $\sigma$". Caves argues that this means that the measurement determines the position immediately after the measurement to be within roughly a distance $\sigma$ of the measured value $X$. Therefore, just after the first measurement it should be

$$\Delta X(0) \leq \sigma. \tag{3.123}$$

This assumption permits that the immediate repetition would yield the same result within the resolution $\sigma$. Consequently, the measured variance in $t = \tau$ is given as

$$\begin{aligned} \Delta^2(\tau) &= \sigma^2 + \Delta X^2(\tau) \\ &\geq \Delta X^2(0) + \Delta X^2(\tau) \qquad (3.124) \\ &\geq 2\Delta X(0)\Delta X(\tau) \geq \frac{\hbar\tau}{m} . \qquad (3.125) \end{aligned}$$

In the derivation of the inequality (3.125) the uncertainty principle

$$\Delta X(0)\Delta X(\tau) \geq \frac{1}{2}|\langle[\hat{X}(0), \hat{X}(\tau)]\rangle| = \frac{\hbar\tau}{2m}$$

was used. The existence of the contractive states obviously does not change anything on the argumentation and the standard quantum limit is relevant to the measured accuracy of the position, provided that measurement process is involved in the description, too. More detailed discussion can be found in the original literature [Yuen (1983), Caves (1985)].

Nevertheless, there is a way how to beat the limit of demolition measurements, as we have already learned in Sec. 3.5. The measured quantity should be isolated from the back-action of the performed measurement. Let us illustrate this on the example of $\hat{X}$ and $\hat{P}$ variables of the free particle. There is no symmetry between these two canonically conjugated operators if one assumes the complemetary situation of the precise measurement of $\hat{P}$. An initial precise measurement of $\hat{P}$ perturbs $\hat{X}$, but since $\hat{X}$ does not feed into $\hat{P}$ during free evolution, the subsequent measurement of $\hat{P}$ in the later time may be arbitrarily accurate. In the next section we will generalize this property of $\hat{P}$ introducing the notion of *nondemolition* measurement.

Especially, we will show that instead of $\hat{X}$, the time-dependent quadrature component may be introduced as a nondemolition variable. Let us therefore notice what is the next physical limitation put on the resolution of the coordinate, in the spirit of the motivation of this section. This can be deduced as the Compton wavelength of the particle — a limitation coming from the relativistic quantum mechanics. This yields the value $\lambda_C = h/mc$, which for the mass $10^3$ kg is about $10^{-45}$ m. For macroscopic systems the relativistic limitation is therefore irrelevant, whereas the standard quantum limit (3.121) characterized by the numerical value of the order $10^{-20}$ m should be taken into account (see Sec. 3.7).

## Nondemolition measurements

Quantum nondemolition (back-action-evading) measurements in which the observable is measured without disturbing its free motion were proposed in order to improve the sensitivity of detectors of gravitational waves [Braginsky and Vorontsov (1974), Unruh (1978, 1979), Hollenhorst (1979), Braginsky, Vorontsov and Thorne (1980), Caves et al. (1980), Braginskij and Khalili (1992)]. In this connection the measurement of classical force weakly coupled to the free particle and harmonic oscillator were considered and the schemes how to increase the sensitivity beyond the limit of demolition measurements were proposed. Significantly, even if the theory of quantum nondemolition measurement was originally designed for mechanical harmonic oscillator realized as a Weber bar, the ideas are directly applicable to the photon field consisting of ensemble of harmonic oscillators. Thus, besides detection of gravitational waves, quantum nondemolition techniques may improve the signal-to-noise ratio of existing classical devices.

Let us provide a simplified theory for the description of these methods [Caves et al. (1980), Imoto, Haus and Yamamoto (1985)] using advantageously the Heisenberg operator picture. In a general quantum measurement the observable of the signal system $\hat{A}_s$ is observed by detecting the changes in the probe system observable $\hat{A}_p$, as is pointed out in Sec. 3.2. Interaction between signal and probe fields is described by the interaction hamiltonian $\hat{H}_I$ and the full hamiltonian then reads

$$\hat{H} = \hat{H}_s + \hat{H}_p + \hat{H}_I\,, \tag{3.126}$$

where $\hat{H}_s$ and $\hat{H}_p$ are the hamiltonians of unperturbed signal and probe systems, respectively. The Heisenberg equations of motion are

$$-i\hbar\frac{d}{dt}\hat{A}_s = [\hat{H}_s, \hat{A}_s] + [\hat{H}_I, \hat{A}_s]\,, \tag{3.127}$$

$$-i\hbar\frac{d}{dt}\hat{A}_p = [\hat{H}_p, \hat{A}_p] + [\hat{H}_I, \hat{A}_p]\,, \tag{3.128}$$

assuming that the measured observables are explicitly time-independent. The first commutators on the right-hand sides of these equations contribute to the free time evolution of observables, whereas the second ones represent the mutual coupling between the signal and probe observables. In general, the quantum nondemolition measurement may now be defined as a sequence of precise measurements of $\hat{A}_s$ such that the result of each measurement (after the first one) is completely predictable from the result of the preceding one. The quantum variable $\hat{A}_s$ is then called nondemolition observable. This will be fulfilled in the model described by the interaction (3.126) if the following conditions are valid:

i) $H_I$ is a function of $\hat{A}_s$ and moreover $[\hat{H}_I, \hat{A}_p] \neq \hat{0}$. This means that the probe variable is coupled to the system and the measurement could bring some information about the system.

ii) $[\hat{H}_I, \hat{A}_s] = \hat{0}$, i.e. the free evolution of the nondemolition variable is not influenced by the interaction with the apparatus described by $\hat{A}_p$.

iii) $\hat{H}_s$ is not a function of the conjugate observable of $\hat{A}_s$, i.e. the free evolution of the quantum nondemolition observable is predictable (since $\hat{A}_s$ is a constant of motion).

An exhaustive discussion how these conditions may be modified and generalized was given for example by Caves et al. (1980).

## Optical back-action-evading measurements

Let us illustrate the properties of nondemolition measurements on physically relevant example of a free harmonic oscillator. Of course, to obtain some useful information the oscillator should be coupled to the observed system and the observable of interest should be a quantum nondemolition observable of the whole hamiltonian. Nevertheless, it is beyond the scope of our explanation which touches the main points only; further details can be found in existing literature [Unruh (1978, 1979), Hollenhorst (1979), Caves et al. (1980), Braginsky and Khalili (1992)]. The hamiltonian of the single-mode harmonic oscillator reads

$$\hat{H}_0 = \frac{\hat{P}^2}{2m} + \frac{m\omega^2\hat{X}^2}{2} = \hbar\omega\hat{N} + \frac{\hbar\omega}{2}, \tag{3.129}$$

where $\hat{N} = \hat{a}^\dagger\hat{a}$ is the operator of the particle number and the annihilation operator is $\hat{a} = \sqrt{m\omega/2\hbar}\left(\hat{X} - i\hat{P}/m\omega\right)$. All constants of motion represent the quantum nondemolition variables. Obviously, $\hat{N}$ fulfils this condition and may be measured by means of quantum counting.

Let us demonstrate this technique by the example of detection of weak classical force [Braginsky and Vorontsov (1974), Hollenhorst (1979), Caves et al. (1980)]. The motivation comes from the analysis of gravitational waves detection, but the solution may easily be treated within the quantum optical formalism. Quantum optical model of nondemolition measurement of photon number using Kerr nonlinearity was formulated by Imoto, Haus and Yamamoto (1985). Assuming the hamiltonian of oscillator driven by the classical force,

$$\hat{H} = \hat{H}_0 - \hat{X}F(t)\,, \tag{3.130}$$

the classical force $F(t)$ (c-number) plays the role of the nondemolition observable, whereas operator $\hat{N}$ serves as a probe system. It is straightforward to show that evolution operator which governs the time evolution of an arbitrary initial state $|\psi\rangle$,

$$|\psi(t)\rangle = \hat{U}(t,t_0)\,|\psi\rangle\,, \tag{3.131}$$

may be specified as

$$\hat{U}(t,t_0) = \exp(-i\beta)\exp\left[-\frac{i(t-t_0)\hat{H}_0}{\hbar}\right]\hat{D}(\alpha)\,, \tag{3.132}$$

where the complex amplitude of the displacement operator $\hat{D}$ is

$$\alpha(t,t_0) = \frac{i}{\sqrt{2m\omega\hbar}}\int_{t_0}^{t} F(t')\exp\left[i\omega(t'-t_0)\right]\,dt'$$

and the (unimportant) overall phase parameter $\beta(t,t_0) = (i/2)\int_{t_0}^{t}(\alpha^*\dot{\alpha} - \alpha\dot{\alpha}^*)\,dt$, $\dot{\alpha}$ being the time derivative of $\alpha$ and the asterisk denotes the complex conjugation. Assume now that the exact measurement at time $t=0$ prepares the oscillator in the eigenstate $|N\rangle$ with $N$ quanta. The system evolves for time interval $\tau$ and then the quantum counting is repeated yielding the value $M$ with the probability

$$P(N\to M,\tau) = |\langle M|\hat{U}(\tau,0)|N\rangle|^2 = \frac{r!}{s!}\left[L_r^{s-r}\left(|\alpha|^2\right)\right]^2\exp\left(-|\alpha|^2\right)|\alpha|^{2(s-r)}\,, \tag{3.133}$$

where $s=\max(N,M)$, $r=\min(N,M)$ and $L_k^l$ denotes the generalized Laguerre polynomials. Consequently the probability that the force induced no change of the quantum state may be expressd as

$$1 - P(N\to N,\tau) = 1 - \exp\left(-|\alpha|^2\right)\left[L_N\left(|\alpha|^2\right)\right]^2. \tag{3.134}$$

This probability is significant if and only if

$$|\alpha(\tau,0)| \geq \frac{1}{\sqrt{N+1}} \tag{3.135}$$

and this condition determines the criterion for the measurement of the weak force by the quantum counting technique. Let us emphasize that no matter how weak the force is and no matter how short the time interval between subsequent countings is, one can detect the force by preparing the oscillator in a sufficiently energetic initial state. The quantum limit of measurement of weak forces does not therefore exist. This explicit example of quantum counting nondemolition measurement beats the *standard quantum limit* for detection of forces

$$\delta F_{SQL} \approx \sqrt{\frac{\hbar m}{\tau}}, \tag{3.136}$$

which could be easily derived from (3.121) assuming the relation $\delta x = \delta F \tau^2 / 2m$.

The time-dependent quadrature operators $\hat{X}_1$, $\hat{X}_2$ defined by the relations

$$\hat{X}_1 = \hat{X}\cos(\omega t) - \frac{\hat{P}}{m\omega}\sin(\omega t), \tag{3.137}$$

$$\hat{X}_2 = \hat{X}\sin(\omega t) + \frac{\hat{P}}{m\omega}\cos(\omega t), \tag{3.138}$$

represent another example of nondemolition technique important in quantum optics. It is straightforward to show that

$$\frac{d}{dt}\hat{X}_j = \frac{\partial}{\partial t}\hat{X}_j - \frac{i}{\hbar}[\hat{X}_j, \hat{H}] = 0. \tag{3.139}$$

The measurement of one quadrature operator disturbs the remaining one, but this influence does not appear in the subsequently repeated measurement of the same quadrature operator. This property uses the phase sensitive back-action-evading amplifier proposed by Yurke (1985). In the input-output formalism the general scheme may be described as the transformation of the input modes $\hat{a}_I$, $\hat{b}_I$ into the respective output modes

$$\hat{a}_O = \exp(i\theta)\hat{a}_I + G\exp\left[i(\theta+\psi)\right]\left[\exp(-i\phi)\hat{b}_I + \exp(i\phi)\hat{b}_I^\dagger\right], \tag{3.140}$$

$$\hat{b}_O = \exp(i\theta')\hat{b}_I + G\exp\left[i(\theta'+\phi)\right]\left[\exp(-i\psi)\hat{a}_I + \exp(i\psi)\hat{a}_I^\dagger\right], \tag{3.141}$$

where the phase parameters are $\theta$, $\theta'$, $\phi$ and $\psi$, and $G$ represents the gain of the amplifiers. Introducing the phase sensitive quadrature operators $\hat{X}(\theta) = (1/2)\left[\exp(-i\theta)\hat{a} + \exp(i\theta)\hat{a}^\dagger\right]$, $\hat{Y}(\theta) = (i/2)\left[\exp(-i\theta)\hat{a} - \exp(i\theta)\hat{a}^\dagger\right]$ for all possible annihilation operators and phases, we can specify the back-action-evading transformation as

$$\hat{X}_a^O(\theta+\psi) = \hat{X}_a^I(\psi) + 2G\hat{X}_b^I(\phi), \tag{3.142}$$

$$\hat{Y}_a^O(\theta+\psi) = \hat{Y}_a^I(\psi), \tag{3.143}$$

$$\hat{X}_b^O(\theta'+\phi) = \hat{X}_b^I(\phi), \tag{3.144}$$

$$\hat{Y}_b^O(\theta'+\phi) = \hat{Y}_b^I(\phi) - 2G\hat{Y}_a^I(\psi). \tag{3.145}$$

The nondemolition character of this detection scheme can be seen from these relations. For example, assuming the input states with decreased noise of $\hat{Y}_b^I(\phi)$ and $\hat{Y}_a^I(\psi)$ components, then the component $\hat{Y}_b^O(\theta' + \phi)$ carries information about $\hat{Y}_a^I(\psi)$ and the noise is delivered to the $\hat{X}_a^O(\theta + \psi)$ component.

## 3.7 High-precision measurement and detection of gravitational waves

As was pointed out in Sec. 3.4, the optical phase measurement belongs to most accurate measurements currently available. Laser interferometry is also designated as a tool for detection of gravitational waves, which could provide new point of view on the Universe. The existence of the gravitational waves was forecasted by Einstein's theory of relativity but it has not been proved experimentally yet. The difficulty of detecting gravitational waves may appear surprising taking into account everyday experience with the static gravity field. Nevertheless, neither falling objects nor orbiting planets involve masses large enough and fast enough to generate appreciable waves. The first detected signals are likely to come from distant astrophysical objects that are more massive than sun and move at nearly speed of light. Gravitational waves are emitted when the massive objects undergo rapid changes in its quadrupole moment. The gravitational waves should distort the shape of any region of space through which they travel. This effect may be used for their detection. The gravitational wave going through the system of two massive bodies suspended in a plane will cause the oscillation of the relative distance $\Delta l$ between them ($l$ being the distance between the two bodies). The strength of the gravitational wave is determined by the relative amount of the separation $\Delta l/l$. The effects induced by the waves are transverse to the direction of propagation and the frequency is determined by the frequency of the source. A system of linked gravitational observatories located at different places on the Earth could therefore determine the orientation of coming gravitational wave. For the concreteness, let us assume the numerical parameters for such a detection. The ideal measurement of the position of two bodies of equal masses with $m = 10^3$ kg, which is repeated with the frequency $10^3$ Hz, is limited by the standard quantum limit (3.121) equal to $\Delta z = 10^{-20}$ m. The values of displacement measured experimentally are reaching the value $10^{-18}$ m [Jeffries, Saulson, Spero and Zucker (1987), Ruthen (1992)] so that the relative displacement reaches the ratio $\Delta l/l = 10^{-20}$. This ratio is identical with the minimum value of estimation of such events as stellar collapse and interaction of two black holes. The question of detection of gravitational waves becomes therefore topical from the point of view of current technology.

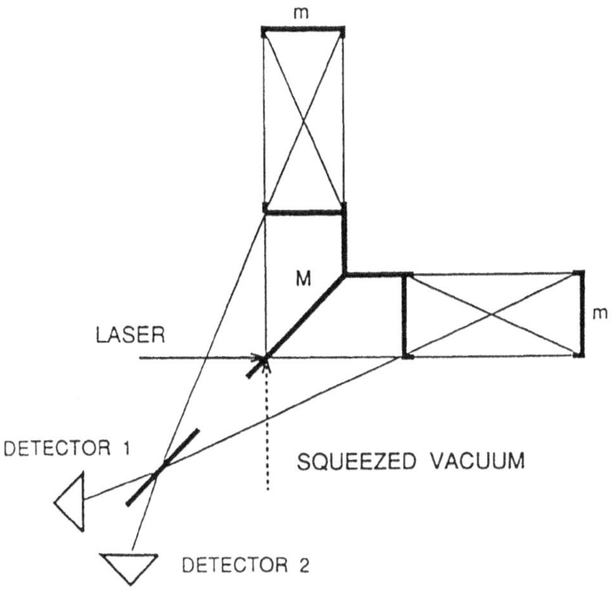


Figure 3.6: Setup of Michelson interferometer designated for detection of gravitational waves.

The analysis of the optical noises associated with the measurement by means of the Michelson interferometr was first given by Caves (1981) and the theory was further developed by Bondurant and Shapiro (1984). In our explanation we will follow the pioneering work by Caves. Adopting his approach let us analyse the noises influencing the resolution of the gravitational wave interferometer. On the contrary to the description of Mach-Zehnder interferometer in Sec. 3.4.3, the analysis of conditions influencing the detection of gravitational waves will be focused instead of description of phase resolution. The scheme sketched in Fig. 3.6 shows the multireflection Michelson interferometer designed for detection of gravitational waves. In the following the idealized lossless model will be assumed only. Laser light enters the input port, splits on an ideal lossless 50/50 beam splitter, bounces back and forth many times between perfectly reflected mirrors and finally it is recombined on the beam splitter. The equal test bodies of masses $m$ are attached to the both reflecting end mirrors. The beam splitter is rigidly attached to the mass $M \gg m$, so that the motion of the beam splitter will be neglected in our model and the beam splitter can be regarded as at rest. Both arms of the interferometer have the same fiducial length $l$ and the induced displacement of the end mirrors from this position will be denoted as $z_1$ and $z_2$. The interferometric pattern is observed by two (ideal) detectors placed in the output ports yielding the information about the difference $z_1 - z_2$. Nevertheless,

due to the finite speed of light the detectors provide information about $z$ as an average value over the storage time $\tau_s = 2bl/c$, where $b$ is number of bounces between the end mirror and the beam splitter. For the simplicity, the measurement time over which each detector averages the signal intensity is assumed to be of the order of the storage time. This is reasonable approximation since for $l = 1000$ m and $b = 200$ it is $\tau_s = 10^{-3}$ s. The main goal of the Caves analysis is the determination of the quantum mechanical limits of the performance of the interferometer. In the simplified model the following three sources of the quantum mechanical uncertainty influence the resolution of the interferometer:
i) the intrinsic quantum-mechanical uncertainties at the end mirrors positions and momenta;
ii) the perturbation of the end mirrors caused by the radiation-pressure fluctuations (the so-called *radiation pressure error*); and
iii) fluctuations in the number of photons at the output ports (the so-called *photon-counting error*).

These effects are manifested in the process of measurement simultaneously. Nevertheless, for the purpose of our analysis this mutual correlation of the noises will be disregarded and the total noise will be assumed as the quadrature sum of separate (independent) errors. The measurement represents the demolition technique discussed in the previous Sec. 3.6 and the uncertainties i) enforce at least the minimum error given as standard quantum limit (3.121). Consequently, the contribution i) to fluctuations coming from the intrinsic quantum–mechanical uncertainties may be eliminated from further discussion and the end mirrors may be treated as classical objects. In the following we will focus on the contributions of remaining two noises influencing the resolution of interferometer.

### Radiation-pressure error

The radiation-pressure error may be obtained in two steps using the following simplification. In the first step the momentum transfer of reflected photons will be calculated under the assumption that the end mirror is at rest and in the second step the perturbation of $z$ position results from the Newton laws. This approximation neglects the contribution to the momentum transfer due to the Doppler shift, since the end mirror is in fact moving for all the time. Nevertheless, this difference is negligible for the cases of interest. In the following the analysis will be performed for one-mode input and output fields, neglecting also the angular deviation of the beams propagating along the arms of interferometer. Abbreviating the electric fields entering the beam splitter as $E_{I,1}$ and $E_{I,2}$ and analogously the respective output fields as $E_{O,1}$ and $E_{O,2}$, the electromagnetic field may be quantized introducing the annihilation operators for the input fields as $\hat{a}_1$ and $\hat{a}_2$ and for the output fields as $\hat{b}_1$ and $\hat{b}_2$. The beam splitter divides the (electromagnetic) fields as

$$\begin{aligned}\hat{b}_1 &= \frac{1}{\sqrt{2}}\left[\hat{a}_1 + \exp(i\mu)\hat{a}_2\right] ,\\ \hat{b}_2 &= \frac{1}{\sqrt{2}}\left[\hat{a}_2 - \exp(-i\mu)\hat{a}_1\right] ,\end{aligned} \tag{3.146}$$

neglecting the overall phase shift induced by the beam splitter. The operator of difference between the momenta transferred to the end mirrors is then proportional to the difference between number of photons in modes 1 and 2

$$\begin{aligned}\mathcal{P} &= \frac{2b\hbar\omega}{c}\left(\hat{b}_2^\dagger\hat{b}_2 - \hat{b}_1^\dagger\hat{b}_1\right)\\ &= -\frac{2b\hbar\omega}{c}\left[\exp(i\mu)\hat{a}_1^\dagger\hat{a}_2 + \exp(-i\mu)\hat{a}_2^\dagger\hat{a}_1\right] .\end{aligned} \tag{3.147}$$

Assuming now the state of the electromagnetic field as a coherent state on the input port 1 and the squeezed state on the port 2, then

$$|\psi\rangle_{\rm in} = \hat{S}_2(\xi)\hat{D}_1(\alpha)|0\rangle_{1,2} , \tag{3.148}$$

where the squeezing and the displacement operators will be discussed in Chap. 4 in greater detail [$\hat{S}(\xi) = \exp[(1/2)\xi^*\hat{a}^2 - (1/2)\xi\hat{a}^{\dagger 2}]$, $\xi = r\exp(i\theta)$ being the squeeze parameter, and $\hat{D}(\alpha) = \exp(\alpha\hat{a}^\dagger - \alpha^*\hat{a})$]. The expectation value and the variance of $\mathcal{P}$ may be simply evaluated as

$$\langle\mathcal{P}\rangle = 0 , \tag{3.149}$$

$$(\Delta\mathcal{P})^2 = \left(\frac{2b\hbar\omega l}{c}\right)^2\left[|\alpha|^2\exp(2r) + \sinh^2(r)\right] . \tag{3.150}$$

Consequently, this disturbance causes the error in determination of the difference between the end mirrors during the time $\tau$ of measurement

$$(\Delta z)_{rp} \approx \frac{(\Delta\mathcal{P})\tau}{2m} = \frac{b\hbar\omega l\tau}{mc}\left[|\alpha|^2\exp(2r) + \sinh^2(r)\right]^{1/2} . \tag{3.151}$$

This is the desired result for *radiation-pressure error.*

### Photon-counting error

Similar analysis determines the signal detected on the output ports of interferometer. Taking into account the propagation of beams in both the arms of interferometer and mixing of light on the beam splitter, the transformation between input modes $\hat{a}_1$, $\hat{a}_2$ and output modes $\hat{c}_1$, $\hat{c}_2$ of the interferometer reaching the detectors can be expressed as

$$\begin{aligned}\hat{c}_1 &= \exp(i\Phi)\left[-i\exp(-i\mu)\hat{a}_1\sin\left(\frac{\phi}{2}\right)+\hat{a}_2\cos\left(\frac{\phi}{2}\right)\right],\\ \hat{c}_2 &= \exp(i\Phi)\left[\hat{a}_2\cos\left(\frac{\phi}{2}\right)-i\exp(i\mu)\hat{a}_2\sin\left(\frac{\phi}{2}\right)\right], \qquad (3.152)\end{aligned}$$

where the phase difference between both the arms $\phi = 2b\omega z/c + \pi - 2\mu$, $z = z_1 - z_2$ and the overall phase shift $\Phi = 2b\omega(z_1 + z_2)/c$. Assuming again the input quantum state (3.148), the expectation value and the variance of the difference between photon numbers $\hat{n} = \hat{c}_2^\dagger\hat{c}_2 - \hat{c}_1^\dagger\hat{c}_1$, to which the photocurrents are proportional, are given as

$$\langle\hat{n}\rangle = \cos(\phi)\left[|\alpha|^2 - \sinh^2(r)\right], \qquad (3.153)$$

$$\begin{aligned}(\Delta n)^2 &= |\alpha|^2\cos^2(\phi) + 2\cos^2(\phi)\cosh^2(r)\sinh^2(r)\\ &\quad + \sin^2(\phi)\left[|\alpha|^2\exp(-2r) + \sinh^2(r)\right]. \qquad (3.154)\end{aligned}$$

The resulting expression may be obtained by comparing (3.154) with the differential version of the dominant term in equation (3.153),

$$\delta n = -\frac{2b\omega}{c}|\alpha|^2\delta z\sin(\phi).$$

Consequently, the *photon-counting error* is given by

$$(\Delta z)_{pc} \approx \frac{c}{2b\omega}\left[\frac{\cot^2(\phi)}{|\alpha|^2} + \frac{2\cot^2(\phi)\cosh^2(r)\sinh^2(r)}{|\alpha|^4} + \frac{\exp(-2r)}{|\alpha|^2} + \frac{\sinh^2(r)}{|\alpha|^4}\right]^{-1/2}. \qquad (3.155)$$

### Optimization of the total error

In the realistic limit of strong field and weak squeezing the dominant contributions to the photon-counting and radiation-pressure errors are

$$(\Delta z)_{pc} \approx \frac{c}{2b\omega}\frac{\exp(-r)}{|\alpha|}, \qquad (3.156)$$

$$(\Delta z)_{rp} \approx \frac{2b\hbar\omega\tau}{mc}|\alpha|\exp(r). \qquad (3.157)$$

The total error

$$(\Delta z) = \left[(\Delta z)_{rp}^2 + (\Delta z)_{pc}^2\right]^{1/2}$$

may be minimized respective to the strong coherent amplitude $|\alpha|$ yielding the optimal regime with the resolution

$$(\Delta z)_{\text{opt}} \approx \sqrt{\frac{\hbar\tau}{m}} \tag{3.158}$$

corresponding to the standard quantum limit (3.121). The optimum amplitude of the laser $|\alpha|^2$ is given as

$$|\alpha|^2_{\text{opt}} \approx |\alpha_0|^2 \exp(-2r)\,, \tag{3.159}$$

where $|\alpha_0|^2 = (1/2)(mc^2/\hbar\omega)(1/\omega\tau)(1/b^2)$ is the complex amplitude needed for optimum resolution with closed (vacuum) input port. The respective optimum power is given as $P_0 = |\alpha_0|^2\hbar\omega/\tau$ leading to the numerical value 8 kW, provided that the parameters are specified as $m = 10^3$ kg, $\tau = 2 \times 10^{-3}$ s, $\omega = 4 \times 10^{15}$ rad s$^{-1}$, $b = 200$.

We conclude that application of squeezed states could improve the sensitivity of interferometers designed for detection of gravitational waves. The noise of existing interferometers is determined by the photon–counting error.

# Chapter 4

# Coherent states

In this chapter we provide the basic properties of the coherent-state technique and its application to the interaction of radiation with matter. Greater details about this technique can be found in books by Klauder and Sudarshan (1968), Malkin and Man'ko (1979), Klauder and Skagerstam (1985), Perelomov (1986), Peřina (1991), in reviews by Zhang, Feng and Gilmore (1990), Kim and Noz (1991) and in proceedings on squeezed states and uncertainty relations edited by Han, Kim and Man'ko (1993) and on harmonic oscillators edited by Han, Kim and Zachary (1993).

## 4.1 The general notion of the coherent state

The name *coherent states* is usually applied to a wide class of objects. Nevertheless all such definitions contain (more or less explicitly) two common properties, which can be used as the minimum set of requirements for a set of states to be referred as coherent states. Here we shall adopt this point of view and follow the nice exposition of [Klauder and Skagerstam (1985)]. Let $\mathcal{H}$ be a separable Hilbert space and $\mathcal{L}$ be some appropriate set (label space) endowed with a notion of continuity. The state of $\mathcal{H}$ labelled by an element $l \in \mathcal{L}$ will be denoted in the Dirac's notation as $|l\rangle$. Then the above mentioned properties can be formulated as follows:

i) **Continuity:** The vector $|l\rangle$ depends strongly continuously on the label $l$.

ii) **Completeness (resolution of unity):** There exists a positive measure $dl$ on $\mathcal{L}$ such that the unit operator $\hat{1}$ admits the resolution of unity

$$\hat{1} = \int_{\mathcal{L}} |l\rangle\langle l|\, dl \tag{4.1}$$

in the weak sense. This means that arbitrary matrix elements of both the sides of the equality coincide. So

$$\langle\psi|\hat{1}|\varphi\rangle = \int_{\mathcal{L}} \langle\psi|l\rangle\langle l|\varphi\rangle\, dl$$

holds for any $|\psi\rangle, |\varphi\rangle \in \mathcal{H}$.
The coherent states $|l\rangle$ are not assumed to be necessarily of the unit norm. Of course, as follows from (4.1), they can always be normalized and this will be assumed implicitly in all what follows.

The completeness gives rise to a functional representation of vectors in $\mathcal{H}$ and operators acting on $\mathcal{H}$. We have an isometrical embedding of the Hilbert space $\mathcal{H}$ into the space $L^2(\mathcal{L}, dl)$ given by the equality

$$\psi(l) = \langle l|\psi\rangle \in L^2(\mathcal{L}, dl)\,. \tag{4.2}$$

We have also

$$\langle\varphi|\psi\rangle = \int \varphi^*(l)\psi(l)\, dl\,, \tag{4.3}$$

$$(\hat{A}\psi)(l) = \int A(l,l')\psi(l')\langle l|l'\rangle\, dl'\,. \tag{4.4}$$

Here

$$A(l,l') = \frac{\langle l|\hat{A}|l'\rangle}{\langle l|l'\rangle}\,. \tag{4.5}$$

If $l = l'$ the above expression is called the *covariant symbol* of the operator $\hat{A}$ and will be denoted as $A(l)$.

On the other hand it is convenient to introduce also the so-called *contravariant symbol* of an operator $\hat{A}$. Let $A^\circ(l)$ be a measurable function on $\mathcal{L}$ and let us assume the operator

$$\hat{A} = \int A^\circ(l)|l\rangle\langle l|\, dl\,, \tag{4.6}$$

where the integral is understood again in the weak sense. The function $A^\circ(l)$ is then called the contravariant symbol of the operator $\hat{A}$. The represenation of the operator in the form (4.6) is called the *diagonal representation.* The relation between the covariant and contravariant symbols of the operator $\hat{A}$, providing that both exist, is easily seen,

$$A(l) = \int A^\circ(l')\frac{\langle l|l'\rangle\langle l'|l\rangle}{\langle l|l\rangle}\, dl'\,. \tag{4.7}$$

The overlap of two coherent states $\langle l|l'\rangle$ is due to its property

$$\langle l|\psi\rangle = \int \langle l|l'\rangle\langle l'|\psi\rangle\, dl'$$

referred as the *reproducing kernel.*

## 4.2 Coherent states associated with Lie groups

The coherent states associated with Lie groups provide us with an unifying approach to all kinds of coherent states which will appear in the subsequent sections. Our exposition will follow the excellent book of Perelomov (1986) [see also Klauder and Skagerstam (1985) and Zhang, Feng and Gilmore (1990)].

Let $G$ be an arbitrary Lie group and $\hat{T}(g), g \in G$ its strongly continuous unitary irreducible representation acting on a separable Hilbert space $\mathcal{H}$. Let $|\psi_0\rangle \in \mathcal{H}$ be an arbitrary fixed unit vector (*fiducial vector*).

**Definition 1** *The set of states*

$$\hat{T}(g)|\psi_0\rangle\,, \qquad g \in G \tag{4.8}$$

*is called the system of coherent states associated with the unitary irreducible representation $\hat{T}$ of the Lie group $G$ and the fiducial vector $|\psi_0\rangle$.*

Some more general definitions are posssible. Hovewer, the definition presented here is the most convenient for our purposes. To be able to handle also with non-compact groups it is sometimes assumed that representations entering the definition of the coherent states should be the so-called square-integrable ones. Let now $H$ denote the set

$$\{h \in G;\; \hat{T}(h)|\psi_0\rangle = \exp[i\alpha(h)]|\psi_0\rangle\}\,,$$

where $\alpha(h) \in [0, 2\pi)$. $H$ is then a soubgroup of $G$, it is the *stationary subgroup* of the vector $|\psi_0\rangle$. Two coherent states $\hat{T}(g_1)|\psi_0\rangle$ and $\hat{T}(g_2)|\psi_0\rangle$ describe the same state (are equivalent) if and only if they differ by a phase factor, i.e. $g_1 g_2^{-1} \in H$. We can pick up one representative in any equivalence class, so that the set of different coherent states will be parametrized naturally by the points in the left coset space $X = G/H$. The coherent state corresponding to the poit $x \in X$ will be denoted then as $|x\rangle$. If we use the notations $x(g) \in X$ for the image of $g \in G$ under the projection $G \to G/H = X$, we will have

$$\hat{T}(g)|\psi_0\rangle = \exp[i\alpha(g)]|x(g)\rangle\,.$$

Now it easily follows that the scalar product $\langle x_1|x_2\rangle$ of two coherent states $|x_1\rangle$ and $|x_2\rangle$ does not depent on the choice of representatives $g_1$ ang $g_2$ in $G$. It is a consequence of a simple property $\alpha(gh) = \alpha(g) + \alpha(h)$ for $g \in G$ and $h \in H$ of the function $\alpha(g)$ and the equality

$$\langle x_1|x_2\rangle = \exp\{i[\alpha(g_1) - \alpha(g_2)]\}\langle\psi_0|\hat{T}(g_1^{-1}g_2)|\psi_0\rangle\,.$$

Let us note that the continuity of the set of coherent states is a consequence of the strong continuity of the assumed representation $\hat{T}$.

Before giving the resolution of unity we have to stress an additional property of the group-related coherent states, namely, the invariance of the set of group-related coherent states under the group action,

$$\hat{T}(g_1)|x\rangle = \exp[-i\alpha(g)]\hat{T}(g_1)\hat{T}(g)|\psi_0\rangle = \exp[i\beta(g_1,g)]|g_1 \cdot x\rangle \,, \tag{4.9}$$

where $x = x(g)$,

$$\beta(g_1, g) = \alpha(g_1 g) - \alpha(g)$$

and $g_1 \cdot x = x(g_1 g)$ is just the action of the group $G$ on its homogeneous space $X = G/H$.

In many cases, particulary in those we will be interested in, the homogeneous space $X = G/H$ possesses a group-invariant measure $dx$. This means that for any $g \in G$ it holds that $d(g \cdot x) = dx$. The completeness of the set of coherent states in the form

$$\hat{1} = \int_X |x\rangle\langle x|\, dx \tag{4.10}$$

is now a consequence of the irreducibility of the assumed representation $\hat{T}$. From the invariance of the measure $dx$ and equation (4.9) we have immediately that the operator defined by the right-hand side of (4.10) commute with all operators $\hat{T}(g), g \in G$. Really

$$\hat{T}(g) \int_X |x\rangle\langle x|\, dx\, \hat{T}^{-1}(g) = \int_X |g \cdot x\rangle\langle g \cdot x|\, d(g \cdot x) = \int_X |x\rangle\langle x|\, dx \,.$$

According to the Schur lemma it is proportional to the unit operator $\hat{1}$. The equality can then be reached by a suitable normalization of the invariant measure $dx$.

One important application of the coherent states described here is the case when the quantum system under consideration admits a group of the dynamical symmetry. In this case the quantum system is assumed to be described by the Hilbert space $\mathcal{H}$ which is a carrier space of a unitary irreducible (strongly continuous) representation $\hat{T}$ of the finite-dimensional Lie group $G$. Let $\mathcal{G}$ be the corresponding Lie algebra. The hamiltonian $\hat{H}$ is some possibly time dependent linear combination of the representatives $\hat{T}(X_i)$ of the generators $X_i \in \mathcal{G}$,

$$\hat{H} = \sum_i h_i(t)\hat{T}(X_i) \,. \tag{4.11}$$

In the following we shall not distinguish between the generators and their representatives and we shall use the notation $X_i$ for both. The operators $X_i$ fulfil the standard commutation relation

$$[X_i, X_j] = c_{ij}^k X_k \,, \tag{4.12}$$

with $c_{ij}^k$ being the structure constants of the Lie algebra $\mathcal{G}$. The evolution operator $\hat{U}(t)$ is then a representative $\hat{T}(g(t))$ of a group element $g(t)$. It follows that if the state of the system $|\psi\rangle$ in some fixed moment $t_0$ was a coherent state $|x_0\rangle = \exp[-i\alpha(g_0)]\hat{T}(g_0)|\psi_0\rangle$, then it will remain to be a coherent state for all values of $t$ as follows from the equality

$$|\psi(t)\rangle = \hat{T}(g(t))|x_0\rangle = \exp\left[i\beta(g(t), g_0)\right]|g(t)\cdot x\rangle \tag{4.13}$$

and from the definition of the coherent states. Moreover the argument of the coherent state as well as the phase in the last expression can be obtained as solutions of some ordinary differential equations which are naturally interpreted in terms of the solution of equations of motion for the underlying classical system.

## 4.3 Coherent states of the harmonic oscillator

### 4.3.1 Basic properties

A simple application of the general construction from the previous section permits to construct the coherent states of the harmonic oscillator (*canonical coherent states*). We shall discuss only the one-dimensional case. The finite-dimensional generalization is straightforward. Let us assume, as in Sec. 2.6, the Hilbert space $\mathcal{H} = L^2(R, dx)$ and the self-adjoint position and momentum operators $\hat{x}$ and $\hat{p}$ satisfying (on some proper dense domain in $L^2(R, dx)$) the canonical commutation relation

$$[\hat{x}, \hat{p}] = i\hat{1}\,, \qquad [\hat{1}, \hat{x}] = [\hat{1}, \hat{p}] = \hat{0}\,. \tag{4.14}$$

We have set $\hbar = 1$ for simplicity. Equivalently, we can assume the Fock representation with creation and annihilation operators $\hat{a}^\dagger$ and $\hat{a}$ with the commutation relation

$$[\hat{a}, \hat{a}^\dagger] = \hat{1}\,, \qquad [\hat{1}, \hat{a}] = [\hat{1}, \hat{a}^\dagger] = \hat{0}\,, \tag{4.15}$$

again on some proper dense domain in $\mathcal{H}$. Let as assume the bounded operators

$$\hat{T}(p, x, t) = \exp[i(p\hat{x} - x\hat{p} + t\hat{1})] = \exp(\alpha\hat{a}^\dagger - \alpha^*\hat{a} + it\hat{1}) = \exp(it\hat{1})\hat{D}(\alpha)\,, \tag{4.16}$$

where

$$\hat{D}(\alpha) = \exp(\alpha\hat{a}^\dagger - \alpha^*\hat{a})\,, \tag{4.17}$$

$x$, $p$ and $t$ are real numbers and $\alpha = (1/\sqrt{2})(x + ip)$. The operator $\hat{D}(\alpha)$ is called the *displacement operator.* From the special form of the Baker-Hausdorff identity

$$\exp(\hat{A})\exp(\hat{B}) = \exp\left\{\frac{1}{2}[\hat{A}, \hat{B}]\right\}\exp\left(\hat{A} + \hat{B}\right), \tag{4.18}$$

which is valid in the case that $[\hat{A}, [\hat{A}, \hat{B}]] = [\hat{B}, [\hat{A}, \hat{B}]] = \hat{0}$, we have immediately the relation

$$\hat{D}(\alpha)\hat{D}(\beta) = \exp\left[i\mathrm{Im}(\alpha\beta^*)\right]\hat{D}(\alpha + \beta). \tag{4.19}$$

The action of the operators $\hat{T}(p, x, t)$ in the Hilbert space $\mathcal{H}$ defines then the unitary irreducible representation of the *Heisenberg-Weyl* group, i.e. the Lie group with elements $(p, x, t)$ parametrized with three real parameters and the composition law

$$(p_1, x_1, t_1) \cdot (p_2, x_2, t_2) = (p_1 + p_2, x_1 + x_2, t_1 + t_2 + p_1x_2 - p_2x_1). \tag{4.20}$$

The operators $\hat{x}$, $\hat{p}$ and $\hat{1}$ define a representation of the corresponding Heisenberg-Weyl algebra, which is a real Lie algebra with three generators $\hat{e}_1$, $\hat{e}_2$, $\hat{e}_3$ and commutation relations

$$[\hat{e}_1, \hat{e}_2] = \hat{e}_3, \qquad [\hat{e}_1, \hat{e}_3] = [\hat{e}_2, \hat{e}_3] = \hat{0}. \tag{4.21}$$

In our representation, which corresponds to the fixed value of the central element $\hat{e}_3 = i\hat{1}$, we have

$$\hat{e}_1 = i\hat{p}, \qquad \hat{e}_2 = i\hat{x}, \qquad \hat{e}_3 = i\hat{1}.$$

Non-equivalent unitary irreducible representations correspond, according to the Schur lemma, to the different values of the central element $\hat{e}_3 = i\lambda$, $\lambda \in R$. The corresponding unitary irreducible representations $\hat{T}^\lambda$ for $\lambda \neq 0$ are then obtained from $\hat{T}^{\lambda=1}$ simply by rescaling the generators. From the Stone-von Neumann theorem (Sec. 2.6) it follows then that for fixed $\lambda \neq 0$ any unitary irreducible representation of the Heisenberg-Weyl group (in separable Hilbert space $\mathcal{H}$) is equivalent to the $\hat{T}^\lambda$ described above. For $\lambda = 0$ all irreducible unitary representations are one-dimensional representations parametrized by two real numbers $\mu$ and $\nu$ in the form

$$\hat{T}^{\mu,\nu}(p, x, t) = \exp[i(\mu p + \nu x)].$$

The usual (canonical) coherent states are associated with the representation $\hat{T}^\lambda$, for $\lambda = 1$ and the choice of the vacuum $|0\rangle$ for the fiducial vector

$|\psi_0\rangle$. As follows from the previous discussion the corresponding set of coherent states is of the form

$$|\alpha\rangle = \hat{D}(\alpha)|0\rangle\,, \qquad \alpha \in C\,. \tag{4.22}$$

Using the Baker-Hausdorff identity (4.18) and the property of the vacuum $\hat{a}|0\rangle = 0$, we can rewrite (4.22) to the form

$$\begin{aligned} |\alpha\rangle &= \exp\left(-\frac{|\alpha|^2}{2}\right)\exp(\alpha\hat{a}^\dagger)\exp(-\alpha^*\hat{a})|0\rangle \\ &= \exp\left(-\frac{|\alpha|^2}{2}\right)\sum_{n=0}^{\infty}\frac{\alpha^n}{\sqrt{n!}}|n\rangle\,. \end{aligned} \tag{4.23}$$

In the last equality we have used the Taylor expansion of the exponentials and the definition (2.40) of the Fock states $|n\rangle$. It follows that the probability that there are $n$ photons in the coherent state labelled by $\alpha$ is given by the Poisson distribution

$$|\langle n|\alpha\rangle|^2 = \frac{|\alpha|^{2n}}{n!}\exp\left(-|\alpha|^2\right)\,, \tag{4.24}$$

with the expectation value $\langle n\rangle = |\alpha|^2$. The scalar product of two coherent states is

$$\langle\alpha|\beta\rangle = \exp\left(-\frac{|\alpha|^2}{2} - \frac{|\beta|^2}{2} + \alpha^*\beta\right)\,, \tag{4.25}$$

and so

$$|\langle\alpha|\beta\rangle|^2 = \exp\left(-|\alpha-\beta|^2\right)\,. \tag{4.26}$$

This shows that the coherent states are not orthogonal.

Here we should mention two specific properties of canonical coherent states. The first one is that the coherent states (4.22) are the eigenstates of the annihilation operator $\hat{a}$

$$\hat{a}|\alpha\rangle = \alpha|\alpha\rangle\,, \tag{4.27}$$

as can be easily verified using (4.23) and (2.37). It is more or less obvious that there are no other eigenstates of the annihilation operator $\hat{a}$.

Normalizing the combinations $(|\alpha\rangle \pm |-\alpha\rangle)$ of coherent states $|\alpha\rangle$ and $|-\alpha\rangle$, we obtain even and odd coherent states, respectively [see e.g. Peřina (1991), Sec. 4.1.9], having strongly nonclassical behaviour. They are eigenstates of $\hat{a}^2$ with the eigenvalue $\alpha^2$. Other generalizations have been reviewed by Knight and Bužek (1995).

The second property is that the coherent states minimalize the uncertainty relation (2.25) applied to operators $\hat{x}$ and $\hat{p}$. It means that if the mean square deviations $\delta x$ and $\delta p$ are evaluated for the system in coherent state, then the equality

$$\delta x \delta p = \frac{1}{2} \tag{4.28}$$

holds. This is easily seen if we express the operators $\hat{x}$ and $\hat{p}$ with the help of $\hat{a}$ and $\hat{a}^\dagger$ and use the fact that the coherent states are eigenstates of $\hat{a}$, which gives $(\delta x)^2 = (\delta p)^2 = 1/2$. Let us remember that we have set $\hbar = 1$. We will return to this property of canonical coherent states later on.

Let us discuss now the completeness of the canonical coherent states. From the group multiplication law (4.20) for the Heisenberg-Weyl group we can see immediately that the action of the group element $g = (p, x, t)$ on the label $\alpha \in C$ is given by the formula

$$g \cdot \alpha = \alpha + \frac{1}{\sqrt{2}}(x + ip),$$

so that the invariant measure on $C$ has to be proportional to $d(\mathrm{Re}\alpha)\, d(\mathrm{Im}\alpha)$ which will be denoted as $d^2\alpha$. The proportionality factor can then be easily computed from the equality

$$\int \langle 0|\alpha\rangle\langle\alpha|0\rangle\, d^2\alpha = \int \exp\left(-|\alpha|^2\right)\, d^2\alpha = \pi .$$

This finally leads to the resolution of unity in the form

$$\frac{1}{\pi}\int |\alpha\rangle\langle\alpha|\, d^2\alpha = \hat{1} . \tag{4.29}$$

The action of the displacement operator $\hat{D}(\alpha)$ transforms one coherent state into another coherent state in accordance with the general result of the previous section,

$$\hat{D}(\alpha)|\beta\rangle = \exp\left[i\mathrm{Im}(\alpha\beta^*)\right] |\alpha + \beta\rangle . \tag{4.30}$$

## 4.3.2 Bargmann representation

As follows from the resolution of unity any pure state $|\psi\rangle \in \mathcal{H}$ is uniquely determined by the function $\langle\alpha|\psi\rangle$,

$$|\psi\rangle = \frac{1}{\pi}\int \langle\alpha|\psi\rangle|\alpha\rangle\, d^2\alpha . \tag{4.31}$$

Let us express the state $|\psi\rangle$ with the help of the Fock states

$$|\psi\rangle = \sum_{n=0}^{\infty} c_n\, |n\rangle , \qquad \langle\psi|\psi\rangle = \sum_{n=0}^{\infty} |c_n|^2 = 1 . \tag{4.32}$$

Using (4.23) we have

$$\langle\alpha|\psi\rangle = \exp\left(-\frac{|\alpha|^2}{2}\right)\psi(\alpha^*), \tag{4.33}$$

where the function

$$\psi(z) = \sum_{n=0}^{\infty} c_n\varphi_n(z), \qquad \varphi_n(z) = \frac{z^n}{\sqrt{n!}} \tag{4.34}$$

is an entire analytic function of the complex variable $z$ because of the convergence of (4.34) for any $z \in C$, which is guaranteed by the normalization condition $\sum_{n=0}^{\infty}|c_n|^2 = 1$. We have

$$\langle\psi|\psi\rangle = \frac{1}{\pi}\int \exp\left(-|z|^2\right)|\psi(z)|^2\, d^2z \ < \infty . \tag{4.35}$$

It can be shown that for any two entire analytic functions $\psi_1(z)$ and $\psi_2(z)$ fulfilling the condition (4.35) the expression

$$\langle\psi_1|\psi_2\rangle = \frac{1}{\pi}\int \exp\left(-|z|^2\right)\psi_1^*(z)\psi_2(z)\, d^2z \tag{4.36}$$

has all properties of the scalar product. Thus we have arrived to a concrete realization of the Fock space $\mathcal{H}$ as a space of all entire analytic functions of the complex variable $z$ fulfilling the condition (4.35). This representation called *Bargmann representation* has many interesting features. Many formulas and expressions take simpler form in this representation in comparison with the Schrödinger representation. For instance, the Fock states $|n\rangle$ are represented by the monomials $\varphi_n(z)$ given by (4.34),

$$|n\rangle \to \varphi_n(z) = \frac{z^n}{\sqrt{n!}} . \tag{4.37}$$

Also the creation and annihilation operators take a simple form

$$\hat{a} = \frac{d}{dz}, \qquad \hat{a}^\dagger = z \tag{4.38}$$

in this representation.

### 4.3.3 Operators and their symbols, orderings

Let us summarize now the general definitions of covariant and contravariant symbols introduced in Sec. 4.1. When specialized to the case of canonical coherent states (4.22) there are different names usually used to refer to them. They are sometimes called *Wick* and *anti-Wick*, or *normal* and *antinormal*, or $Q$- and $P$- symbols, respectively. Explicitly, the covariant symbol $A(\alpha^*, \alpha)$ of the operator $\hat{A}$ is defined as

$$A(\alpha^*, \alpha) = \langle \alpha | \hat{A} | \alpha \rangle . \tag{4.39}$$

The covariant symbol exists e.g. in the cases where the operator $\hat{A}$ is a bounded one or it is a polynomial in creation and annihilation operators $\hat{a}^\dagger$ and $\hat{a}$. Let us assume the following two functions $A(\alpha^*, \beta)$ and $\tilde{A}(\alpha^*, \beta)$ of two complex variables $\alpha$, $\beta$

$$A(\alpha^*, \beta) = \langle \alpha | \hat{A} | \beta \rangle = \exp\left[-\frac{1}{2}\left(|\alpha|^2 + |\beta|^2\right)\right] \tilde{A}(\alpha^*, \beta) . \tag{4.40}$$

The covariant symbol is then the diagonal value at $\alpha = \beta$ of the function $A(\alpha^*, \beta)$ and the function $\tilde{A}(\alpha^*, \beta)$ is an analytic function of two complex variables $\alpha$ and $\beta$ and it determines the operator $\hat{A}$ uniquely owing to the relation

$$\langle \alpha | \hat{A} | \psi \rangle = \frac{1}{\pi} \int \exp\left[-\frac{1}{2}\left(|\alpha|^2 + |\beta|^2\right)\right] \tilde{A}(\alpha^*, \beta) \langle \beta | \psi \rangle \, d^2\beta , \tag{4.41}$$

following immediately from the completeness of coherent states. Moreover, it is an entire function of two complex variables (this is easily seen e.g. expressing the function $A(\alpha^*, \beta)$ in terms of the matrix elements $\langle m | \hat{A} | n \rangle$ of the operator $\hat{A}$ in Fock states) and such a function is uniquely determined by its values on the diagonal $\alpha = \beta$. It follows that the operator $\hat{A}$ is uniquely determined by its covariant symbol $\langle \alpha | \hat{A} | \alpha \rangle$.

For the contravariant symbols the following result is known. If $A^\circ(\alpha^*, \alpha)$ is a function such that for $r > 2$

$$\int |A^\circ(\alpha^*, \alpha)|^r \exp\left(-|\alpha|^2\right) d^2\alpha < \infty$$

holds, then this function is a contravariant symbol of an operator with a dense domain. For any bounded operator or a polynomial in creation and annihilation operators $\hat{a}^\dagger$, $\hat{a}$ the contravariant symbol exists as a generalized function, which is a Fourier transform of some compactly supported distribution. This fact follows easily from the relation (4.7) between the covariant and contravariant symbols

$$A(\alpha^*, \alpha) = \frac{1}{\pi} \int A^\circ(\alpha^*, \alpha) \exp\left(-|\alpha - \beta|^2\right) d^2\beta . \tag{4.42}$$

More precisely we have to use the Fourier transform of (4.42).

The kernel entering the integral relation (4.42) is a smoothing kernel and so the covariant symbol on the left-hand side is defined for any contravariant symbol entering the right-hand side. In general this convolution integral cannot be inverted. There are operators having covariant symbols but not admitting the diagonal representation with contravariant symbol as a usual function. The diagonal representation of an operator in terms of coherent states is called also the *Glauber-Sudarshan* representation.

The covariant and contravariant symbols are in some sense dual each other. To explain this let us note that both give a simple formula for the trace of an operator

$$\mathrm{Tr}(\hat{A}) = \frac{1}{\pi}\int A(\alpha^*,\alpha)\, d^2\alpha = \frac{1}{\pi}\int A^\circ(\alpha^*,\alpha)\, d^2\alpha\,. \tag{4.43}$$

For the trace of a product of operators $\hat{A}_1$ and $\hat{A}_2$ we easily obtain

$$\mathrm{Tr}(\hat{A}_1\hat{A}_2) = \frac{1}{\pi}\int A_1(\alpha^*,\alpha)A_2^\circ(\alpha^*,\alpha)\, d^2\alpha = \frac{1}{\pi}\int A_1^\circ(\alpha^*,\alpha)A_2(\alpha^*,\alpha)\, d^2\alpha\,. \tag{4.44}$$

The operators $\hat{A}$ and $\hat{B}$ are said to be orthogonal if $\mathrm{Tr}(\hat{A}^\dagger\hat{B}) = 0$. Using the above formula for the trace of a product of two operators it can be shown (in the framework of general formulations of Sec. 4.1) that the set of all operators with vanishing covariant symbols is orthogonal to the set of all operators admitting the diagonal representation.

Finally it can be shown that any operator can be viewed as a limit of a sequence of operators admitting the diagonal representation with smooth contravariant symbols.

The covariant symbol as well as the contravariant symbol are very simple for polynomials in $\hat{a}^\dagger$ and $\hat{a}$. Let us assume a *normally ordered* monomial $\hat{a}^{\dagger m}\hat{a}^n$. Normal ordering means that all annihilation operators stand to the right of all creation operators. It follows from the canonical commutation relation that any operator in $\hat{a}^\dagger$ and $\hat{a}$ can be normally ordered. Then we have

$$\langle\alpha|\hat{a}^{\dagger m}\hat{a}^n|\alpha\rangle = \alpha^{*m}\alpha^n\,. \tag{4.45}$$

Similarly for the *antinormally ordered* monomial $\hat{A} = \hat{a}^m\hat{a}^{\dagger n}$ (all creation operators are to the right of all annihilation operators; any operator can be antinormally ordered) we obtain, inserting the resolution of unity between $\hat{a}^m$ and $\hat{a}^{\dagger n}$,

$$A^\circ(\alpha^*,\alpha) = \alpha^m\alpha^{*n}\,.$$

We should also mention that there is, of course, very close relation between the properties of the functions defining the covariant and contravariant symbols and the properties of the corresponding operator. The reader interested in greater details can consult the book by Perelomov (1986).

Now we can introduce the third frequently used symbol, i.e. the so-called *Weyl* symbol. Using (4.18) we can order the displacement operator $\hat{D}(\gamma)$

normally or antinormally and we can easily compute its covariant or contravariant symbols. Then using the formula (4.43) we obtain immediately

$$\mathrm{Tr}\{\hat{D}(\gamma)\} = \pi\delta(\gamma) \tag{4.46}$$

and consequently

$$\mathrm{Tr}\{\hat{D}(\alpha)\hat{D}^\dagger(\beta)\} = \pi\delta(\alpha - \beta)\,. \tag{4.47}$$

Thus we obtain for a "good" operator $\hat{A}$ (e.g. any bounded operator or polynomial in $\hat{a}^\dagger$ and $\hat{a}$) the representation in the form

$$\hat{A} = \frac{1}{\pi}\int \mathrm{Tr}\{\hat{D}^\dagger(\alpha)\hat{A}\}\hat{D}(\alpha)\,d^2\alpha\,. \tag{4.48}$$

The Fourier transform $W_A(\alpha)$ of the function $\mathrm{Tr}\{\hat{D}^\dagger(\beta)\hat{A}\}$,

$$W_A(\alpha) = \frac{1}{\pi}\int \mathrm{Tr}\{\hat{D}^\dagger(\beta)\hat{A}\}\exp(\alpha^*\beta - \beta^*\alpha)\,d^2\beta\,, \tag{4.49}$$

then defines the *Weyl symbol* of the operator $\hat{A}$. The operator $\hat{A}$ is uniquely determined by its Weyl symbol $W_A(\alpha)$ by the formula

$$\hat{A} = \frac{1}{\pi}\int W_A(\alpha)\hat{\Delta}(\alpha)\,d^2\alpha\,, \tag{4.50}$$

where

$$\hat{\Delta}(\alpha) = \frac{1}{\pi}\int \exp(\alpha\beta^* - \alpha^*\beta)\hat{D}(\beta)\,d^2\beta \tag{4.51}$$

and

$$W_A(\alpha) = \mathrm{Tr}\{\hat{\Delta}(\alpha)\hat{A}\}\,. \tag{4.52}$$

The Weyl symbol corresponds to the *symmetric* ordering. This is most easily to see comparing the relations (4.48) and (4.49). It follows that the operator is obtained from the Fourier expansion of its Weyl symbol just by replacing $\alpha \to \hat{a}$ and $\alpha^* \to \hat{a}^\dagger$ in the exponential function.

So the operator corresponding to a monomial $\alpha^{*m}\alpha^n$ is the symmetrized form $\{\hat{a}^{\dagger m}\hat{a}^n\}$ of the operator $\hat{a}^{\dagger m}\hat{a}^n$,

$$\{\hat{a}^{\dagger m}\hat{a}^n\} = \frac{1}{(m+n)!}S[\hat{a}^{\dagger m}\hat{a}^n]\,,$$

where $S$ represents the sum over all $(m+n)!$ permutations of all factors.

We have the following relations between the Weyl, covariant and contravariant symbols

$$W_A(\alpha) = \frac{2}{\pi}\int A^\circ(\alpha^*,\alpha)\exp\left(-2|\alpha-\beta|^2\right)\,d^2\beta\,, \tag{4.53}$$

$$A(\alpha^*, \alpha) = \frac{2}{\pi} \int W_A(\alpha) \exp\left(-2|\alpha - \beta|^2\right) d^2\beta \,. \qquad (4.54)$$

The covariant symbol is a smoothened version of the Weyl symbol and of contravariant symbol as well. The Weyl symbol is then a smoothened version of the contravariant symbol. When specialized to the statistical operator $\hat{\varrho}$ the above symbols are denoted as follows: the covariant symbol as $\pi\Phi_{\mathcal{A}}(\alpha)$, the symmetric symbol as $\pi\Phi_{\mathcal{S}}(\alpha)$, and the contravariant symbol as $\pi\Phi_{\mathcal{N}}(\alpha)$. While the covariant and symmetric symbols, as follows from the previous discussion, exist as usual functions for all statistical operators, the contravariant symbol exists only in the sense of distributions as already explained above, e.g. the contravariant symbol of the statistical operator $\hat{\varrho} = |\beta\rangle\langle\beta|$ representing the coherent state $|\beta\rangle$ is the $\delta$-function $\delta(\alpha - \beta)$ and in general the distribution $\Phi_{\mathcal{N}}$ (Glauber-Sudarshan quasidistribution) may have singularities even stronger than the $\delta$-function. The function $\Phi_{\mathcal{A}}$ is a non-negative, regular, bounded and integrable function. So it possesses many properties of the probability density, nevertheless it cannot be measured directly. The function $\Phi_{\mathcal{S}}$, which is called also *Wigner function*, need not be even non-negative. For the differences in comparison with the probability densities the symbols of the statistical operator are called *quasiprobability functions* or *quasidistributions*.

All quasidistributions are normalized owing to the unit trace of the statistical operator,

$$\int \Phi_{\mathcal{X}}(\alpha)\, d^2\alpha = 1 \,, \qquad (4.55)$$

where the index $\mathcal{X}$ represents any of $\mathcal{A}$, $\mathcal{S}$ and $\mathcal{N}$.

The corresponding *characteristic functions* $C_{\mathcal{X}}$ are then introduced as

$$C_{\mathcal{X}}(\beta) = \mathrm{Tr}\{\hat{\varrho}\hat{D}(\beta, s_{\mathcal{X}})\} \,, \qquad \hat{D}(\beta, s_{\mathcal{X}}) = \exp\left(\frac{s_{\mathcal{X}}|\beta|^2}{2}\right) \hat{D}(\beta) \,, \qquad (4.56)$$

where the value of the parameter $s_{\mathcal{X}}$ in the definition of the operator $\hat{D}(\beta, s_{\mathcal{X}})$ is $s_{\mathcal{X}} = 1$ for $\mathcal{X} = \mathcal{N}$, $s_{\mathcal{X}} = 0$ for $\mathcal{X} = \mathcal{S}$ and $s_{\mathcal{X}} = -1$ for $\mathcal{X} = \mathcal{A}$.

Making use of the Baker-Hausdorff identity we have

$$C_{\mathcal{N}}(\beta) = \exp\left(\frac{|\beta|^2}{2}\right) C_{\mathcal{S}}(\beta) = \exp\left(|\beta|^2\right) C_{\mathcal{A}}(\beta) \,. \qquad (4.57)$$

Here we have to comment on the terminology and notation used. Let us write out more explicitly the formulas for the characteristic functions,

$$C_{\mathcal{N}}(\beta) = \mathrm{Tr}\{\hat{\varrho} \exp(\beta\hat{a}^\dagger) \exp(-\beta^*\hat{a})\} \,, \qquad (4.58)$$

$$C_{\mathcal{A}}(\beta) = \mathrm{Tr}\{\hat{\varrho} \exp(-\beta^*\hat{a}) \exp(\beta\hat{a}^\dagger)\} \,, \qquad (4.59)$$

$$C_{\mathcal{S}}(\beta) = \mathrm{Tr}\{\hat{\varrho} \exp(\beta\hat{a}^\dagger - \beta^*\hat{a})\} \,. \qquad (4.60)$$

The index of the characteristic function refers to the ordering of the exponentials in these definition expressions; in the case of the corresponding quasidistributions $\Phi_{\mathcal{A}}$ and $\Phi_{\mathcal{N}}$ the suffix expresses the ordering of operators which are to be averaged by the corresponding quasidistribution, whereas they are obtained with the help of the opposite ordering.

The characteristic functions can easily be recognized as Fourier transforms of the corresponding quasidistributions $\Phi_{\mathcal{X}}$

$$\begin{aligned} C_{\mathcal{X}}(\beta) &= \int \Phi_{\mathcal{X}}(\alpha) \exp(\alpha^* \beta - \alpha \beta^*)\, d^2\alpha\,, \\ \Phi_{\mathcal{X}}(\alpha) &= \frac{1}{\pi^2} \int C_{\mathcal{X}}(\beta) \exp(\alpha \beta^* - \alpha^* \beta)\, d^2\beta\,. \end{aligned} \tag{4.61}$$

They are suitable for the computations of the moments

$$\langle \{\hat{a}^{\dagger m} \hat{a}^n\}_{\mathcal{X}} \rangle_{\hat{\varrho}} = \frac{\partial^{m+n}}{\partial \beta^m \partial(-\beta^*)^n} C_{\mathcal{X}} \bigg|_{\beta=\beta^*=0}\,, \tag{4.62}$$

where $\{\hat{a}^{\dagger m} \hat{a}^n\}_{\mathcal{X}}$ has been used for $\hat{a}^{\dagger m} \hat{a}^n$ in the case of normal ordering, for $\hat{a}^n \hat{a}^{\dagger m}$ in the case of antinormal ordering and for the $\{\hat{a}^{\dagger m} a^n\}$ in the case of symmetric ordering.

### 4.3.4 $\Omega$-ordering

The formula (4.57) related the characteristic functions for various types of orderings suggests a straightforward generalization unifying all kinds of already introduced symbols. The following discussion is valid for a sufficiently large class of operators and we will proceed formally without precise specification of the operators. Let us start from the Weyl symbol $W_A$ and its Fourier transform $\mathcal{F}W_A$

$$(\mathcal{F}W_A)(\beta) = \int \Phi_{\mathcal{X}}(\alpha) \exp(\alpha^* \beta - \alpha \beta^*)\, d^2\alpha\,.$$

The same relation as (4.57) holds for the Fourier transforms of the covariant, contravariant and Weyl symbols. The idea is to introduce the *$\Omega$-symbol* $a(\alpha, \Omega)$ of the operator $\hat{A}$ as a symbol with the Fourier transform $(\mathcal{F}a)(\beta, \Omega)$ of the form

$$(\mathcal{F}a)(\beta, \Omega) = \Omega(\beta^*, \beta)\, (\mathcal{F}W_A)(\beta)\,, \tag{4.63}$$

where $\Omega(\beta^*, \beta)$ is some function of the following properties: $\Omega(\beta^*, \beta)$ is a diagonal value of some entire analytic function $\Omega(\alpha, \beta)$ of two complex variables and it has no zeros, $\Omega(0,0) = 1$ and it is symmetric, $\Omega(\alpha, \beta) = \Omega(-\alpha, -\beta)$. In the spirit of the relation between the symbols and orderings it is natural to speak in this relation about $\Omega$-*ordering*. The relation between the symbol $a(\alpha, \Omega)$ and the Weyl symbol $W_A(\alpha)$ is then given by the inverse Fourier transform,

$$a(\alpha, \Omega) = \int W_A(\beta) K(\alpha - \beta)\, d^2\beta \,, \tag{4.64}$$

where the integration kernel is

$$K(\beta) = \frac{1}{\pi} \int \Omega(\gamma^*, \gamma) \exp(\beta\gamma^* - \beta^*\gamma)\, d^2\gamma \,. \tag{4.65}$$

Let us denote the ordering with the function $\Omega^{-1}$ as $\tilde{\Omega}$-ordering. Then it is easy to verify with the use of (4.50) that

$$\hat{A} = \frac{1}{\pi} \int a(\beta, \tilde{\Omega}) \hat{\Delta}(\beta, \Omega)\, d^2\beta \,, \tag{4.66}$$

$$\hat{\Delta}(\alpha, \Omega) = \frac{1}{\pi} \int \Omega(\beta^*, \beta) \hat{D}(\beta) \exp(\alpha\beta^* - \alpha^*\beta)\, d^2\beta \,, \tag{4.67}$$

which provides the $\Omega$-ordered form of the operator $\hat{A}$. In particular (4.66) can be applied to the statistical operator $\hat{\varrho}$ with the quasidistribution $\Phi(\beta, \tilde{\Omega}) \equiv a(\beta, \tilde{\Omega})/\pi$. Operators $\hat{\Delta}(\alpha, \Omega)$ are complete in the following sense

$$\mathrm{Tr}\{\hat{\Delta}(\alpha, \Omega) \hat{\Delta}(\alpha, \tilde{\Omega})\} = \pi\delta(\alpha - \beta) \,, \tag{4.68}$$

which is a direct consequence of the completeness (4.47) of the displacement operators $\hat{D}(\alpha)$. Hence the $\Omega$-symbol $a(\alpha, \Omega)$ of the operator $\hat{A}$ can be expressed as

$$a(\alpha, \Omega) = \mathrm{Tr}\{\hat{\Delta}(\alpha, \Omega) \hat{A}\} \,. \tag{4.69}$$

The covariant, contravariant and symmetric symbols correspond in view of (4.57) to the choice

$$\Omega(\beta^*, \beta) = \exp\left(\frac{s|\beta|^2}{2}\right) \,, \tag{4.70}$$

with $s = 1$ for the normal ordering, $s = 0$ for symmetric ordering and $s = -1$ for the antinormal ordering. So for $s = 0$ we have from (4.66) the symmetrized form (4.50) of the operator $\hat{A}$ and for $s = -1$ we have the diagonal representation of $\hat{A}$ ( $\hat{\Delta}(\beta, -1) = |\beta\rangle\langle\beta|$ ). It follows from (4.47) [(4.68)] that

$$\mathrm{Tr}\{\hat{A}\hat{B}\} = \frac{1}{\pi} \int a(\alpha, \Omega)\, b(\alpha, \tilde{\Omega})\, d^2\alpha \tag{4.71}$$

and because the $\Omega$-symbol of the unit operator $\hat{1}$ is equal to the unity we have also

$$\mathrm{Tr}\{\hat{A}\} = \frac{1}{\pi} \int a(\alpha, \Omega)\, d^2\alpha \,. \tag{4.72}$$

Let us now denote as $\{\hat{A}\}_\Omega$ the operator related to the operator $\hat{A}$ by the prescription

$$\{\hat{A}\}_\Omega = \frac{1}{\pi}\int A(\alpha^*,\alpha)\hat{\Delta}(\alpha,\Omega)\, d^2\alpha\,. \tag{4.73}$$

The correspoding characteristic fuction $C(\beta,\Omega)$ and quasidistribution $\Phi(\alpha,\Omega)$ are now related as

$$C(\beta,\Omega) = \mathrm{Tr}\{\hat{\varrho}\Omega(\beta^*,\beta)\hat{D}(\beta)\} = \int \Phi(\alpha,\Omega)\exp\left(\alpha^*\beta - \alpha\beta^*\right)\, d^2\alpha \tag{4.74}$$

and $\Phi_{\mathcal{N}}(\alpha) = \Phi(\alpha,1)$, $\Phi_{\mathcal{S}}(\alpha) = \Phi(\alpha,0)$, $\Phi_{\mathcal{A}}(\alpha) = \Phi(\alpha,-1)$. Now using the definition of the operator $\{\hat{a}^{\dagger m}\hat{a}^n\}_\Omega$ and the completeness of the operators $\hat{\Delta}(\alpha,\Omega)$ we can compute the $\Omega$-ordered moments of $\Phi(\alpha,\Omega)$ with the help of the characteristic function $C(\beta,\Omega)$ as

$$\begin{aligned}\langle \hat{a}^{\dagger m}\hat{a}^n\rangle_\Omega &\equiv \mathrm{Tr}\left[\hat{\varrho}\left\{\hat{a}^{\dagger m}\hat{a}^n\right\}_\Omega\right] = \int \Phi(\alpha,\Omega)\,\alpha^{*m}\alpha^n\, d^2\alpha \\ &= \left.\frac{\partial^{m+n}}{\partial\beta^m\partial(-\beta^*)^n}C(\beta,\Omega)\right|_{\beta=\beta^*=0}.\end{aligned} \tag{4.75}$$

If we now consider two orderings $\Omega_1$ and $\Omega_2$, all connecting relations for all corresponding quantities can be given starting from the obvious relation for the characteristic functions $C(\alpha,\Omega_1)$ and $C(\alpha,\Omega_2)$,

$$C(\alpha,\Omega_2) = C(\alpha,\Omega_1)\tilde{\Omega}_1(\alpha^*,\alpha)\Omega_2(\alpha^*,\alpha)\,. \tag{4.76}$$

Taking the inverse Fourier transform we get the connecting relation between the corresponding quasidistributions

$$\Phi(\alpha,\Omega_2) = \int \Phi(\beta,\Omega_1)K_{21}(\alpha-\beta)\, d^2\beta\,, \tag{4.77}$$

where the integration kernel $K_{21}$ is given by

$$K_{21}(\beta) = \frac{1}{\pi}\int \tilde{\Omega}_1(\gamma^*,\gamma)\,\Omega_2(\gamma^*,\gamma)\exp\left(\beta\gamma^* - \beta^*\gamma\right)\, d^2\gamma\,. \tag{4.78}$$

The same connecting relation for the operators $\hat{\Delta}(\alpha,\Omega_1)$ and $\hat{\Delta}(\alpha,\Omega_2)$

$$\hat{\Delta}(\alpha,\Omega_2) = \int \hat{\Delta}(\beta,\Omega_1)K_{21}(\alpha-\beta)\, d^2\beta \tag{4.79}$$

holds as follows form the definition (4.67) of the operator $\hat{\Delta}(\alpha,\Omega)$. Starting from (4.76) and remembering the properties of the Fourier transform, we see immediately that the above integral relations possess also differential form

$$\Phi(\alpha,\Omega_2) = \tilde{\Omega}_1\left(\frac{\partial}{\partial\alpha},-\frac{\partial}{\partial\alpha^*}\right)\Omega_2\left(\frac{\partial}{\partial\alpha},-\frac{\partial}{\partial\alpha^*}\right)\Phi(\alpha,\Omega_1)\,. \tag{4.80}$$

The same relation takes place, of course, also for the corresponding operators $\hat{\Delta}(\alpha,\Omega_i)$, $i=1,2$.

## 4.4 Squeezed states

Let us now assume such pure states for which the equality in the Heisenberg uncertainty relation (2.25), i.e. equation (4.28), is valid. This is possible as follows from the discussion leading to (2.25) if and only if the expectation value of the operator

$$(\hat{b}^\dagger - \beta^*)(\hat{b} - \beta), \tag{4.81}$$

where

$$\hat{b} = \frac{\hat{x}}{2\delta x} + i\hat{p}\delta x, \qquad \beta = \langle \hat{b} \rangle, \tag{4.82}$$

is zero. For the pure state $|\beta\rangle_m$ this means equivalently that there is an eigenstate

$$\hat{b}|\beta\rangle_m = \beta|\beta\rangle_m \tag{4.83}$$

of the operator $\hat{b}$ with the eigenvalue $\beta = \langle \hat{b} \rangle$. As can be shown [Thirring (1983)], for the mixed states the equality in the uncertainty relation can never occur and we have obtained the most general states for which (4.28) holds.

The operators $\hat{b}$ and $\hat{b}^\dagger$ commute as

$$[\hat{b}, \hat{b}^\dagger] = \hat{1} \tag{4.84}$$

and they can be expressed simply as proper linear combinations of creation and annihilation operators $\hat{a}^\dagger$ and $\hat{a}$. Let us now assume the general homogeneous linear transformation of creation and annihilation operators $\hat{a}^\dagger$ and $\hat{a}$ of the form

$$\begin{aligned} \hat{b} &= \mu\hat{a} + \nu\hat{a}^\dagger \\ \hat{b}^\dagger &= \mu^*\hat{a}^\dagger + \nu^*\hat{a}\,. \end{aligned} \tag{4.85}$$

We shall also demand that the transformations are canonical so that (4.84) holds. This implies that

$$|\mu|^2 - |\nu|^2 = 1\,. \tag{4.86}$$

Such transformations form a group, which can be identified with symplectic group Sp$(2, R)$ of $2 \times 2$ matrices $g(\mu, \nu)$ of the form

$$g(\mu,\nu) = \begin{bmatrix} \mu & \nu \\ \nu^* & \mu^* \end{bmatrix},$$

fulfilling (4.86). This group is isomorphic with the group SU$(1,1)$. As a consequence of the Stone-von Neumann theorem of Sec. 2.6 there exists a unitary transformation $\hat{U}(\mu,\nu) = \exp[i\hat{H}(\mu,\nu)]$ with the Hermitian infinitesimal generator $\hat{H}(\mu,\nu)$ such that

$$\hat{b} = \hat{U}(\mu,\nu)\hat{a}\hat{U}^\dagger(\mu,\nu)\,. \tag{4.87}$$

Now we shall omit the dependence of the operators $\hat{U}$ and $\hat{H}$ on the parameters $\mu$ and $\nu$ for convenience. It is easy to describe the general form of the generator $\hat{H}$. It follows from (4.85) that the result of the commutator $[\hat{H}, \hat{a}]$ should be linear in creation and annihilation operators $\hat{a}^\dagger$ and $\hat{a}$, so that $\hat{H}$ is a linear combination of terms at most quadratic in operators $\hat{a}^\dagger$ and $\hat{a}$. Consequently

$$\hat{U} = \exp\left[i\frac{\gamma}{2}(\hat{a}^\dagger\hat{a} + \hat{a}\hat{a}^\dagger) + \frac{\zeta}{2}\hat{a}^{\dagger 2} - \frac{\zeta^*}{2}\hat{a}^2\right] . \tag{4.88}$$

The real parameter $\gamma$ and complex parameter $\zeta$ are of course depending on the parameters $\mu$ and $\nu$. We shall not need this dependence explicitly in the following. The operators then describe a unitary representation of the symplectic group $\mathrm{Sp}(2, R)$ acting on the Fock space $\mathcal{H}$. The operators

$$\hat{X} = \hat{a}^2 , \qquad \hat{X}^\dagger = \hat{a}^{\dagger 2} , \qquad \hat{Y} = \frac{1}{2}(\hat{a}\hat{a}^\dagger + \hat{a}^\dagger\hat{a}) \tag{4.89}$$

describe the unitary representation of the corresponding Lie algebra $\mathrm{sp}(2, R)$ (or equivalently $\mathrm{su}(1,1)$). This representation is not irreducible because the subspaces of the Fock space $\mathcal{H}$ corresponding to the odd and even particle numbers are invariant under the action of the generators.

Now we can easy describe the eigenvectors of the operator $\hat{b}$. It follows from (4.87) that all eigenvectors $|\beta\rangle_{\mu,\nu}$ of the operator $\hat{b}$ are of the form

$$\begin{aligned} |\beta\rangle_{\mu,\nu} &= \hat{U}|\beta\rangle , \\ \hat{b}|\beta\rangle_{\mu,\nu} &= \beta|\beta\rangle_{\mu,\nu} , \end{aligned} \tag{4.90}$$

where $|\beta\rangle = \hat{D}(\beta)|0\rangle$ is the canonical coherent state.

From the equations (4.85) and (4.87) we also deduce that

$$\hat{U}\hat{D}(\alpha)\hat{U}^\dagger = \hat{D}(\tilde{\alpha}) , \qquad \tilde{\alpha} = \mu\alpha - \nu^*\alpha^* .$$

In this way the eigenstates of the operator $\hat{b}$ can be understood as coherent states associated with the representation of the Weyl-Heisenberg group in the Fock space and with the fiducial vector $\hat{U}|0\rangle$. Let us note that the operator $\hat{U}$ can always be written in the form

$$\hat{U} = N\exp\left(\frac{\xi}{2}\hat{X}^\dagger\right)\exp\left(\alpha\hat{Y}\right)\exp\left(\frac{\eta}{2}\hat{X}\right) , \tag{4.91}$$

where $N$ is a normalization factor, which will be explicitly given later on. This can be shown using the fundamental representation of the Lie algebra $\mathrm{sp}(2, R)$ $(\mathrm{su}(1,1))$ given with the help of Pauli matrices $\hat{\sigma}_i$, $i = 1, 2, 3$

$$\hat{X}^\dagger = -(\hat{\sigma}_1 + i\hat{\sigma}_2) , \qquad \hat{X} = -(\hat{\sigma}_1 - i\hat{\sigma}_2) , \qquad \hat{Y} = i\hat{\sigma}_3 .$$

So up to some non-substantional phase factor, the fiducial vector can be written in the form

$$N \exp\left(\frac{\xi}{2}\hat{X}^\dagger\right)|0\rangle\,. \tag{4.92}$$

Because of its dependence on the parameter $\xi$ it will be denoted as $|\xi\rangle$. The fiducial vector $|\xi\rangle$ is of course the eigenvector of the operator $\hat{b} = \mu\hat{a} + \nu\hat{a}^\dagger$ with the zero eigenvalue. From this we conclude that $\xi = -\nu/\mu$ and that $|\mu|^2 - |\nu|^2 = 1$ so that we have directly $\xi\xi^* < 1$. The normalization factor $N$ then is

$$N = (1 - \xi\xi^*)^{1/4}\,.$$

So as follows from our discussion only the relative phase of the complex numbers $\mu$ and $\nu$ is essential and we might assume without loss of generality the $\mu$ to be real from the beginning.

The states just described (the eigenstates of the operator $\hat{b}$ given in (4.85) are called *squeezed states.* The corresponding fiducial vector $|\xi\rangle$ determined by (4.92) is referred to as the *squeezed vacuum.* It is evident that if expressed in terms of the Fock states only, the squeezed vacuum will be a superposition of even-number Fock states. For this reason the squeezed states are called also *two-photon coherent states.* If both $\mu$ and $\nu$ are real, then we obtain the minimum uncertainty states $|\beta\rangle_m$ discussed above.

Let us compute the mean square deviations $\delta x$ and $\delta p$ of the position and momentum operators $\hat{x}$ and $\hat{p}$ in the squeezed state $|\beta\rangle_{\mu,\nu}$. The result is

$$\begin{aligned} \delta x &= \frac{1}{\sqrt{2}}|\mu - \nu|\,, \\ \delta p &= \frac{1}{\sqrt{2}}|\mu + \nu|\,. \end{aligned} \tag{4.93}$$

It can happen that one of these quantities may be less then $1/\sqrt{2}$, the value which corresponds to the canonical coherent states. This is e.g. the case of any minimum uncertainty state which is not a canonical coherent state. If in this case $\xi \to -1$ or $\xi \to 1$ then $\delta x \to 0$ or $\delta p \to 0$, respectively. This explaines the name squeezed states. Unlike the canonical coherent state $|\alpha\rangle$, which satisfies

$$\langle\alpha|(\hat{a} - \langle\hat{a}\rangle)^\dagger(\hat{a} - \langle\hat{a}\rangle)|\alpha\rangle = 0\,, \tag{4.94}$$

the squeezed state $|\beta\rangle_{\mu,\nu}$ involves the quantum noise

$$_{\mu,\nu}\langle\beta|(\hat{a} - \langle\hat{a}\rangle)^\dagger(\hat{a} - \langle\hat{a}\rangle)|\beta\rangle_{\mu,\nu} = |\nu|^2\,. \tag{4.95}$$

A direct computation shows that the squeezed states minimalize the more general uncertainty relation (2.26). This can be understood taking into account the fact that the equality in the uncertainty relation (2.26) is covariant under the linear transformations (4.85) (see Sec. 5.1 and relation (5.10)).

Another useful interpretation of the squeezed states is given in terms of the inhomogeneous symplectic Lie group ISp(2, $R$) which is a subgroup of the symplectic Lie group Sp(4). More concretely we have a unitary irreducible representation of the isp(2, $R$) algebra in the Fock space $\mathcal{H}$ in which the generators are represented by the following set of operators:

$$\left\{\hat{n} = \hat{a}^\dagger \hat{a} + \frac{1}{2}, \hat{a}^{\dagger 2}, \hat{a}^2, \hat{a}^\dagger, \hat{a}, \hat{1}\right\} . \tag{4.96}$$

It is easy to see that the commutator of any two linear combinations of the generators (4.96) gives again a linear combination of the generators. So the linear space spanned by these generators is really a Lie algebra. The Lie sub-algebras generated by elements $\left\{\hat{a}^\dagger\hat{a} + 1/2, \hat{a}^{\dagger 2}, \hat{a}^2\right\}$, $\left\{\hat{a}^\dagger\hat{a} + 1/2, \hat{a}^\dagger, \hat{a}, \hat{1}\right\}$ and $\left\{\hat{a}^\dagger, \hat{a}, \hat{1}\right\}$ are the sp(2, $R$) (su(1, 1)) algebra, the *oscillator* algebra and the Weyl-Heisenberg algebra, respectively. Let us now assume the states in $\mathcal{H}$ of the form

$$\exp\left(i\hat{H}\right)|0\rangle , \tag{4.97}$$

where $\hat{H} \in$ isp(2, $R$). We can use the $4 \times 4$ faithful matrix representation of isp(2, $R$) to prove [Zhang, Feng and Gilmore (1990)] that the operator $\exp(i\hat{H})$ can always be decomposend to

$$\exp(i\hat{H}) = \exp\left(\alpha\hat{a}^\dagger - \alpha^*\hat{a}\right)\exp\left(\zeta\hat{a}^{\dagger 2} - \zeta^*\hat{a}^2\right)\exp\left(\eta\hat{a}^\dagger\hat{a} + \delta\hat{1}\right) . \tag{4.98}$$

So the states (4.97) can always be written in the form

$$\exp\left(\alpha\hat{a}^\dagger - \alpha^*\hat{a}\right)\exp\left(\zeta\hat{a}^{\dagger 2} - \zeta^*\hat{a}^2\right)|0\rangle . \tag{4.99}$$

Using already mentioned decomposition property (4.91) of the Sp(2, $R$) (SU(1, 1)) group, we can see that the state (4.99) is identical with the squeezed state $|\xi\rangle$ of (4.92) if we identify $\xi = \exp(i\phi)\tanh(|\zeta|)$, where $\phi$ is the phase of $\zeta$.

From the above discussion we see that the squeezed states can also be viewed as coherent states associated with the unitary irreducible representation of the inhomogeneous symplectic group ISp(2, $R$) in the Fock space and the Fock vacuum $|0\rangle$ as the fiducial vector. So in accordance with the general result of Sec. 4.2 the set of squeezed states remains invariant under the time evolution generated by the hamiltonian $\hat{H}$ of the form

$$\hat{H} = \omega(t)\hat{a}^\dagger\hat{a} + f(t)\hat{a}^{\dagger 2} + f^*(t)\hat{a}^2 + g(t)\hat{a}^\dagger + g^*(t)\hat{a} \tag{4.100}$$

with arbitrary deterministic functions $\omega(t)$, $f(t)$ and $g(t)$. The subset of canonical coherent states is then invariant under the time evolution governed by the hamiltonian (4.100) with $f(t) = 0$.

## 4.5 Spin coherent states

The *spin coherent states*, which are sometimes called also *atomic coherent states* are a further particular example of the general construction of the group-related coherent states. Let us assume the compact Lie group SO(3) of the rotations of the three-dimensional space. For our purpose it will be more convenient to work with the group SU(2) which is closely related to the group SO(3). The group SO(3) is a factor group of the group SU(2) by its center (the set of its elements commuting with all group elements) $Z_2 = \{1, -1\}$. The compact Lie group SU(2) is by definition the group of $2 \times 2$ unitary matrices $g$ with the unit determinant. So we can write

$$g = \begin{bmatrix} a & b \\ -b^* & a^* \end{bmatrix}, \qquad |a|^2 + |b|^2 = 1 . \tag{4.101}$$

The matrices $g$ given by (4.101) define the fundamental $2 \times 2$ representation of the Lie group SU(2). The corresponding Lie algebra su(2) is generated by three elements $\hat{J} = (\hat{J}_1, \hat{J}_2, \hat{J}_3)$, with familiar commutation relations

$$[\hat{J}_i, \hat{J}_j] = i\varepsilon_{ijk}\hat{J}_k , \tag{4.102}$$

where $\varepsilon_{ijk}$ is the Levi-Civita symbol (summation is performed over the repeated indices). The raising and lowering operators $\hat{J}_+$ and $\hat{J}_-$ are introduced as

$$\hat{J}_+ = \hat{J}_1 + i\hat{J}_2 , \qquad \hat{J}_- = \hat{J}_1 - i\hat{J}_2 ,$$
$$[\hat{J}_3, \hat{J}_-] = \hat{J}_+ , \qquad [\hat{J}_3, \hat{J}_+] = -\hat{J}_- , \qquad [\hat{J}_+, \hat{J}_-] = 2\hat{J}_3 . \tag{4.103}$$

All non-equivalent unitary irreducible representations $\hat{T}^j$ of the Lie group (Lie algebra) SU(2) (su(2)) are as well known and labelled by distinct half-integer values $j = 0, 1/2, 1, \ldots$ of the Casimir operator

$$\hat{J}_1^2 + \hat{J}_2^2 + \hat{J}_3^2 = j(j+1) . \tag{4.104}$$

For the fixed value $j$ the correspoding $(2j+1)$-dimensional representation Hilbert space $\mathcal{H}_j$ is then spanned on the vectors (Dicke states)

$$|m, j\rangle = \left(\frac{(j-m)!}{(2j)!\,(j+m)!}\right)^{1/2} \hat{J}_+^{m+j}|-j, j\rangle , \tag{4.105}$$

where $m = -j, -j+1, \ldots, j-1, j$. The generators are then acting on $\mathcal{H}_j$ as

$$\begin{aligned} \hat{J}_3|m, j\rangle &= m|m, j\rangle , \\ \hat{J}_+|m, j\rangle &= [(j-m)(j+m+1)]^{1/2}|m+1, j\rangle , \\ \hat{J}_-|m, j\rangle &= [(j+m)(j-m+1)]^{1/2}|m-1, j\rangle . \end{aligned} \tag{4.106}$$

The representation with $j = 0$ is the trivial one-dimensional representation. The simplest non-trivial representation with $j = 1/2$ is the $2 \times 2$ representation given by Pauli matrices $\hat{J}_i = (1/2)\hat{\sigma}_i$ corresponding to the definition of fundamental representation of the Lie group SU(2).

The most natural fiducial state for the construction of the set of coherent states is the lowest (highest) weight vector $|-j,j\rangle$ ($|j,j\rangle$). Let us assume the first possibility because the second choice leads to the same set of coherent states. From (4.105) we see immediately that the stability subgroup of the fiducial state $|-j,j\rangle$ is the one parameter subgroup U(1) of diagonal $2 \times 2$ matrices with the unit determinant,

$$\begin{bmatrix} \exp(i\psi) & 0 \\ 0 & \exp(-i\psi) \end{bmatrix}, \tag{4.107}$$

generated by the generator $\hat{J}_3$. The factor space $G/H$ can then be identified with the set of elements of SU(2) in the form

$$\tilde{g} = \begin{bmatrix} \alpha & \beta \\ -\beta^* & \alpha \end{bmatrix}, \qquad \alpha^2 + |\beta|^2 = 1, \qquad \alpha \geq 0, \qquad \alpha \in R. \tag{4.108}$$

The convenient parametrization of the coset space $G/H$ which can be identified with the two-dimensional unit sphere $S^2$ (the set of unit vectors $\mathbf{n} = (\sin\theta\cos\psi, \sin\theta\sin\psi, \cos\theta)$) is the following

$$\alpha = \cos\left(\frac{\theta}{2}\right), \qquad \beta = -\exp(-i\psi)\sin\left(\frac{\theta}{2}\right), \tag{4.109}$$

with $\theta \in [0,\pi]$ and $\psi \in [0,2\pi)$. Using the properties of the Pauli matrices it is easy to obtain that any element $\tilde{g}$ of the coset space $G/H$ given by (4.108) can be written as

$$\tilde{g} = \exp\left(i\frac{\theta}{2}\left[\hat{\sigma}_1\sin(\psi) - \hat{\sigma}_2\cos(\psi)\right]\right). \tag{4.110}$$

So we have

$$\hat{T}^j(\tilde{g}) \equiv \hat{D}(\mathbf{n}) = \exp\left(i\theta\left[\hat{J}_1\sin(\psi) - \hat{J}_2\cos(\psi)\right]\right). \tag{4.111}$$

In the last formula we did not distinguish graphically between the generators $\hat{J}_i$ and their representatives in the representation $\hat{T}^j$. So the set of coherent states associated with the representation $\hat{T}^j$ is of the form

$$\{\hat{T}^j(\tilde{g})|-j,j\rangle\,;\ \tilde{g} \in G/H\}. \tag{4.112}$$

To get a more explicit formula for the coherent states we can use a convenient disentangling formula which is sufficient to be used in the fundamental representation

$$\exp\left(\xi\hat{J}_+ - \xi^*\hat{J}_-\right) = \exp\left(\zeta\hat{J}_+\right)\exp\left(\eta\hat{J}_3\right)\exp\left(-\zeta^*\hat{J}_-\right),$$

$$\xi = -\frac{\theta}{2}\exp(-i\psi)\,, \qquad \zeta = -\tan\left(\frac{\theta}{2}\right)\exp(-i\psi)\,, \qquad \eta = \ln\left(1+|\zeta|^2\right). \tag{4.113}$$

According to the above formulas we can write the coherent state corresponding to the point of the coset space $G/H$ parametrized by the coordinates $(\theta, \psi)$, or equivalently by the coordinate $\zeta$, as (atomic coherent states — Bloch states)

$$\begin{aligned} |\theta,\psi\rangle &= |\zeta\rangle = \left(1+|\zeta|^2\right)^{-j}\exp\left(\zeta\hat{J}_+\right)|-j,j\rangle \\ &= \left(1+|\zeta|^2\right)^{-j}\sum_{m=-j}^{m=j}\left(\frac{(2j)!}{(j-m)!\,(j+m)!}\right)^{1/2}\zeta^{j+m}|m,j\rangle\,. \end{aligned} \tag{4.114}$$

Using the parametrization of the homogeneous space $G/H$ with the help of the coordinate $\zeta$, we can write any element $\tilde{g}$ in (4.108) in the form

$$\tilde{g} = \left(1+|\zeta|^2\right)^{-1/2}\begin{bmatrix} 1 & \zeta \\ -\zeta^* & 1 \end{bmatrix}. \tag{4.115}$$

It is easy to see that the group action in the coordinate $\zeta$ is given by

$$\tilde{g}\ :\ \zeta \to \frac{a\zeta + b}{-b^*\zeta + a^*} \tag{4.116}$$

and any invariant measure on the homogeneous space $G/H$ is proportional to the measure $d\mu$

$$d\mu(\zeta) = \frac{d^2\zeta}{\pi(1+|\zeta|^2)^2}\,. \tag{4.117}$$

In variables $\theta$, $\psi$ we have

$$d\mu(\theta,\psi) = \frac{\sin\theta}{4\pi}\,d\theta\,d\psi\,. \tag{4.118}$$

The relation between the two types of coordinates introduced on $G/H$ is just the stereographic projection of the two-dimensional sphere from its south pole on to the complex plane $\zeta$. The sphere without its south pole is then isomorphic to the complex plane via the stereographic projection. The south pole itself is of the measure zero with respect to the invariant measure (4.118), so it does not contribute to the resolution of unity which takes the form

$$\hat{1} = (2j+1)\int |\zeta\rangle\langle\zeta|\, d\mu(\zeta) = (2j+1)\int |\theta,\psi\rangle\langle\theta,\psi|\, d\mu(\theta,\psi)\,. \qquad (4.119)$$

The spin coherent states have the following eigenvalue property

$$(\mathbf{n}\cdot\hat{J})\,|\theta,\psi\rangle = -j|\theta,\psi\rangle\,, \qquad (4.120)$$

where $\mathbf{n} = (\sin\theta\cos\psi, \sin\theta\sin\psi, \cos\theta)$ is the unit vector in the direction $(\theta,\psi)$. It holds also that

$$(\hat{J}_- + 2\zeta\hat{J}_3 - \zeta^2\hat{J}_+)|\zeta\rangle = 0\,. \qquad (4.121)$$

Unlike the canonical coherent states, the spin coherent states cannot be all simultaneously the minimum uncertainty states for any choice of the fiducial vector. The sets of states which do fulfil the equality in the Heisenberg uncertainty relation concerning the operators $\hat{J}_i$ are called *intelligent* spin states. Nevertheless the spin coherent state $|\theta,\psi\rangle$ minimalizes the uncertainty relation

$$\delta\hat{J}_1'\delta\hat{J}_2' \geq \frac{1}{2}\,|\langle\hat{J}_3'\rangle|\,, \qquad (4.122)$$

where

$$\hat{J}_i' = \hat{D}(\mathbf{n})\hat{J}_i\hat{D}^{-1}(\mathbf{n})\,. \qquad (4.123)$$

As in the case of canonical coherent states we have, using the spin coherent states, the realization of the represenation space $\mathcal{H}_j$ as the space of polynomials of the order less than or equal to $2j$ in the complex variable $\zeta$ . Really, for any state $|\phi\rangle \in \mathcal{H}_j$ the expression $\langle\zeta^*|\phi\rangle$ is such a polynomial appart from a common factor $(1+|\zeta|)^{-j}$ and it determines the state $|\phi\rangle$ uniquely. The resolution of unity (4.119) then gives the scalar product for two polynomials $f_1(\zeta)$ and $f_2(\zeta)$ in the form

$$\langle f_1|f_2\rangle = \frac{2j+1}{\pi}\int f_1^*(\zeta)f_2(\zeta)\,\frac{d^2\zeta}{(1+|\zeta|^2)^{2+2j}}\,. \qquad (4.124)$$

The group representation $\hat{T}^j$ is given by

$$(\hat{T}^j(g)f)(\zeta) = (b\zeta + a^*)^{2j} f\left(\frac{a\zeta - b^*}{b\zeta + a^*}\right) \qquad (4.125)$$

and the generators $\hat{J}_\pm$ and $\hat{J}_3$ are realized as the first-order differential operators

$$\hat{J}_+ = -\zeta^2\frac{d}{d\zeta} + 2j\zeta\,, \qquad \hat{J}_- = \frac{d}{d\zeta}\,, \qquad \hat{J}_3 = \zeta\frac{d}{d\zeta} - j\,. \qquad (4.126)$$

The states $|m,j\rangle$ are represented by

$$|m,j\rangle = \left(\frac{(2j)!}{(j+m)!\,(j-m)!}\right)^{1/2} \zeta^{j+m} . \tag{4.127}$$

We should also mention some relations of the spin coherent states with the canonical coherent states. For the first relation let us note that there is a unitary reducible representation of the group SU(2) in the Fock space corresponding to two different sorts of bosons. Let $\hat{a}_1$, $\hat{a}_2$ be the corresponding annihilation operators. Then we have

$$\hat{J}_+ = \hat{a}_1^\dagger \hat{a}_2 , \qquad \hat{J}_- = \hat{J}_+^\dagger , \qquad \hat{J}_3 = \frac{1}{2}(\hat{a}_1^\dagger \hat{a}_1 - \hat{a}_2^\dagger a_2) . \tag{4.128}$$

The irreducible representation $\hat{T}^j$ is obtained by the restriction to the invariant subspace $\mathcal{H}_j$ of the Fock space on which the total number of particles $n_1 + n_2 = 2j$ is fixed. Let us assume the canonical coherent state $|\alpha_1, \alpha_2\rangle$ for two degrees of freedom. The spin coherent state $|\zeta\rangle$ with $\zeta = \alpha_1/\alpha_2$ is then obtained up to some phase factor from the canonical coherent state $|\alpha_1, \alpha_2\rangle$ by the projection to the subspace $\mathcal{H}_j$ and by subsequent normalization.

Secondly, for the values of $j \to \infty$ the spin coherent states go to the canonical coherent states. For this it is necessary to take the contraction of the Lie algebra su(2) and its representation $\hat{T}^j$ yielding the Heisenberg-Weyl Lie algebra

$$(2j)^{-1/2}\hat{J}_+ \to \hat{a}^\dagger , \qquad (2j)^{-1/2}\hat{J}_- \to \hat{a} , \qquad j^{-1}\hat{J}_3 \to -1 . \tag{4.129}$$

If we denote $\lim_{j\to\infty}(2j)^{1/2}\zeta = \alpha$ then we have

$$|\zeta\rangle \to |\alpha\rangle ,$$

where $|\alpha\rangle$ is the canonical coherent state.

## 4.6 Heisenberg-Langevin approach to quantum statistics

At the end of this chapter we will review briefly the Heisenberg-Langevin and Schrödinger approaches to quantum statistics of interacting systems of radiation and matter. We will do that from the point of view of the coherent-state technique. More detailed treatments can be found in the books [Lax (1968), Haken (1970), Scully and Whitney (1972), Louisell (1973), Agarwal (1973), Haake (1973), Van Kampen (1981), Risken (1984), Meystre and Sargent III (1991), Gardiner (1991)].

We will consider $N$ radiation modes described by the annihilation and creation photon operators $\hat{a}_j$ and $\hat{a}_j^\dagger$, respectively, with photon frequencies $\omega_j$ in the interaction with large reservoirs described by the boson annihilation (creation) operators $\hat{b}_l^{(j)}$ ($\hat{b}_l^{(j)\dagger}$) having frequencies $\psi_l^{(j)}$. We assume a smooth broad dense spectrum of oscillator frequencies and a linear interaction between radiation and the reservoir oscillators with smooth coupling constants $\kappa_l^{(j)}$ as functions of frequencies. We take also into account rotating terms (squeezed reservoirs). The hamiltonian of this system can be written as

$$\hat{H} = \hat{H}_{0,\mathrm{rad}} + \hat{H}_{0,\mathrm{res}} + \hat{H}_{\mathrm{int}}\,, \tag{4.130}$$

where

$$\hat{H}_{0,\mathrm{rad}} = \sum_j \hbar\omega_j \left(\hat{a}_j^\dagger \hat{a}_j + \frac{1}{2}\right) \tag{4.131}$$

is the free radiation hamiltonian,

$$\hat{H}_{0,\mathrm{res}} = \sum_{j,l} \hbar\psi_l^{(j)} \left(\hat{b}_j^{(l)\dagger}\hat{b}_j^{(l)} + \frac{1}{2}\right) \tag{4.132}$$

is the free reservoir hamiltonian, and

$$\hat{H}_{\mathrm{int}} = \sum_{j,l} \hbar\kappa_l^{(j)}\hat{b}_l^{(j)}\left(\hat{a}_j^\dagger + \eta\hat{a}_j\right) + \mathrm{h.c.} \tag{4.133}$$

is the interaction hamiltonian among the radiation modes and the infinite reservoir system $\{\hat{b}_l^{(j)}\}$. Here $\eta = 0$ for the standard reservoir neglecting the rotating-wave terms, and $\eta = 1$ for the squeezed reservoir.

We can write down the Heisenberg equations of motion for $\hat{a}_j$ and $\hat{b}_l^{(j)}$ as follows

$$\frac{d\hat{a}_j}{dt} = -i\omega_j\hat{a}_j - i\sum_l \kappa_l^{(j)}\hat{b}_l^{(j)} - i\eta\sum_l \kappa_l^{(j)*}\hat{b}_l^{(j)\dagger}\,, \tag{4.134a}$$

$$\frac{d\hat{b}_l^{(j)}}{dt} = -i\psi_l^{(j)}\hat{b}_l^{(j)} - i\kappa_l^{(j)*}\left(\hat{a}_j + \eta\hat{a}_j^\dagger\right)\,. \tag{4.134b}$$

The solution of the last equation (4.134b) can be written in the form

$$\begin{aligned}
\hat{b}_l^{(j)}(t) &= \hat{b}_l^{(j)}(0)\exp\left(-i\psi_l^{(j)}t\right) \\
&\quad - i\kappa_l^{(j)*}\int_0^t \left[\hat{a}_j(t') + \eta\hat{a}_j^\dagger(t')\right]\exp\left[-i\psi_l^{(j)}(t-t')\right]dt' \\
&\approx \hat{b}_l^{(j)}(0)\exp\left(-i\psi_l^{(j)}t\right) \\
&\quad - i\kappa_l^{(j)*}\hat{a}_j(t)\int_0^\infty \exp\left[-i(\psi_l^{(j)} - \omega_j)t'\right]dt'\,,
\end{aligned} \tag{4.135}$$

assuming the quasimonochromatic radiation, $\hat{a}_j(t') \approx \hat{a}_j(t)\exp[-i\omega_j(t-t')]$; the first term in (4.135) represents the free solution for the reservoir operator in the absence of the interaction, the second one expresses that the radiation field is a source of $\hat{b}_l^{(j)}$ in the interaction. Performing the integration approximately over the interval $(0,\infty)$ and using the symbolic identity

$$\int_0^{\infty} \exp\left[i(\omega_j - \psi_l^{(j)})t\right] dt = \lim_{\varepsilon\to+0} \frac{1}{-i(\omega_j-\psi_l^{(j)})+\varepsilon} = \frac{i\mathcal{P}}{\omega_j-\psi_l^{(j)}} + \pi\delta(\omega_j-\psi_l^{(j)}), \quad (4.136)$$

$\mathcal{P}$ denoting the Cauchy principal value of the integral, we obtain from equation (4.134a)

$$\frac{d\hat{a}_j}{dt} = -\left(i\omega_j' + \frac{\gamma_j}{2}\right)\hat{a}_j + \eta\left(-i\Delta\omega + \frac{\gamma_j}{2}\right)\hat{a}_j^{\dagger} + \hat{L}_j, \quad (4.137)$$

where $\omega_j' = \omega_j + \Delta\omega_j$, the frequency shifts $\Delta\omega_j$ and the damping constants $\gamma_j$ are defined as

$$\Delta\omega_j = \mathcal{P}\int_{-\infty}^{+\infty} \frac{|\kappa^{(j)}(\psi^{(j)})|^2 \varrho_j(\psi^{(j)})}{\omega_j - \psi^{(j)}}\, d\psi^{(j)}, \quad (4.138a)$$

$$\gamma_j = 2\pi|\kappa^{(j)}(\omega_j)|^2 \varrho_j(\omega_j), \quad (4.138b)$$

respectively, $\varrho_j(\omega_j)$ being the density functions of the reservoir oscillators, the sum over $l$ has been replaced by the integration over $\psi^{(j)}$ with respect to the high reservoir oscillator density assumed and the Langevin noise forces $\hat{L}_j$ are obtained in the form

$$\hat{L}_j(t) = -i\sum_l \left[\kappa_l^{(j)}\hat{b}_l^{(j)}(0)\exp\left(-i\psi_l^{(j)}t\right) + \eta\kappa_l^{(j)*}\hat{b}_l^{(j)\dagger}(0)\exp\left(i\psi_l^{(j)}t\right)\right]. \quad (4.139)$$

We see that the squeezed reservoir involving the rotating terms stimulates the occurrence of creation operators in equation (4.137), which must be solved together with its adjoint. In such a way squeezing of vacuum fluctuations in radiation and other nonclassical effects may arise.

Additionally we assume Gaussian distribution of the initial reservoir oscillators $\{\hat{b}_l^{(j)}(0)\}$, i.e. for the Glauber-Sudarshan quasidistribution of the reservoir coherent-state amplitudes $\{b_l^{(j)}\}$ it holds that

$$\Phi_{\mathcal{N}}(\{b_l^{(j)}\}) = \prod_{j,l} \frac{1}{\pi\langle n_{Rl}^{(j)}\rangle} \exp\left(-\frac{|b_l^{(j)}|^2}{\langle n_{Rl}^{(j)}\rangle}\right), \quad (4.140)$$

where $\langle n_{Rl}^{(j)}\rangle = \langle \hat{b}_l^{(j)\dagger}(0)\hat{b}_l^{(j)}(0)\rangle_R$ represents the mean number of the reservoir oscillators in the reservoir mode $l$ coupled to the radiation mode $j$.

The Langevin forces (4.130) fulfil the following Markov properties as a consequence of the above assumptions:
i) squeezed reservoir ($\eta = 1$)

$$\begin{aligned}
\langle \hat{L}_j(t)\rangle_R &= \langle \hat{L}_j^\dagger(t)\rangle_R = 0\,, \\
\langle \hat{L}_j^\dagger(t)\hat{L}_k(t')\rangle_R &= \langle \hat{L}_k(t')\hat{L}_j^\dagger(t)\rangle_R = -\langle \hat{L}_j(t)\hat{L}_k(t')\rangle_R = -\langle \hat{L}_j^\dagger(t)\hat{L}_k^\dagger(t')\rangle_R \\
&= \gamma_j\,(2\langle n_{Rj}\rangle + 1)\,\delta_{jk}\delta(t-t')\,,
\end{aligned} \tag{4.141}$$

ii) standard reservoir ($\eta = 0$)

$$\begin{aligned}
\langle \hat{L}_j(t)\rangle_R &= \langle \hat{L}_j^\dagger(t)\rangle_R = 0\,, \\
\langle \hat{L}_j^\dagger(t)\hat{L}_k(t')\rangle_R &= \gamma_j\langle n_{Rj}\rangle\delta_{jk}\delta(t-t')\,, \\
\langle \hat{L}_k(t')\hat{L}_j^\dagger(t)\rangle_R &= \gamma_j\,(\langle n_{Rj}\rangle + 1)\,\delta_{jk}\delta(t-t')\,, \\
\langle \hat{L}_j(t)\hat{L}_k(t')\rangle_R &= \langle \hat{L}_j^\dagger(t)\hat{L}_k^\dagger(t')\rangle_R = 0\,.
\end{aligned} \tag{4.142}$$

Here $\langle\ \rangle_R$ means the average over the reservoir with respect to (4.140) and the broad-band reservoir spectrum has been used ($\langle n_{Rj}\rangle \equiv \langle n_{Rl}^{(j)}\rangle$ independently of $l$; at temperature $T$ it equals $\langle n_{Rj}\rangle = [\exp(\hbar\psi^{(j)}/KT) - 1]^{-1}$, $K$ being the Boltzmann constant).

The solution of equation (4.137) together with its adjoint is of the form

$$\hat{a}_j(t) = u_j(t)\hat{a}_j(0) + v_j(t)\hat{a}_j^\dagger(0) + \sum_l w_{jl}(t)\hat{b}_l^{(j)}(0) + \sum_l z_{jl}(t)\hat{b}_l^{(j)\dagger}(0)\,, \tag{4.143}$$

where the time-dependent functions $u_j(t)$, $v_j(t)$, $w_{jl}(t)$ and $z_{jl}(t)$ must fulfil some identities as a consequence of the commutation rules for $\hat{a}_j(t)$ and $\hat{a}_j^\dagger(t)$, e.g. it follows that

$$|u_j(t)|^2 - |v_j(t)|^2 + \sum_l |w_{jl}(t)|^2 - \sum_l |z_{jl}(t)|^2 = 1 \tag{4.144}$$

from the commutation rule $[\hat{a}_j(t), \hat{a}_j^\dagger(t)] = \hat{1}$. The presence of the initial creation operators for the radiation and the reservoir in (4.143) describes the self-effects of the system if the initial fields are zero, as it is seen from the vacuum mean values of the photon number operator $\langle 0|\hat{a}_j^\dagger(t)\hat{a}_j(t)|0\rangle = |v_j(t)|^2 + \sum_l |z_{jl}(t)|^2$. Such terms are also sources of nonclassical behaviour (squeezing of vacuum fluctuations, oscillations in photocount distribution, quantum phase, etc.). These functions can easily be derived for the case of the standard reservoir ($\eta = 0$):

$$u_j(t) = \exp\left(-i\omega_j' t - \frac{\gamma_j}{2}t\right),$$

$$\begin{aligned} v_j(t) &= 0, \\ w_{jl}(t) &= \kappa_l^{(j)} \exp(-i\psi_l^{(j)} t) \frac{1 - \exp\left[i(\psi_l^{(j)} - \omega_j')t - \gamma_j t/2\right]}{\psi_l^{(j)} - \omega_j' + i\gamma_j/2}, \\ z_{jl}(t) &= 0; \end{aligned} \tag{4.145}$$

from the commutation rule it holds that $|u_j(t)|^2 + \sum_l |w_{jl}(t)|^2 = 1$ and

$$\sum_l |w_{jl}(t)|^2 = 1 - \exp(-\gamma t). \tag{4.146}$$

In this case ($\eta = 0$) we obtain from (4.143)

$$\langle \hat{n}_j(t) \rangle_R = \langle \hat{a}_j^\dagger(t) \hat{a}_j(t) \rangle_R = \exp(-\gamma_j t) \langle \hat{n}_j(0) \rangle + \langle n_{Rj} \rangle \left[1 - \exp(-\gamma_j t)\right]. \tag{4.147}$$

The first term represents the time evolution of the initial radiation system whereas the second one is the contribution of the reservoir. This expression obeys the following equation of motion

$$\frac{d\langle \hat{n}_j(t) \rangle_R}{dt} = -\gamma_j \langle \hat{n}_j(t) \rangle_R + \gamma_j \langle n_{Rj} \rangle. \tag{4.148}$$

Making use of the equation of motion (4.137) with $\eta = 0$, we arrive at

$$\begin{aligned} \frac{d\langle \hat{a}_j^\dagger \hat{a}_j \rangle_R}{dt} &= \left\langle \frac{d\hat{a}_j^\dagger}{dt} \hat{a}_j \right\rangle_R + \left\langle \hat{a}_j^\dagger \frac{d\hat{a}_j}{dt} \right\rangle_R \\ &= -\gamma_j \langle \hat{n}_j(t) \rangle_R + \langle \hat{L}_j^\dagger(t) \hat{a}_j(t) \rangle_R + \langle \hat{a}_j^\dagger(t) \hat{L}_j(t) \rangle_R, \end{aligned} \tag{4.149}$$

and comparing both the equations, we have the important identity

$$\langle \hat{L}_j^\dagger(t) \hat{a}_j(t) + \hat{a}_j^\dagger(t) \hat{L}_j(t) \rangle_R = \gamma_j \langle n_{Rj} \rangle. \tag{4.150}$$

The properties of the Langevin force operator ensure clearly the validity of the boson commutation rules for all times and makes the description fully quantum consistent. If they are neglected, the solution $\hat{a}_j(t) = \hat{a}_j(0) \exp(-i\omega_j' t - \gamma_j t/2)$ violates the commutation rule since $[\hat{a}_j(t), \hat{a}_j^\dagger(t)] = \exp(-\gamma_j t) \to 0$ for $t \to \infty$ and the inclusion of the damping factor $\exp(-\gamma_j t)$ can mostly provide only classical results because such an approach approximately agrees with the quantum description for $\gamma_j t \ll 1$ only.

To characterize fully the quantum statistical properties of the system in the framework of the Heisenberg-Langevin approach [Peřina (1991)], we use the normal quantum characteristic function

$$C_{\mathcal{N}}(\{\beta_j\}, t) = \mathrm{Tr} \left\{ \hat{\varrho}(0) \prod_j \exp\left[\beta_j \hat{a}_j^\dagger(t)\right] \exp\left[-\beta_j^* \hat{a}_j(t)\right] \right\}$$

$$= \left\langle \prod_j \exp\left[ -\langle n_{Rj} \rangle \left[ 1 - \exp(-\gamma_j t) \right] |\beta_j|^2 + \beta_j \xi_j^*(t) - \beta_j^* \xi_j(t) \right] \right\rangle , \tag{4.151}$$

where $\beta_j$ are the parameters of the characteristic function, the reservoir quasiprobability (4.140) has been used, after substituting from (4.143), together with the integrals

$$\int \beta^n \beta^{*m} \exp(-s|\beta|^2)\, d^2\beta = \frac{\pi n!}{s^{n+1}} \delta_{nm} , \qquad \mathrm{Re}\, s > 0 , \tag{4.152a}$$

$$\int \exp(-s|\beta|^2 + \alpha\beta^* + \alpha'\beta)\, d^2\beta = \frac{\pi}{s} \exp\left( \frac{\alpha\alpha'}{s} \right) , \quad \mathrm{Re}\, s > 0 , \tag{4.152b}$$

$\xi_j(t) = \xi_j \exp(-i\omega_j' t - \gamma_j t/2)$ and the brackets $\langle\ \rangle$ mean the average over the initial radiation complex amplitudes $\{\xi_j\}$ described by a quasidistribution $\Phi_{\mathcal{N}}(\{\xi_j\}, 0)$. The time development of the radiation system is determined by the statistical operator

$$\hat{\varrho}(t) = \int \Phi_{\mathcal{N}}(\{\alpha_j\}, t) |\{\alpha_j\}\rangle\langle\{\alpha_j\}|\, d^2\{\alpha_j\} , \tag{4.153}$$

where, performing the Fourier transformation of (4.151), we arrive at

$$\Phi_{\mathcal{N}}(\{\alpha_j\}, t) = \int C_{\mathcal{N}}(\{\beta_j\}, t) \prod_j \exp(\alpha_j \beta_j^* - \alpha_j^* \beta_j) \frac{d^2\beta_j}{\pi^2}$$

$$= \prod_j \frac{1}{\pi \langle n_{Rj} \rangle [1 - \exp(-\gamma_j t)]} \left\langle \exp\left[ - \frac{|\alpha_j - \xi_j(t)|^2}{\langle n_{Rj} \rangle [1 - \exp(-\gamma_j t)]} \right] \right\rangle . \tag{4.154}$$

If the initial radiation is in a coherent state $|\{\xi_j\}\rangle$, then the brackets in (4.154) can be omitted and the quasidistribution moves along the spiral $\xi_j(t) = \xi_j \exp(-i\omega_j' t - \gamma_j t/2)$ with increasing noise $\langle n_{Rj} \rangle [1 - \exp(-\gamma_j t)]$. The photocount distribution $p(n,t)$ and its factorial moments $\langle W^k(t) \rangle_{\mathcal{N}}$ can be expressed with the help of the superposition of signal $\xi_j(t)$ and noise $\langle n_{Rj} \rangle [1 - \exp(-\gamma_j t)]$ in terms of the Laguerre polynomials [Peřina (1991)].

## 4.7 Schrödinger approach to quantum statistics — master equation and generalized Fokker-Planck equation

Sometimes it is more convenient to write down directly an equation of motion for the statistical operator traced over non-interesting variables (e.g.

reservoir variables), which provides the master equation, and the use of the coherent-state technique makes it possible to obtain the generalized Fokker-Planck equation. Of course, both the Heisenberg-Langevin and Schrödinger approaches are fully equivalent.

This approach is based on the equation of motion for the statistical operator $\hat{\varrho}(t)$,

$$i\hbar\frac{\partial\hat{\varrho}}{\partial t} = [\hat{H}_{0,\mathrm{rad}} + \hat{H}_{0,\mathrm{res}} + \hat{H}_{\mathrm{int}}, \hat{\varrho}]\,. \tag{4.155}$$

Up to the second iteration appropriate to the Markov approximation, it can be solved as

$$\begin{aligned}\hat{\varrho}(t) &= \hat{\varrho}(0) - \frac{i}{\hbar}\int_0^t dt'[\hat{H}(t'), \hat{\varrho}(0)] \\ &\quad - \frac{1}{\hbar^2}\int_0^t dt'\int_0^{t'} dt''[\hat{H}(t'), [\hat{H}(t''), \hat{\varrho}(t'')]]\,. \end{aligned} \tag{4.156}$$

Differentiating with respect to $t$ we obtain

$$\begin{aligned}\frac{d\hat{\varrho}(t)}{dt} &= -\frac{i}{\hbar}[\hat{H}(t), \hat{\varrho}(0)] - \frac{1}{\hbar^2}\int_0^t [\hat{H}(t), [\hat{H}(t'), \hat{\varrho}(t')]]\,dt' \\ &\approx -\frac{i}{\hbar}[\hat{H}(t), \hat{\varrho}(0)] - \frac{1}{\hbar^2}t[\hat{H}(t), [\hat{H}(t), \hat{\varrho}(t)]]\,, \end{aligned} \tag{4.157}$$

assuming that $t$ is much larger than a typical correlation time of the reservoir and $t \ll 1/\gamma$, $\gamma$ being a typical damping constant for the radiation. Now we can perform the trace over the reservoir variables (the reduced statistical operator) and denoting $\hat{\varrho}(t)$ as $\mathrm{Tr}_{\mathrm{res}}\hat{\varrho}(t) = \hat{\varrho}_{\mathrm{rad}}(t)$ again, we obtain from (4.157), under the assumption of the initial independence of the radiation and reservoir, $\hat{\varrho}(0) = \hat{\varrho}_{\mathrm{res}}(0)\hat{\varrho}_{\mathrm{rad}}(0)$, that $\mathrm{Tr}_{\mathrm{res}}[\hat{H}_{\mathrm{int}}, \hat{\varrho}(0)] = 0$ and we have the master equation for single mode

$$\begin{aligned}\frac{d\hat{\varrho}}{dt} &= -i\omega\left([\hat{a}, \hat{\varrho}\hat{a}^\dagger] - [\hat{a}\hat{\varrho}, \hat{a}^\dagger]\right) \\ &\quad + \frac{\gamma}{2}(\langle n_R\rangle + 1)\left([\hat{a}\hat{\varrho}, \hat{a}^\dagger] + [\hat{a}, \hat{\varrho}\hat{a}^\dagger]\right) \\ &\quad + \frac{\gamma}{2}\langle n_R\rangle\left([\hat{a}^\dagger\hat{\varrho}, \hat{a}] + [\hat{a}^\dagger, \hat{\varrho}\hat{a}]\right) \\ &= \left(\frac{\gamma}{2} - i\omega\right)[\hat{a}, \hat{\varrho}\hat{a}^\dagger] + \left(\frac{\gamma}{2} + i\omega\right)[\hat{a}\hat{\varrho}, \hat{a}^\dagger] \\ &\quad + \frac{\gamma}{2}\langle n_R\rangle\left([[\hat{a}, \hat{\varrho}], \hat{a}^\dagger] + [\hat{a}, [\hat{\varrho}, \hat{a}^\dagger]]\right)\,. \end{aligned} \tag{4.158}$$

This can also be written as

$$\begin{aligned}\frac{d\hat{\varrho}}{dt} &= -i\omega\left(\hat{a}^\dagger\hat{a}\hat{\varrho} - \hat{\varrho}\hat{a}^\dagger\hat{a}\right) + \frac{\gamma}{2}(\langle n_R\rangle + 1)\left(2\hat{a}\hat{\varrho}\hat{a}^\dagger - \hat{a}^\dagger\hat{a}\hat{\varrho}\right. \\ &\quad \left. - \hat{\varrho}\hat{a}^\dagger\hat{a}\right) + \frac{\gamma}{2}\langle n_R\rangle\left(2\hat{a}^\dagger\hat{\varrho}\hat{a} - \hat{a}\hat{a}^\dagger\hat{\varrho} - \hat{\varrho}\hat{a}\hat{a}^\dagger\right)\,, \end{aligned} \tag{4.159}$$

or more generally

$$\begin{aligned}\frac{d\hat{\varrho}}{dt} &= -\frac{i}{\hbar}[\hat{H}_{\text{rad}}, \hat{\varrho}] + \frac{\gamma}{2}\left(\langle n_R\rangle + 1\right)\left(2\hat{a}\hat{\varrho}\hat{a}^\dagger - \hat{a}^\dagger\hat{a}\hat{\varrho}\right.\\ &\quad \left. - \hat{\varrho}\hat{a}^\dagger\hat{a}\right) + \frac{\gamma}{2}\langle n_R\rangle\left(2\hat{a}^\dagger\hat{\varrho}\hat{a} - \hat{a}\hat{a}^\dagger\hat{\varrho} - \hat{\varrho}\hat{a}\hat{a}^\dagger\right). \qquad (4.160)\end{aligned}$$

Omitting the free oscillations included in the first term (using the interaction picture), we can obtain for the diagonal Fock matrix elements $p_n = \langle n|\hat{\varrho}|n\rangle$

$$\frac{dp_n}{dt} = -\gamma\left(\langle n_R\rangle + 1\right)\left[np_n - (n+1)p_{n+1}\right] - \gamma\langle n_R\rangle\left[(n+1)p_n - np_{n-1}\right], \qquad (4.161)$$

which can be schematically derived from the following scheme of flows of number probabilities:

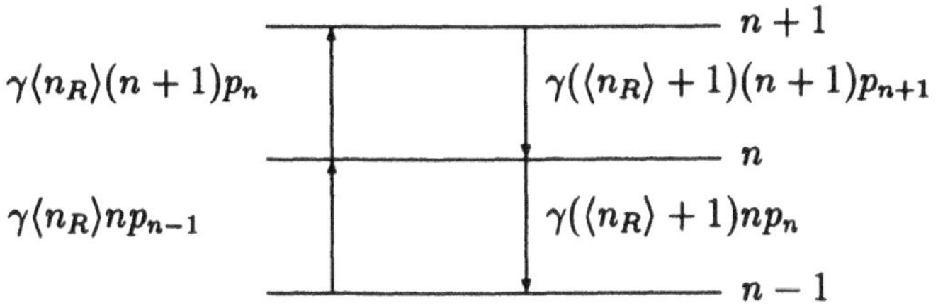


A steady state ($dp_n/dt = 0$) occurs when a detailed balance takes place between the absorption and emission processes,

$$\gamma\left(\langle n_R\rangle + 1\right)np_n = \gamma\langle n_R\rangle np_{n-1}. \qquad (4.162)$$

This recursive relation leads to the thermal equilibrium distribution

$$\begin{aligned}p_n &= \frac{\langle n_R\rangle}{\langle n_R\rangle + 1}p_{n-1} = \left(\frac{\langle n_R\rangle}{\langle n_R\rangle + 1}\right)^n p_0\\ &= \left[1 - \exp\left(-\frac{\hbar\omega}{kT}\right)\right]\exp\left(-\frac{n\hbar\omega}{kT}\right),\\ p_0 &= \left(\langle n_R\rangle + 1\right)^{-1}. \qquad (4.163)\end{aligned}$$

Using the q-c-number correspondence of the coherent-state technique (see e.g. Peřina (1991)), we can adopt the Glauber-Sudarshan quasidistribution $\Phi_{\mathcal{N}}(\alpha) = \hat{\varrho}^{(\mathcal{A})}(\hat{a} \to \alpha, \hat{a}^\dagger \to \alpha^*)/\pi$ to obtain the generalized Fokker-Planck equation applying the antinormal or normal ordering to the master equation (4.159)

$$\begin{aligned}\frac{\partial \Phi_s}{\partial t} &= -\frac{\partial}{\partial \alpha}\left(A_\alpha \Phi_s\right) - \frac{\partial}{\partial \alpha^*}\left(A_{\alpha^*}\Phi_s\right) + 2\frac{\partial^2}{\partial\alpha\partial\alpha^*}\left(D^{(s)}_{\alpha^*\alpha}\Phi_s\right) \\ &= \left(\frac{\gamma}{2}+i\omega\right)\frac{\partial}{\partial\alpha}\left(\alpha\Phi_s\right) + \left(\frac{\gamma}{2}-i\omega\right)\frac{\partial}{\partial\alpha^*}\left(\alpha^*\Phi_s\right) \\ &\quad +\gamma\left(\langle n_R\rangle + \frac{1-s}{2}\right)\frac{\partial^2\Phi_s}{\partial\alpha\partial\alpha^*}, \end{aligned} \tag{4.164}$$

where $s = 1$ for $\Phi_{\mathcal{N}}$ and $s = -1$ for $\Phi_{\mathcal{A}}$. For $s = 0$ we have the equation for the Wigner function $\Phi_{\mathcal{S}}(\alpha)$. The first two terms are related to the dynamical equations of motion giving for the drift vectors

$$\begin{aligned} A_\alpha &= \left\langle \frac{d\alpha}{dt}\right\rangle_R = -\left(i\omega + \frac{\gamma}{2}\right)\alpha, \\ A_{\alpha^*} &= \left\langle \frac{d\alpha^*}{dt}\right\rangle_R = \left(i\omega - \frac{\gamma}{2}\right)\alpha^*, \end{aligned} \tag{4.165}$$

and for the diffusion coefficient

$$2D_{\alpha^*\alpha} = \left\langle \frac{d(\alpha^*\alpha)}{dt}\right\rangle_R - \left\langle \alpha^*\frac{d\alpha}{dt}\right\rangle_R - \left\langle \frac{d\alpha^*}{dt}\alpha\right\rangle_R = \gamma\langle n_R\rangle \tag{4.166}$$

for the normal order, whereas $2D_{\alpha\alpha^*} = \gamma\left(\langle n_R\rangle + 1\right)$ for the antinormal order. The first two drift terms involving the first order derivatives are responsible for a movement of the peak of quasidistributions, the third diffusion second-order derivative term causes a broadening of the quasidistribution and an increase of uncertainty.

The Fourier fransformation provides the equation of motion for the characteristic functions

$$\frac{\partial C_s}{\partial t} = -\left(-i\omega + \frac{\gamma}{2}\right)\beta\frac{\partial C_s}{\partial\beta} - \left(i\omega + \frac{\gamma}{2}\right)\beta^*\frac{\partial C_s}{\partial\beta^*} - \gamma\left(\langle n_R\rangle + \frac{1-s}{2}\right)|\beta|^2 C_s. \tag{4.167}$$

More general derivation of the generalized Fokker-Planck equation can be obtained as follows. Consider any operator $\hat{M}(\hat{a},\hat{a}^\dagger)$ and decompose it using the Taylor series to construct

$$\begin{aligned}\left\langle \frac{d\hat{M}}{dt}\right\rangle &\approx \left\langle \sum_j \frac{\partial\hat{M}}{\partial\hat{a}_j}\frac{\Delta\hat{a}_j}{\Delta t} + \sum_{j>k}\frac{\partial^2\hat{M}}{\partial\hat{a}_j\partial\hat{a}_k}\frac{\Delta\hat{a}_j\Delta\hat{a}_k}{\Delta t}\right. \\ &\quad \left. +\sum_j \frac{1}{2}\frac{\partial^2\hat{M}}{\partial\hat{a}_j^2}\frac{(\Delta\hat{a}_j)^2}{\Delta t}\right\rangle. \end{aligned} \tag{4.168}$$

Using the coherent-state technique and integrating by parts we obtain for a quasidistribution $\Phi$

$$\frac{\partial \Phi}{\partial t} = \sum_j \left[ -\frac{\partial}{\partial \alpha_j} (A_j \Phi) + \sum_{j,k} \frac{\partial^2}{\partial \alpha_j \partial \alpha_k} (D_{jk} \Phi) \right], \tag{4.169}$$

where $A_j = \langle \Delta \hat{a}_j / \Delta t \rangle$, $2D_{jk} = \langle \Delta \hat{a}_j \Delta \hat{a}_k / \Delta t \rangle$ and using the identity $\Delta(\hat{a}_j \hat{a}_k) = \hat{a}_j(t+\Delta t)\hat{a}_k(t+\Delta t) - \hat{a}_j(t)\hat{a}_k(t) = \Delta \hat{a}_j \Delta \hat{a}_k + (\Delta \hat{a}_j)\hat{a}_k + \hat{a}_j(\Delta \hat{a}_k)$, we arrive at the definition of the drift vector and diffusion coefficients in agreement with (4.165) and (4.166),

$$\begin{aligned} A_j &= \left\langle \frac{d\hat{a}_j}{dt} \right\rangle, \\ 2D_{jk} &= \left\langle \frac{d(\hat{a}_j \hat{a}_k)}{dt} \right\rangle - \left\langle \hat{a}_j \frac{d\hat{a}_k}{dt} \right\rangle - \left\langle \frac{d\hat{a}_j}{dt} \hat{a}_k \right\rangle . \end{aligned} \tag{4.170}$$

With respect to the above treatment in the complex variables we can denote, for instance, $\alpha_1 = \alpha$, $\alpha_2 = \alpha^*$, $A_\alpha = \langle \Delta \hat{a} / \Delta t \rangle$, $A_{\alpha^*} = \langle \Delta \hat{a}^\dagger / \Delta t \rangle$, $2D_{\alpha\alpha} = \langle (\Delta \hat{a})^2 / \Delta t \rangle$, $2D_{\alpha^*\alpha^*} = \langle (\Delta \hat{a}^\dagger)^2 / \Delta t \rangle$, $D_{\alpha^*\alpha} = \langle \Delta \hat{a}^\dagger \Delta \hat{a} / \Delta t \rangle \neq D_{\alpha\alpha^*} = \langle \Delta \hat{a} \Delta \hat{a}^\dagger / \Delta t \rangle$ and we have the generalized Fokker-Planck equation

$$\begin{aligned} \frac{\partial \Phi}{\partial t} &= -\frac{\partial}{\partial \alpha} (A_\alpha \Phi) - \frac{\partial}{\partial \alpha^*} (A_{\alpha^*} \Phi) + \frac{\partial^2}{\partial \alpha^2} (D_{\alpha\alpha} \Phi) \\ &+ \frac{\partial^2}{\partial \alpha^{*2}} (D_{\alpha^*\alpha^*} \Phi) + \frac{\partial^2}{\partial \alpha \partial \alpha^*} \left[ (D_{\alpha\alpha^*} + D_{\alpha^*\alpha}) \Phi \right], \end{aligned} \tag{4.171}$$

which includes the rotating term $\partial^2 (D_{\alpha\alpha} \Phi) / \partial \alpha^2$ and its complex conjugate.

It is also possible to write the Fokker-Planck equation in the form

$$\frac{\partial \Phi}{\partial t} = -\sum_j \frac{\partial S_j}{\partial \alpha_j}, \tag{4.172}$$

where the probability current $S_j$ is defined as

$$S_j = A_j \Phi - \sum_k \frac{\partial}{\partial \alpha_k} (D_{jk} \Phi) . \tag{4.173}$$

Consider more general case of the interaction of radiation with a nonlinear medium composed of two-level atoms, taking into account multiphoton absorptions and emissions. Using the interaction picture we can write the interaction hamiltonian in the form [McNeil and Walls (1974)]

$$\begin{aligned} \hat{H}_{\text{int}} &= \sum_j \hbar \mu^{(n)}(\mathbf{x}_j) \hat{c}_{2j}^\dagger \hat{c}_{1j} \hat{O}^{(n)} + \text{h.c.}, \\ \hat{O}^{(n)} &= \prod_{l=1}^{m} \hat{a}_l^\dagger \prod_{k=m+1}^{n} \hat{a}_k , \end{aligned} \tag{4.174}$$

where $\hat{a}_l$ and $\hat{a}_l^\dagger$ are again the annihilation and creation operators of a photon in the radiation mode, $\hat{c}_{\lambda j}$ and $\hat{c}_{\lambda j}^\dagger$ are the annihilation and creation operators of the $\lambda$th level of the $j$th atom and $\mu^{(n)}(\mathbf{x}_j)$ is the coupling coefficient proportional to the $n$-photon transition matrix element of the atom situated at the point $\mathbf{x}_j$. This hamiltonian describes $m$ emission and $n-m$ absorption events during one atomic transition. If we are mostly interested in the properties of radiation, as is the case in quantum optics, the atomic variables may be eliminated in the same way as the reservoir variables and we can obtain the Heisenberg-Langevin equations or the generalized Fokker-Planck equation. If only virtual electronic transitions are taken into account whereas real transitions are neglected, an effective hamiltonian can be derived (Peřina, Jr. (1993) and references therein) and the Heisenberg-Langevin and generalized Fokker-Planck equations directly follow. In the Markov approximation, the master equation for the reduced statistical operator can be derived in the same way as above

$$\begin{aligned}\frac{\partial\hat{\varrho}}{\partial t} &= K\Big\{N_1\left([\hat{O}^{(n)}\hat{\varrho},\hat{O}^{(n)\dagger}]+[\hat{O}^{(n)},\hat{\varrho}\hat{O}^{(n)\dagger}]\right)\\ &\quad -N_2\left([\hat{O}^{(n)},\hat{O}^{(n)\dagger}\hat{\varrho}]+[\hat{\varrho}\hat{O}^{(n)},\hat{O}^{(n)\dagger}]\right)\Big\}, \qquad (4.175)\end{aligned}$$

where $K$ is a constant related to $\mu^{(n)}$ and $N_1$ and $N_2$ are the occupation numbers of the atomic levels 1 and 2 respectively, under the condition of the thermal equilibrium. Considering more general interaction of the radiation modes symbolicly described by the expression ($\hat{G}^\dagger\hat{O}^{(n)}$+h.c.) in the hamiltonian $\hat{H}_{\text{int}}$, $\hat{G}$ being a random noise operator, we obtain the equation of motion for the reduced statistical operator in the interaction picture in the same way as above

$$\begin{aligned}\frac{\partial\hat{\varrho}}{\partial t} &= -\frac{i}{\hbar}[\hat{H}_{\text{int}},\hat{\varrho}]+D_{G^*G^*}\left[2\hat{O}^{(n)}\hat{\varrho}\hat{O}^{(n)}-\hat{O}^{(n)2}\hat{\varrho}-\hat{\varrho}\hat{O}^{(n)2}\right]\\ &\quad +D_{G^*G}\left[2\hat{O}^{(n)\dagger}\hat{\varrho}\hat{O}^{(n)}-\hat{O}^{(n)}\hat{O}^{(n)\dagger}\hat{\varrho}-\hat{\varrho}\hat{O}^{(n)}\hat{O}^{(n)\dagger}\right]\\ &\quad +D_{GG^*}\left[2\hat{O}^{(n)}\hat{\varrho}\hat{O}^{(n)\dagger}-\hat{O}^{(n)\dagger}\hat{O}^{(n)}\hat{\varrho}-\hat{\varrho}\hat{O}^{(n)\dagger}\hat{O}^{(n)}\right]\\ &\quad +D_{GG}\left[2\hat{O}^{(n)\dagger}\hat{\varrho}\hat{O}^{(n)\dagger}-\left(\hat{O}^{(n)\dagger}\right)^2\hat{\varrho}-\hat{\varrho}\left(\hat{O}^{(n)\dagger}\right)^2\right],\end{aligned} \qquad (4.176)$$

where diffusion coefficients are defined as

$$D_{jk}=\frac{1}{2\hbar^2}\lim_{\Delta t\to 0}\left\langle\int_t^{t+\Delta t}\int_t^{t'}G_j(t')G_k(t'')\,dt'dt''\right\rangle_R \qquad (4.177)$$

and $j,k=G,G^*$, and $G_G=G$ and $G_{G^*}=G^*$. It is seen that $D_{GG^*}$ and $D_{G^*G}$ correspond to $KN_1$ and $KN_2$ in (4.175), respectively. For the above case of a one-photon absorption $\hat{O}^{(n)} = \hat{a}$ ($m = 0$, $n = 1$) in a single mode, we have $K = \gamma/2$, $N_1 = \langle n_R \rangle + 1$, $N_2 = \langle n_R \rangle$ and we arrive at the result for the damped harmonic oscillator.

## 4.8 Generalized superposition of coherent fields and quantum noise

For the purpose of applications in next chapters we introduce basic definitions and main results for the generalized superposition of coherent fields and quantum noise, which is equivalent to the two-photon coherent state technique. Such a description can also approximately describe higher-order interaction conserving nonclassical effects in optical fields. Even if the Glauber-Sudarshan quasidistribution $\Phi_{\mathcal{N}}$ does not exist generally, the quasidistribution $\Phi_{\mathcal{A}}$ related to antinormal ordering, or the Wigner quasidistribution $\Phi_{\mathcal{S}}$ are Gaussian, which is convenient for calculations. Greater details concerning this quantum description can be found in the book by Peřina (1991) (Secs. 8.5, 9.3 and 9.4) where an extended treatment of nonlinear optical processes, such as optical parametric processes, Raman and hyper-Raman scattering, four-wave mixing, phase conjugation, multiphoton absorption and emission, anharmonic oscillations, etc., is presented in Chap. 10 of that book.

Describing interaction of optical fields with matter, e.g. with nonlinear media, we want to include also initial nonclassical light states into our considerations and consequently we adopt the antinormal ordering of field operators in order to describe the nonclassical states simply and to avoid manipulations with generalized functions appropriate for the Glauber-Sudarshan quasidistribution in this case. If solutions of Heisenberg equations are linear in the initial (input) annihilation and creation operators $\hat{a}_j$ and $\hat{a}_j^\dagger$, respectively, i.e. when the interaction hamiltonian is quadratic in the annihilation and creation operators, we can obtain the antinormal characteristic function for an $M$-mode field in the form

$$
\begin{aligned}
C_{\mathcal{A}}(\{\beta_j\},t) &= \left\langle \prod_{j=1}^{M} \exp\left[-\beta_j^* \hat{a}_j(t)\right] \prod_{k=1}^{M} \exp\left[\beta_k \hat{a}_k^\dagger(t)\right] \right\rangle \\
&= \exp\left\{\frac{1}{2}\hat{\beta}^\dagger \hat{A}(t)\hat{\beta} + \hat{\beta}^\dagger \hat{\xi}\right\} \\
&= \exp\left\{\sum_{j=1}^{M}\left[-B_j(t)|\beta_j|^2 + \left(C_j^*(t)\frac{\beta_j^2}{2} + \text{c.c.}\right)\right]\right. \\
&\quad + \sum_{j<k}^{M}\left[D_{jk}(t)\beta_j^*\beta_k^* + \bar{D}_{jk}(t)\beta_j\beta_k^* + \text{c.c.}\right]
\end{aligned}
$$

$$+\sum_{j=1}^{M}\left[\beta_j\xi_j^*(t) - \text{c.c.}\right]\Bigg\}, \qquad (4.178)$$

where c.c. denotes the complex conjugate terms, $\beta_j$ are complex parameters, and the angle brackets mean the average with the help of the initial statistical operator $\hat{\varrho}(0)$. Here

$$\hat{\beta} = \begin{bmatrix} \beta_1 \\ \beta_1^* \\ \vdots \\ \beta_M \\ \beta_M^* \end{bmatrix}, \quad \hat{\xi} = \begin{bmatrix} -\xi_1 \\ \xi_1^* \\ \vdots \\ -\xi_M \\ \xi_M^* \end{bmatrix},$$

$$\hat{A} = \begin{bmatrix} -B_1 & C_1 & \bar{D}_{12}^* & D_{12} & \dots & \bar{D}_{1M}^* & D_{1M} \\ C_1^* & -B_1 & D_{12}^* & \bar{D}_{12} & \dots & D_{1M}^* & \bar{D}_{1M} \\ \bar{D}_{12} & D_{12} & -B_2 & C_2 & \dots & \bar{D}_{2M}^* & D_{2M} \\ D_{12}^* & \bar{D}_{12}^* & C_2^* & -B_2 & \dots & D_{2M}^* & \bar{D}_{2M} \\ \dots & \dots & \dots & \dots & \dots & \dots & \dots \\ \bar{D}_{1M} & D_{1M} & \bar{D}_{2M} & D_{2M} & \dots & -B_M & C_M \\ D_{1M}^* & \bar{D}_{1M}^* & D_{2M}^* & \bar{D}_{2M}^* & \dots & C_M^* & -B_M \end{bmatrix}, \qquad (4.179)$$

where the time-dependent coefficients $B_j$, $C_j$, $D_{jk}$, $\bar{D}_{jk}$ are defined in terms of fluctuations $\Delta\hat{a}_j = \hat{a}_j - \langle\hat{a}_j\rangle$ as follows:

$$B_j = \langle\Delta\hat{a}_j\Delta\hat{a}_j^\dagger\rangle, \qquad C_j = \langle(\Delta\hat{a}_j)^2\rangle,$$

$$D_{jk} = \langle\Delta\hat{a}_j\Delta\hat{a}_k\rangle, \qquad \bar{D}_{jk} = -\langle\Delta\hat{a}_j^\dagger\Delta\hat{a}_k\rangle, \qquad (4.180)$$

$j,k = 1,2,\dots,M$, $j \neq k$, and $\xi_j(t)$ are the time-dependent complex amplitudes determined by the interaction. If the normal characteristic function $C_{\mathcal{N}}(\{\beta_j\},t)$ is to be calculated, one must substitute $B_j = \langle\Delta\hat{a}_j^\dagger\Delta\hat{a}_j\rangle = \langle\Delta\hat{a}_j\Delta\hat{a}_j^\dagger\rangle - 1$ instead of the original $B_j = \langle\Delta\hat{a}_j\Delta\hat{a}_j^\dagger\rangle$. Note that derivatives of (4.178) with respect to $\beta_j^*$ and $\beta_k$ at $\beta_1 = \dots = \beta_M = \beta_1^* = \dots = \beta_M^* = 0$ provide general antinormal moments $\langle\hat{a}_1^{k_1}\dots\hat{a}_M^{k_M}\,\hat{a}_M^{\dagger l_M}\dots\hat{a}_1^{\dagger l_1}\rangle$ in the ways discussed in Sec. 2.9. The corresponding quasidistribution related to normal ordering, provided that the coupling of modes is neglected, is

$$\begin{aligned}\Phi_{\mathcal{A}}(\{\alpha_j\},t) &= \Bigg\langle\prod_{j=1}^{M}\frac{1}{\pi\sqrt{K_j(t)}}\exp\Bigg\{-\frac{B_j(t)}{K_j(t)}|\alpha_j - \xi_j(t)|^2 \\ &\quad + \left[\frac{C_j^*(t)}{2K_j(t)}(\alpha_j - \xi_j(t))^2 + \text{c.c.}\right]\Bigg\}\Bigg\rangle, \qquad (4.181)\end{aligned}$$

where $K_j = B_j^2 - |C_j|^2$ is the determinant of the Fourier transformation and the angle brackets mean the average over the initial complex amplitudes. The corresponding normal generating function can be obtained directly from the normal characteristic function [Peřina (1991), Sec. 8.5]

$$\begin{aligned}\langle \exp[-\lambda W(t)] \rangle_{\mathcal{N}} &= \frac{1}{(\pi\lambda)^M} \int \exp\left[-\frac{1}{\lambda}\sum_{j=1}^{M} |\beta_j|^2\right] C_{\mathcal{N}}(\{\beta_j\}, t)\, d^2\{\beta_j\} \\ &= \left\langle \prod_{j=1}^{M} [1+\lambda(E_j-1)]^{-1/2} [1+\lambda(F_j-1)]^{-1/2} \right. \\ &\qquad \left. \times \exp\left[-\frac{\lambda A_{1j}}{1+\lambda(E_j-1)} - \frac{\lambda A_{2j}}{1+\lambda(F_j-1)}\right] \right\rangle, \end{aligned} \tag{4.182}$$

where $\lambda = -is$ compared to denotations of Sec. 2.9,

$$E_j = B_j - |C_j|, \quad F_j = B_j + |C_j|,$$

$$A_{1,2j} = \frac{1}{2}\left\{ |\xi_j(t)|^2 \mp \frac{1}{2|C_j|} \left[\xi_j^2(t) C_j^* + \text{c.c.}\right] \right\}. \tag{4.183}$$

The expression in the angle brackets of (4.182) has the form of the product of generating functions for the superposition of signal components $A_{1,2j} \geq 0$ and quantum noise components $(F_j - 1) \geq 0$, $(E_j - 1)$ may be arbitrary. The deviation of the expression within the angle brackets in (4.182) from the Poisson generating function $\exp\left[-\lambda \sum_{j=1}^{M} (A_{1j} + A_{2j})\right] = \exp\left[-\lambda \sum_{j=1}^{M} |\xi_j(t)|^2\right]$ reflects the change of the photon statistics caused by the nonlinear dynamics of the corresponding interaction of light with matter, whereas the average over the initial complex amplitudes of the optical field, represented by the angle brackets, produces an additional change of the photon statistics with respect to the initial (input) state of the field (including nonclassical states). The quantum features of optical fields under the discussion are reflected by negative values of quantum noise components $E_j - 1 = B_j - |C_j| - 1$ and therefore we speak of the generalized superposition of coherent fields and quantum noise.

For discussions of quantum statistical properties of optical fields interacting with matter, cases of single-mode fields and two-mode fields (compound-mode fields) involving effects of coupling of modes are most frequent.

Instead of considering the single-mode case only we can assume $M$ independent modes with the same parameters $E_j$, $F_j$. The corresponding factors in the brackets of (4.182) are also the generating functions for the Laguerre polynomials $L_j^{M/2-1}$ [Peřina (1991), Secs. 5.2 and 5.3]. Consequently, the photon-number distribution and its factorial moments are expressed in the form (omitting the mode index $j$)

$$p(n,t) = \left\langle (EF)^{-M/2}\left(1-\frac{1}{F}\right)^n \exp\left[-\frac{A_1}{E}-\frac{A_2}{F}\right] \right.$$
$$\times \sum_{k=0}^{n} \frac{1}{\Gamma(k+M/2)\Gamma(n-k+M/2)}\left(\frac{1-1/E}{1-1/F}\right)^k$$
$$\left. \times L_k^{M/2-1}\left[-\frac{A_1}{E(E-1)}\right] L_{n-k}^{M/2-1}\left[-\frac{A_2}{F(F-1)}\right]\right\rangle, \quad (4.184a)$$

$$\langle W^k\rangle_{\mathcal{N}} = k!\left\langle (F-1)^k \sum_{l=0}^{k} \frac{1}{\Gamma(l+M/2)\Gamma(k-l+M/2)}\left(\frac{E-1}{F-1}\right)^l \right.$$
$$\left. \times L_l^{M/2-1}\left[-\frac{A_1}{E-1}\right] L_{k-l}^{M/2-1}\left[-\frac{A_2}{F-1}\right]\right\rangle, \quad (4.184b)$$

where $\Gamma$ is the gamma function and $A_{1,2} = \sum_{j=1}^{M} A_{1,2j}$. The mean number of photons equals

$$\langle \hat{n}\rangle = \langle W\rangle_{\mathcal{N}} = \left\langle A_1 + A_2 + \frac{M}{2}(E+F-2)\right\rangle, \quad (4.185a)$$

and the photon-number variance is

$$\langle (\Delta W)^2\rangle_{\mathcal{N}} = \left\langle 2A_1(E-1) + 2A_2(F-1) + \frac{M}{2}\left[(E-1)^2 + (F-1)^2\right]\right\rangle. \quad (4.185b)$$

The last terms in (4.185a, b) represent quantum noise contributions related to physical vacuum fluctuations. They are non-zero even if the input light intensity is zero ($A_1 = A_2 = 0$). For nonclassical fields the first interference term between signal $A_1$ and quantum noise $(E-1)$ may be dominant at least for some time intervals (e.g. after the switching on of the interaction). Then $(E-1) < 0$ and $\langle (\Delta W)^2\rangle_{\mathcal{N}} < 0$, $\langle (\Delta \hat{n})^2\rangle < \langle \hat{n}\rangle$ and light is sub-Poissonian and antibunched.

In the framework of this model we can easily derive the expression for the entropy $S = -\mathrm{Tr}\,\{\hat{\varrho}\ln\hat{\varrho}\}$ as a global characteristic of photon statistics of optical fields including nonclassical states [Peřinová, Křepelka and Peřina (1986), Rai (1992)]. It holds that

$$S = -\ln\left[\frac{x^x}{(1+x)^{1+x}}\right], \quad (4.186)$$

where $x = [(B-1/2)^2 - |C|^2]^{1/2} - 1/2$. For a chaotic state $x = \langle \hat{a}^\dagger\hat{a}\rangle = \langle n\rangle$. The value of the entropy remains, of course, unchanged if any additional coherent component is superimposed. The entropy of squeezed coherent states as pure states is zero.

For compound mode case the corresponding generating function reads [Peřinová (1981), Peřinová and Peřina (1981)]

$$\langle \exp(-\lambda W) \rangle_{\mathcal{N}} = \prod_{k=1}^{4} (1+\lambda\lambda_k)^{-1/2} \exp\left[-\sum_{j=1}^{4} \frac{\lambda A_k}{1+\lambda\lambda_k}\right], \tag{4.187}$$

where $\lambda_k$ are roots of certain polynomials and coherent components $A_k$ are constructed from $\lambda_k$ and input field quantities. The procedure of calculations is rather complex in this case and we refer the reader to the mentioned literature. The corresponding photon-number distribution and its factorial moments are then expressed as

$$\begin{aligned} p(n,t) &= \exp\left[-\sum_{l=1}^{4} \frac{A_l}{1+\lambda_l}\right] \sum{}' \prod_{l=1}^{4} \frac{\lambda_l^{k_l}}{\Gamma(k_l+1/2)\,(1+\lambda_l)^{k_l+1/2}} \\ &\quad \times L_{k_l}^{-1/2}\left[-\frac{A_l}{\lambda_l(1+\lambda_l)}\right], \end{aligned} \tag{4.188a}$$

$$\langle W^k(t) \rangle_{\mathcal{N}} = k! \sum{}' \prod_{l=1}^{4} \frac{\lambda_l^{k_l}}{\Gamma(k_l+1/2)} L_{k_l}^{-1/2}\left[-\frac{A_l}{\lambda_l}\right], \tag{4.188b}$$

where $\sum'$ is to be taken under the condition $\sum_{l=1}^{4} k_l = n$ or $k$.

# Chapter 5

# Nonclassical optical phenomena and their relations

The purpose of this chapter is to review the basic concepts leading to definition of nonclassical optical phanomena which have fully quantum origin, have no classical analogues and can be explained only in the framework of quantum theory. Their experimental observations (Chap. 8) represent a complete confirmation of the quantum theory of Nature. The most important nonclassical phenomena are quadrature squeezing of vacuum fluctuations, antibunching of photons, sub-Poissonian photon statistics or photon-number squeezing, oscillations occurring in the photocount distributions, quantum phase, collapses and revivals of atomic populations and violation of various classical inequalities. Interesting measurements of nonclassical quasidistributions, e.g. Wigner quasiprobability function can also be realized. Although these phenomena can occur separately, some relations among them can be outlined.

## 5.1 Quadrature squeezing — squeezing of vacuum fluctuations

Considering first a single mode for simplicity, we can define quadrature operators

$$\begin{aligned}\hat{Q} &= \hat{a} + \hat{a}^{\dagger} = \sqrt{\frac{2\omega}{\hbar}}\,\hat{q}\,, \\ \hat{P} &= -i(\hat{a} - \hat{a}^{\dagger}) = \sqrt{\frac{2}{\hbar\omega}}\,\hat{p} \end{aligned} \tag{5.1}$$

in terms of the generalized coordinate $\hat{q}$ and momentum $\hat{p}$. In classical terms the quadrature operators are proportional to real and imaginary parts of the complex field amplitude. The quadrature operators satisfy the commutation rule

$$[\hat{Q}, \hat{P}] = 2i\hat{1}\,, \tag{5.2}$$

which means that the fluctuations $\Delta\hat{Q} = \hat{Q} - \langle\hat{Q}\rangle$ and $\Delta\hat{P} = \hat{P} - \langle\hat{P}\rangle$ are limited by the Heisenberg uncertainty relation

$$\langle(\Delta\hat{Q})^2\rangle\,\langle(\Delta\hat{P})^2\rangle \geq 1\,. \tag{5.3}$$

As shown before this inequality reduces to equality for the coherent state $|\alpha\rangle$ since $\langle(\Delta\hat{Q})^2\rangle = \langle(\Delta\hat{P})^2\rangle = 1$ in this case. This means that in the complex $\alpha$-plane the coherent state is represented by the radius-vector $\alpha$ with a circle centered at the end point of the radius $1/2$ representing isotropic vacuum fluctuations of the incoherent energy $[\langle(\Delta\hat{p})^2\rangle + \omega^2\langle(\Delta\hat{q})^2\rangle]/2 = \hbar\omega/2$ in every mode. However, this quantum noise isotropy is not necessary to satisfy the inequality (5.3). There are quantum states for which the condition $\langle(\Delta\hat{Q})^2\rangle < 1$ or $\langle(\Delta\hat{P})^2\rangle < 1$, i.e. one of the quadrature fluctuations can be reduced compared to vacuum fluctuations or fluctuations in the coherent state. Of course, this reduction of fluctuations in one quadrature must lead to the increase of fluctuations in the other quadrature to fulfil the uncertainty relation (5.3). Such a quantum state is called squeezed state because the noise isotropy is cancelled and noise circle of the coherent state is transformed e.g. to a noise ellipse. Such a state can be a minimum uncertainty state if the equality in (5.3) is fulfilled (squeezed coherent state), or it may be a mixed state if the sign of inequality is appropriate in (5.3) (more generally, inequality (5.10) can be adopted).

For the quadrature variances we obtain

$$\left\langle\left(\Delta \begin{matrix}\hat{Q}\\ \hat{P}\end{matrix}\right)^2\right\rangle = 1 + 2\left[\langle\Delta\hat{a}^\dagger\Delta\hat{a}\rangle \pm \mathrm{Re}\left(\langle(\Delta\hat{a})^2\rangle\right)\right] \tag{5.4a}$$

and using the Glauber-Sudarshan representation of the statistical operator

$$\begin{aligned}\langle(\Delta\hat{Q})^2\rangle &= 1 + \int \Phi_{\mathcal{N}}(\alpha)\left[(\alpha+\alpha^*) - (\langle\alpha\rangle + \langle\alpha^*\rangle)\right]^2 d^2\alpha\,,\\ \langle(\Delta\hat{P})^2\rangle &= 1 + \int \Phi_{\mathcal{N}}(\alpha)\left[\frac{1}{i}(\alpha-\alpha^*) - \frac{1}{i}(\langle\alpha\rangle - \langle\alpha^*\rangle)\right]^2 d^2\alpha\,.\end{aligned} \tag{5.4b}$$

In order to have $\langle(\Delta\hat{Q})^2\rangle$ or $\langle(\Delta\hat{P})^2\rangle < 1$, $\Phi_{\mathcal{N}}(\alpha)$ must be negative or singular.

As an example we can mention the squeezed states as discussed in Sec. 4.4 produced by a two-photon process or by the process of degenerate subharmonic generation with classical pumping (Sec. 5.4). If an operator $\hat{b} = \mu\hat{a} + \nu\hat{a}^\dagger$, $[\hat{a}, \hat{a}^\dagger] = [\hat{b}, \hat{b}^\dagger] = \hat{1}$ defines the squeezed state $|\beta, \mu, \nu\rangle$ ( $\hat{b}|\beta, \mu, \nu\rangle =$

$\beta|\beta,\mu,\nu\rangle$ ), then $\mu = \cosh(r)$, $\nu = \exp(i\varphi)\sinh(r)$, $r$ being the squeeze parameter and $\varphi$ is a phase of squeezing, $|\mu|^2 - |\nu|^2 = 1$ and

$$\left\langle \left( \Delta \begin{matrix} \hat{Q} \\ \hat{P} \end{matrix} \right)^2 \right\rangle = |\mu \mp \nu|^2 = \exp(\mp 2r) , \tag{5.5a}$$

provided that $\varphi = 0$. More generaly

$$\begin{aligned} \langle (\Delta\hat{Q})^2 \rangle &= \exp(-2r)\cos^2\left(\frac{\varphi}{2}\right) + \exp(2r)\sin^2\left(\frac{\varphi}{2}\right) , \\ \langle (\Delta\hat{P})^2 \rangle &= \exp(-2r)\sin^2\left(\frac{\varphi}{2}\right) + \exp(2r)\cos^2\left(\frac{\varphi}{2}\right) . \end{aligned} \tag{5.5b}$$

The squeezing in $\hat{Q}$ occurs if

$$\cos(\varphi) > \tanh(r) . \tag{5.5c}$$

The mean photon number $\langle \hat{b}^\dagger \hat{b} \rangle = \langle |\alpha|^2 \rangle_{\mathcal{N}} + \sinh^2(r)$ is composed of the stimulated part $\langle |\alpha|^2 \rangle_{\mathcal{N}}$ and spontaneous emission $\sinh^2(r)$ independent of the input field.

Such reduction of quadrature fluctuations is usually detected with the help of homodyne detection, where a coherent component of a local oscillator of the coherent complex amplitude $\eta$ is superimposed on the signal beam using a beam splitter (Secs. 3.3.2 and 8.4). Photodetectors placed beyond the splitter provide the difference of photocurrents as follows

$$\hat{Q} = \frac{1}{2}\left(\hat{a}^\dagger + \eta^*\right)(\hat{a} + \eta) - \frac{1}{2}\left(\hat{a}^\dagger - \eta^*\right)(\hat{a} - \eta) = \eta\hat{a}^\dagger + \eta^*\hat{a} . \tag{5.6}$$

Without loss of generality, we can choose $\gamma = u_1 + iu_2 = \exp(i\theta)$, $u_1^2 + u_2^2 = 1$, where $u_1$, $u_2$ and $\theta$ are real numbers. The measured variance can be written in the form

$$\begin{aligned} V \equiv \langle (\Delta\hat{Q})^2 \rangle &= \gamma^2 \langle (\Delta\hat{a}^\dagger)^2 \rangle + \gamma^{*2} \langle (\Delta\hat{a})^2 \rangle + \langle \Delta\hat{a}^\dagger \Delta\hat{a} \rangle + \langle \Delta\hat{a}\Delta\hat{a}^\dagger \rangle \\ &= u_1^2 \langle (\Delta\hat{Q})^2 \rangle + u_2^2 \langle (\Delta\hat{P})^2 \rangle + u_1^2 u_2^2 \langle \{\Delta\hat{Q}, \Delta\hat{P}\} \rangle , \end{aligned} \tag{5.7}$$

where $\hat{Q} = \hat{a} + \hat{a}^\dagger$, $\hat{P} = (\hat{a} - \hat{a}^\dagger)/i$ again and the anticommutator is defined as $\{\Delta\hat{Q}, \Delta\hat{P}\} = \Delta\hat{Q}\Delta\hat{P} + \Delta\hat{P}\Delta\hat{Q}$. Writing (5.7) in a matrix form

$$V = (u_1, u_2) \begin{pmatrix} \langle (\Delta\hat{Q})^2 \rangle & \frac{1}{2}\langle \{\Delta\hat{Q}, \Delta\hat{P}\} \rangle \\ \frac{1}{2}\langle \{\Delta\hat{Q}, \Delta\hat{P}\} \rangle & \langle (\Delta\hat{P})^2 \rangle \end{pmatrix} \begin{pmatrix} u_1 \\ u_2 \end{pmatrix} = \hat{u}^\dagger \hat{M} \hat{u} , \tag{5.8}$$

we can diagonalize the matrix $\hat{M}$ by a transformation to another vector $\hat{v}$ fulfilling the eigenvalue equation $\hat{M}\hat{v} = \lambda\hat{v}$, with the eigenvalues $\lambda$ determined from the characteristic equation $\mathrm{Det}(\hat{M} - \lambda \hat{1}) = 0$,

$$\begin{aligned}\lambda_{1,2} &= \frac{1}{2}\mathrm{Tr}(\hat{M})\left[1 \pm \left(1 - \frac{4\mathrm{Det}(\hat{M})}{[\mathrm{Tr}(\hat{M})]^2}\right)^{1/2}\right] \\ &= \langle\{\Delta\hat{a}, \Delta\hat{a}^\dagger\}\rangle \pm 2|\langle(\Delta\hat{a})^2\rangle| \\ &= 1 + 2\left[\langle\Delta\hat{a}^\dagger\Delta\hat{a}\rangle \pm |\langle(\Delta\hat{a})^2\rangle|\right], \end{aligned} \qquad (5.9)$$

which provide the maximum and minimum half-axes of the noise ellipse. These results are just analogous to the description of partial polarization [Born and Wolf (1965)]. It is interesting to note that

$$\mathrm{Det}(\hat{M}) = \lambda_1\lambda_2 = \langle(\Delta\hat{Q})^2\rangle\langle(\Delta\hat{P})^2\rangle - \frac{1}{4}\langle\{\Delta\hat{Q}, \Delta\hat{P}\}\rangle^2 \geq 1. \qquad (5.10)$$

Thus the principal squeezing of vacuum fluctuations defined as

$$\lambda_2 \equiv \lambda = 1 + 2\left[\langle\Delta\hat{a}^\dagger\Delta\hat{a}\rangle - |\langle(\Delta\hat{a})^2\rangle|\right] < 1 \qquad (5.11)$$

represents the maximum value of squeezing related to Schrödinger-Robertson inequality (5.10) rather than to standard Heisenberg inequality (5.3) (cf. (2.25) and (2.26)). It holds that $\lambda_1 \geq 1$ (for the maximum half-axis of the noise ellipse) from the geometric-arithmetic mean inequality $\sqrt{\mathrm{Det}(\hat{M})} \leq \mathrm{Tr}(\hat{M})/2$. This definition of principal squeezing involves the anticommutator, which is neglected in the standard definition or it is equal to zero. As the principal squeezing is rotational invariant it is independent of the phase of local oscillator. It involves the cross-correlations $\langle\{\Delta\hat{Q}, \Delta\hat{P}\}\rangle$ playing an important role in nonlinear interactions. It reduces to the standard definition if $\langle\{\Delta\hat{Q}, \Delta\hat{P}\}\rangle = 0$. Another interpretation of the Schrödinger-Robertson inequality can be given in terms of the so-called correlated coherent states [Markov (1991)]. A graphical representation of principal and standard squeezing with the help of the noise ellipse and its lemniscate has been suggested by Loudon (1989).

Principal squeezing for a pair of modes is defined in an analogous way as

$$\begin{aligned}\lambda_{jk} &= 2\Big[1 + \langle\Delta\hat{a}_j^\dagger\Delta\hat{a}_j\rangle + \langle\Delta\hat{a}_k^\dagger\Delta\hat{a}_k\rangle + 2\mathrm{Re}\left(\langle\Delta\hat{a}_j^\dagger\Delta\hat{a}_k\rangle\right) \\ &\quad - |\langle(\Delta\hat{a}_j)^2\rangle + \langle(\Delta\hat{a}_k)^2\rangle + 2\langle\Delta\hat{a}_j\Delta\hat{a}_k\rangle|\Big] < 2, \quad j \neq k, \end{aligned} \qquad (5.12)$$

whereas

$$\begin{aligned}\left\langle\left(\Delta \begin{matrix}\hat{Q}_{jk} \\ \hat{P}_{jk}\end{matrix}\right)^2\right\rangle &= 2\Big[1 + \langle\Delta\hat{a}_j^\dagger\Delta\hat{a}_j\rangle + \langle\Delta\hat{a}_k^\dagger\Delta\hat{a}_k\rangle + 2\mathrm{Re}\left(\langle\Delta\hat{a}_j^\dagger\Delta\hat{a}_k\rangle\right) \\ &\quad \pm\mathrm{Re}\left(\langle(\Delta\hat{a}_j)^2\rangle + \langle(\Delta\hat{a}_k)^2\rangle + 2\langle\Delta\hat{a}_j\Delta\hat{a}_k\rangle\right)\Big]. \end{aligned} \qquad (5.13)$$

Also other kinds of vacuum fluctuations can be defined. Hong and Mandel (1985) defined the higher-order squeezing as follows: a state is squeezed in the $N$-th order if the value of $\langle(\Delta\hat{a}_j)^N\rangle$ is less than its corresponding coherent-state value. Decomposing the Baker-Hausdorff identity

$$\langle\exp(\mu\Delta\hat{Q})\rangle = \langle:\exp(\mu\Delta\hat{Q}):\rangle \exp\left(\frac{\mu^2 C}{2}\right), \tag{5.14}$$

where we assume more generally $[\hat{Q},\hat{P}] = 2iC$, $C$ being a complex number (so that $\langle(\Delta\hat{Q})^2\rangle\langle(\Delta\hat{P})^2\rangle \geq |C|^2$), according to $\mu$ and comparing the coefficients at $\mu^N/N!$, we obtain

$$\begin{aligned}\langle(\Delta\hat{Q})^N\rangle &= \langle:(\Delta\hat{Q})^N:\rangle + N(N-1)\frac{C}{2}\langle:(\Delta\hat{Q})^{N-2}:\rangle \\ &+ \frac{N(N-1)(N-2)(N-3)}{2!}\left(\frac{C}{2}\right)^2\langle:(\Delta\hat{Q})^{N-4}:\rangle + \cdots \\ &+ (N-1)!!\,C^{N/2}, \qquad N \text{ even}.\end{aligned} \tag{5.15}$$

For a coherent atate $\langle:(\Delta\hat{Q})^j:\rangle$, $j \neq 0$, vanish and squeezing of the $N$-th order occurs if

$$\langle(\Delta\hat{Q})^N\rangle < (N-1)!!\,C^{N/2}. \tag{5.16}$$

As this means that some of the normal moments $\langle:(\Delta\hat{Q})^j:\rangle$ for $j \leq N$ have to be negative, the effect can occur only for nonclassical states. A relation between the second-order and higher-order squeezing can be found [Hong and Mandel (1985)]. A convenient parameter for describing the $N$-th-order squeezing is

$$\lambda_N = \frac{\langle(\Delta\hat{Q})^N\rangle}{(N-1)!!\,C^{N/2}} - 1, \tag{5.17}$$

which has to be negative in the case of squeezing. Using the commutation rule (5.2) we have $C = 1$.

Another kind of squeezing of vacuum fluctuations was introduced by Hillery (1987a,b) which can be called amplitude-squared squeezing. Defining

$$\hat{Q} = \hat{a}^2 + \hat{a}^{\dagger 2}, \qquad \hat{P} = -i\left(\hat{a}^2 - \hat{a}^{\dagger 2}\right), \tag{5.18}$$

we obtain the commutation rule

$$[\hat{Q},\hat{P}] = 4i(2\hat{n}+1) \tag{5.19}$$

giving

$$\langle(\Delta\hat{Q})^2\rangle\langle(\Delta\hat{P})^2\rangle \geq 4\langle 2\hat{n}+1\rangle^2 \tag{5.20}$$

and the amplitude-squared squeezing occurs if

$$\langle(\Delta\hat{Q})^2\rangle \text{ or } \langle(\Delta\hat{P})^2\rangle < 2\langle 2\hat{n}+1\rangle. \tag{5.21}$$

Such squeezing is appropriate for the description of quantum effects in the second harmonic generation. It can be shown that the fundamental (subfrequency) mode just exhibits the amplitude-squared squeezing while the second harmonic mode is squeezed in the standard way and both the squeezing effects are directly related. Of course, this concept is more general because the right-hand side in (5.19) is dependent on a state of the field. Also sum- and difference amplitude squeezing can be introduced [Hillery (1989)] suitable for description of non-degenerate optical parametric processes [for a review, see Hillery, Yu and Bergou (1992)].

Squeezing of vacuum fluctuations was reported in a number of reviews, see e.g. Loudon and Knight (1987), Teich and Saleh (1989), Zaheer and Zubairy (1991), Milonni and Singh (1991), Kimble (1992). There are particular proceedings discussed this topic thoroughly from various points of view, e.g. Haake, Narducci and Walls (1986), Pike and Sarkar (1986), Harvey and Walls (1986), Yamamoto (1991), Ehlotzky (1993) and the monograph by Gardiner (1991) deals with various questions of quantum noise.

## 5.2 Antibunching of photons

The property of bunching, unbunching and antibunching of photons is described by the fourth-order correlation function, as discussed in Sec. 2.9. Probability of detecting two photons separated by a time interval $\tau$ by a photodetector is proportional to the correlation function $\Gamma_{\mathcal{N}}^{(2,2)}(t,t+\tau,t+\tau,t) = \langle :\hat{I}(t)\hat{I}(t+\tau):\rangle = \langle \hat{A}^{(-)}(t)\hat{A}^{(-)}(t+\tau)\hat{A}^{(+)}(t+\tau)\hat{A}^{(+)}(t)\rangle = \langle I(t)I(t+\tau)\rangle_{\mathcal{N}}$. We can define the degree of coherence

$$\gamma_{\mathcal{N}}^{(2)}(\tau) = \gamma_{\mathcal{N}}^{(2,2)}(\tau) = \frac{\Gamma_{\mathcal{N}}^{(2,2)}(t,t+\tau,t+\tau,t)}{\left[\Gamma_{\mathcal{N}}^{(1,1)}(t,t)\right]^2} = \frac{\langle I(t)I(t+\tau)\rangle_{\mathcal{N}}}{\langle I(t)\rangle_{\mathcal{N}}^2}. \tag{5.22}$$

From the Schwarz inequality it must hold in the classical description that $\langle I(t)I(t+\tau)\rangle_{\mathcal{N}}^2 \le \langle I^2(t)\rangle_{\mathcal{N}}^2$ because for a stationary field $\langle I^2(t+\tau)\rangle_{\mathcal{N}} = \langle I^2(t)\rangle_{\mathcal{N}}$. This means that for classical fields $\gamma_{\mathcal{N}}^{(2)}(\tau) \le \gamma_{\mathcal{N}}^{(2)}(0)$, which expresses the bunching property of classical photons.

For example for chaotic light from natural sources

$$\Gamma_{\mathcal{N}}^{(m,n)}(x_1,\ldots,x_{m+n}) = \delta_{mn}\sum_{\pi}\prod_{j=1}^{n}\Gamma_{\mathcal{N}}^{(1,1)}(x_1,x_{n+1})\ldots\Gamma_{\mathcal{N}}^{(1,1)}(x_n,x_{2n}). \tag{5.23}$$

For the fourth-order correlations involving partial polarization

$$\langle I_1(t)I_2(t+\tau)\rangle_{\mathcal{N}} = \langle I_1\rangle_{\mathcal{N}}\langle I_2\rangle_{\mathcal{N}}\left[1+|\gamma_{\mathcal{N},12}(\tau)|^2\,\frac{1+P^2}{2}\right] \tag{5.24a}$$

and

$$\begin{aligned}\langle \Delta I_1(t)\Delta I_2(t+\tau)\rangle_{\mathcal{N}} &= \langle I_1(t)I_2(t+\tau)\rangle_{\mathcal{N}} - \langle I_1\rangle_{\mathcal{N}}\langle I_2\rangle_{\mathcal{N}} \\ &= \langle I_1\rangle_{\mathcal{N}}\langle I_2\rangle_{\mathcal{N}} \frac{1+P^2}{2} |\gamma_{\mathcal{N},12}(\tau)|^2 , \qquad (5.24b)\end{aligned}$$

$\gamma_{\mathcal{N},12}(\tau)$ being the second-order degree of coherence $\gamma_{\mathcal{N}}^{(1,1)}(\tau)$. Thus we see from (5.22) that for chaotic light

$$\gamma_{\mathcal{N}}^{(2)}(\tau) = 1 + |\gamma_{\mathcal{N}}(\tau)|^2 \frac{1+P^2}{2} \qquad (5.25)$$

and there is an enhancement of the fourth-order degree of coherence over its coherent state value 1 since $\langle I(t)I(t+\tau)\rangle_{\mathcal{N}} \approx \langle I(t)\rangle_{\mathcal{N}}\langle I(t+\tau)\rangle_{\mathcal{N}} = \langle I(t)\rangle_{\mathcal{N}}^2$ in this case due to the stabilization of the intensity. Therefore, assuming full polarized field ($P = 1$), the probability to detect two close photons of natural light is twice of that for two successive photons being very separated, since $\gamma_{\mathcal{N}}(0) = 1$ and $\lim_{\tau\to\infty} \gamma_{\mathcal{N}}(\tau) = 0$ (e.g. for Lorentzian spectrum of light $\gamma_{\mathcal{N}}(\tau) = \exp(-i\bar{\omega}\tau - \Delta\omega|\tau|)$, $\bar{\omega}$ is the mean frequency of quasimonochromatic light and $\Delta\omega$ its spectral half-width). Therefore photons of natural light are bunched, whereas photons of ideal stabilized laser are unbunched.

We can show that $\langle I^2\rangle_{\mathcal{N}} \geq \langle I\rangle_{\mathcal{N}}^2$ for classical fields (i.e. for optical fields having a classical analogue with the existing Glauber-Sudarshan quasidistribution $\Phi_{\mathcal{N}}$), which means that for classical fields it always holds (putting $\tau = 0$ in (5.25)), that $\gamma_{\mathcal{N}}^{(2)}(0) \geq 1$ and $\gamma_{\mathcal{N}}^{(2)}(0) = 1$ for fully coherent fields. As we have mentioned in Sec. 2.9 $\langle(\Delta W)^2\rangle_{\mathcal{N}} \approx T^2\langle(\Delta I)^2\rangle_{\mathcal{N}} < 0$ can occur for nonclassical fields (this means that the above classical inequality is violated because $\langle I^2\rangle_{\mathcal{N}} < \langle I\rangle_{\mathcal{N}}^2$ in this case). Therefore for such nonclassical optical fields

$$\gamma_{\mathcal{N}}^{(2)}(0) < 1 . \qquad (5.26)$$

In this case we speak of antibunching of photons because they tend to be separated. This is fully quantum phenomenon.

## 5.3 Photon-number squeezing — sub-Poissonian photon statistics

Starting from the fluctuation formula (2.123) we can write for the variance of the number of emitted photoelectrons from the photocathode

$$\langle(\Delta n)^2\rangle = \langle n\rangle + \frac{\langle n\rangle^2}{T^2}\int_0^T\int_0^T \left[\gamma_{\mathcal{N}}^{(2)}(t_2-t_1) - 1\right] dt_1 dt_2 , \qquad (5.27)$$

where again $\langle n \rangle = \eta \langle I \rangle_{\mathcal{N}} T$ is the mean number of emitted photoelectrons, if we assume a stationary field. As we have seen in Sec. 2.9 the Fano factor $F_n = \langle (\Delta n)^2 \rangle / \langle n \rangle$ is a good measure to decide whether photon statistics of an optical field are Poissonian, super-Poissonian or sub-Poissonian. From (5.27) we arrive at the following relation of basic properties of the field [Teich and Saleh (1988)]

$$F_n - 1 = \frac{\langle n \rangle}{M} \left[ \gamma_{\mathcal{N}}^{(2)}(0) - 1 \right] , \tag{5.28}$$

where we have introduced the number of degrees of freedom by the relation

$$\frac{1}{M} = \frac{2}{T} \int_0^T \left( 1 - \frac{\tau}{T} \right) \xi(\tau)\, d\tau , \tag{5.29}$$

where

$$\xi(\tau) = \frac{\gamma_{\mathcal{N}}^{(2)}(\tau) - 1}{\gamma_{\mathcal{N}}^{(2)}(0) - 1} \tag{5.30}$$

and $\int_0^T \int_0^T \xi(t_2 - t_1)\, dt_1 dt_2 = 2 \int_0^T (T - \tau) \xi(\tau)\, d\tau$ has been used. As we have defined in Sec. 2.9 the optical field is Poissonian if $F_n = 1$, super-Poissonian if $F_n > 1$ and sub-Poissonian if $F_n < 1$ (taking into account the statistics conservation law (2.125)). Hence we see from (5.28) that, defining photon bunching if $\gamma_{\mathcal{N}}^{(2)}(0) > 1$, unbunching if $\gamma_{\mathcal{N}}^{(2)}(0) = 1$ and antibunching if $\gamma_{\mathcal{N}}^{(2)}(0) < 1$, there exists one-to-one correspondence between sub-Poissonian, Poissonian and super-Poissonian photon statistics and photon antibunching, unbunching and bunching, respectively, provided that $M > 0$. This represents several physically important cases. If $T \ll \tau_c$, $\gamma_{\mathcal{N}}^{(2)}(\tau)$ behaves monotonically and $\lim_{\tau \to \infty} \gamma_{\mathcal{N}}^{(2)}(\tau) = 1$, it is $\gamma_{\mathcal{N}}^{(2)}(\tau) - 1 \geq 0$ and $M > 0$ in the case of classical fields which are bunched and super-Poissonian, or $\gamma_{\mathcal{N}}^{(2)}(\tau) - 1 \leq 0$ and again $M > 0$ for nonclassical fields which are antibunched and sub-Poissonian. In general, antibunching and sub-Poissonian behaviour need not accompany each other. Bunching of chaotic photons is a consequence of the Bose-Einstein statistics and maximum entropy principle. The bunching effect is completely compensated by correlations in the laser source, which leads to the fact that ideal laser photons are unbunched. Various cases of relation between photon antibunching and sub-Poissonian behaviour are shown in Fig. 5.1. In Fig. 5.1a we illustrate the case of light which is bunched and super-Poissonian, in Fig. 5.1b light is antibunched and super-Poissonian, in Fig. 5.1c it is bunched and sub-Poissonian and finally in Fig. 5.1d light is antibunched and sub-Poissonian. Only in the case in Fig. 5.1a it is classical one, all other cases are nonclassical. We note that photon antibunching is defined in the sense that $\lim_{\tau \to 0+} d\gamma_{\mathcal{N}}^{(2)}(\tau)/d\tau > 0$, which can serve as a weaker definition of photon antibunching.

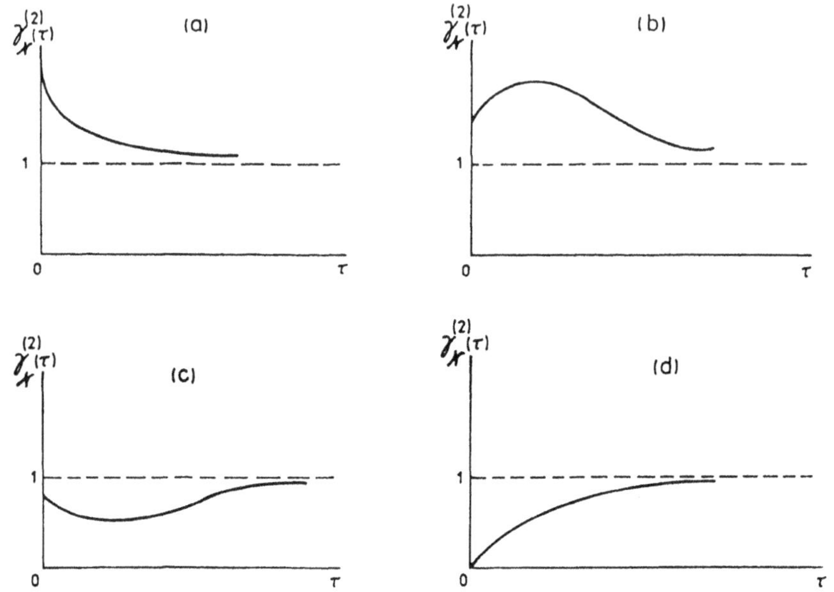


Figure 5.1: Relation between bunching, antibunching, super-Poissonian and sub-Poissonian statistics of photons: (a) bunched super-Poissonian light, (b) antibunched super-Poissonian light, (c) bunched sub-Poissonian light, and (d) antibunched sub-Poissonian light.

Sometimes the quantum effect of sub-Poissonian photon statistics is called photon-number squeezing.

If the detection volume $V$ can be identified with the quantization volume $L^3$ of the field, then $\hat{n}_V = \sum_\lambda \hat{a}_\lambda^\dagger \hat{a}_\lambda$, $W = \sum_\lambda |\alpha_\lambda|^2$ and the difference between photon antibunching and sub-Poissonian statistics is ruled out.

## 5.4 Oscillations in photon distributions

When studying the photon statistics of degenerate and non-degenerate optical parametric processes with classical pumping [Mišta, Peřinová, Peřina and Braunerová (1977), Mišta and Peřina (1978)] it has been found that there are several regimes in the process after the switching on of the interaction. If the initial field is coherent and there is a suitable relation between initial phases of the signal amplitudes and pumping phase, the photon statistics develop from the initial Poissonian statistics to sub-Poissonian statistics and further to super-Poissonian statistics exhibiting oscillations in the photon-number distribution as an expression of the competition of quantum states. We can explain this fully quantum phenomena using more straightforward method of generating function [Peřina and Bajer (1990)] or interferences in phase space [Schleich and Wheeler (1987)], which represents more sophisticated way.

Let us follow the description of the degenerate subharmonic generation with classical pumping. The corresponding hamiltonian reads

$$\hat{H} = \hbar\omega\left(\hat{a}^\dagger\hat{a} + \frac{1}{2}\right) - \left(\frac{1}{2}\hbar g\hat{a}^2 \exp(i2\omega t - i\varphi) + \text{h.c.}\right), \tag{5.31}$$

where $g$ is the coupling constant, $2\omega$ is a frequency of strong classical pumping of the phase $\varphi$. The amplitude of pumping is included in $g$. The solution of the Heisenberg equations for $\hat{a}$ and $\hat{a}^\dagger$ is in the form

$$\hat{a}(t) = u(t)\,\hat{a}(0) + v(t)\,\hat{a}^\dagger(0), \tag{5.32}$$

where

$$\begin{aligned} u(t) &= \exp(-i\omega t)\cosh(gt), \\ v(t) &= i\exp(-i\omega t + i\varphi)\sinh(gt), \\ &|u(t)|^2 - |v(t)|^2 = 1, \end{aligned} \tag{5.33}$$

where we assume that $g$ is real (phase matching occurs). Following Sec. 4.8 we can write for the antinormal characteristic function

$$\begin{aligned} C_{\mathcal{A}}(\beta,t) &= \text{Tr}\left\{\hat{\varrho}\exp\left(-\beta^*\hat{a}(t)\right)\exp\left(\beta\hat{a}^\dagger(t)\right)\right\} \\ &= \exp\left[-B|\beta|^2 + \left(\frac{1}{2}C^*\beta^2 + \text{c.c.}\right) + \beta\xi^* - \beta^*\xi\right], \end{aligned} \tag{5.34}$$

with the corresponding quasidistribution

$$\Phi_{\mathcal{A}}(\beta,t) = \frac{1}{\pi\sqrt{K(t)}}\exp\left[-\frac{B(t)}{K(t)}|\alpha - \xi(t)|^2 + \left(\frac{C^*(t)}{2K(t)}(\alpha - \xi(t))^2 + \text{c.c.}\right)\right], \tag{5.35}$$

where the initial coherent state $|\xi\rangle$ is assumed, $\xi(t) = u(t)\xi + v(t)\xi^*$ and $K = B^2 - |C|^2 = \cosh^2(gt)$ is the determinant of the Fourier transformation. The generating function equals

$$\begin{aligned} \langle\exp(-\lambda W)\rangle_{\mathcal{N}} &= \frac{1}{\pi\lambda}\int \exp\left(-\frac{1}{\lambda}|\beta|^2\right) C_{\mathcal{N}}(\beta,t)\, d^2\beta \\ &= [1 + \lambda(E-1)]^{-1/2}\,[1 + \lambda(F-1)]^{-1/2} \\ &\quad \times \exp\left[-\frac{\lambda A_1}{1 + \lambda(E-1)} - \frac{\lambda A_2}{1 + \lambda(F-1)}\right], \end{aligned} \tag{5.36}$$

where the normal characteristic function equals $C_{\mathcal{N}}(\beta,t) = \exp(|\beta|^2)C_{\mathcal{A}}(\beta,t)$, $W = |\alpha|^2$ and

$$\begin{aligned}
A_{1,2} &= \frac{1}{2}\left[|\xi(t)|^2 \mp \frac{1}{2|C|}\left(C^*\xi^2(t) + \text{c.c.}\right)\right] \\
&= \frac{1}{2}|\xi|^2 \exp(\mp 2gt)\left[1 \mp \sin(2\vartheta - \varphi)\right] \geq 0\,, \\
E - 1 &= B - 1 - |C| = \frac{1}{2}\left[\exp(-2gt) - 1\right] \leq 0\,, \\
F - 1 &= B - 1 + |C| = \frac{1}{2}\left[\exp(2gt) - 1\right] \geq 0\,, \\
B &= \langle \Delta\hat{a}\Delta\hat{a}^\dagger\rangle = |u|^2 = \cosh^2(gt)\,, \\
C &= \langle(\Delta\hat{a})^2\rangle = uv = \frac{i}{2}\sinh(2gt)\exp(i\varphi)\,,
\end{aligned} \tag{5.37}$$

when $\vartheta$ is the phase of the initial coherent amplitude $\xi$. The above expressions are just appropriate for squeezed states (Sec. 4.4) if we choose for the squeeze parameter $r = gt$. We can point out that $A_{1,2}$ are playing the role of "mean numbers of coherent photons" related to the input coherent amplitude $\xi$, whereas $E - 1$ and $F - 1$ are playing the role of "mean numbers of chaotic photons" in the generating function (5.36) for the Laguerre polynomials. This means that in classical terms light generated by this process can be considered as the superposition of signals $A_{1,2}$ and quantum noise components ($E$, $F-1$). However, the quantum properties are reflected by negative values of the noise component $E - 1$, which is impossible in any classical description.

For the mean integrated intensity and its variance we obtain

$$\begin{aligned}
\langle W\rangle_{\mathcal{N}} &= \frac{d}{d(-\lambda)}\langle\exp(-\lambda W)\rangle_{\mathcal{N}}\Big|_{\lambda=0} = A_1 + A_2 + \frac{1}{2}(E + F - 2) \\
&= \left[\cosh(2r) + \sinh(2r)\sin(2\vartheta - \varphi)\right]|\xi|^2 + \sinh^2(r)\,,
\end{aligned} \tag{5.38a}$$

$$\begin{aligned}
\langle(\Delta W)^2\rangle_{\mathcal{N}} &= \frac{d^2}{d(-\lambda)^2}\langle\exp(-\lambda W)\rangle_{\mathcal{N}}\Big|_{\lambda=0} - \left[\frac{d}{d(-\lambda)}\langle\exp(-\lambda W)\rangle_{\mathcal{N}}\Big|_{\lambda=0}\right]^2 \\
&= 2A_1(E-1) + 2A_2(F-1) + \frac{1}{2}\left[(E-1)^2 + (F-1)^2\right] \\
&= \sinh(r)\Bigg[2|\xi|^2\left(\sinh(3r) + \cosh(3r)\sin(2\vartheta - \varphi)\right) \\
&\quad + \frac{\sinh(3r) - \sinh(r)}{2}\Bigg]\,.
\end{aligned} \tag{5.38b}$$

For vacuum state $\xi = 0$ and $\langle(\Delta n)^2\rangle = 2\langle n\rangle(1 + \langle n\rangle)$, $\langle n\rangle = \sinh^2(r)$, i.e. the squeezed vacuum is super-chaotic. In both the expressions (5.38a,b) the first term dependent on the input amplitude is related to stimulated effect whereas the second term represents the effect of vacuum fluctuations.

Making use of the fact that (5.36) is also the generating function for the Laguerre polynomials we can write for the photon-number distribution $p(n,t)$ and its factorial moments $\langle W^k\rangle_{\mathcal{N}}$ using expressions (4.184a,b) of Sec. 4.8 for $M=1$

$$\begin{aligned} p(n) &= \frac{1}{\sqrt{EF}}\left(1-\frac{1}{F}\right)^n \exp\left(-\frac{A_1}{E}-\frac{A_2}{F}\right) \\ &\times \sum_{k=0}^{n} \frac{1}{\Gamma(k+1/2)\Gamma(n-k+1/2)}\left(\frac{1+1/E}{1+1/F}\right)^k \\ &\times L_k^{-1/2}\left(-\frac{A_1}{E(E-1)}\right) L_{n-k}^{-1/2}\left(-\frac{A_2}{F(F-1)}\right), \end{aligned} \tag{5.39a}$$

$$\begin{aligned} \langle W^k\rangle_{\mathcal{N}} &= (F-1)^k \sum_{l=0}^{k} \frac{1}{\Gamma(l+1/2)\Gamma(k-l+1/2)}\left(\frac{E-1}{F-1}\right)^l \\ &\times L_l^{-1/2}\left(-\frac{A_1}{E-1}\right) L_{k-l}^{-1/2}\left(-\frac{A_2}{F-1}\right). \end{aligned} \tag{5.39b}$$

Applying the identity between the Laguerre and Hermite polynomials (Peřina (1991), p. 247) we can rewrite the photon-number distribution (5.39a) for the squeezed state in the form provided by Yuen (1976)

$$\begin{aligned} p(n) &= \frac{1}{2^n n!}\frac{1}{|u|}\left(\frac{|v|}{|u|}\right)^n \exp\left[-|\xi(t)|^2+\frac{v^*}{2u}\xi^2(t)+\frac{v}{2u^*}\xi^{*2}(t)\right] \\ &\times H_n\left(\frac{\xi(t)}{\sqrt{2uv}}\right) H_n\left(\frac{\xi^*(t)}{\sqrt{2u^*v^*}}\right); \end{aligned} \tag{5.40}$$

here $|v|/|u| = \tanh(r)$ and $2|u||v| = \sinh(2r)$. The generating function (5.36) is also the product of two factors representing the generalized superposition of signal $A_1$ with noise $(E-1)$ and that of signal $A_2$ with the noise $(F-1)$. Denoting the corresponding distributions as $p_1(n)$ and $p_2(n)$ respectively, we can also obtain the discrete convolution

$$p(n) = \sum_{k=0}^{n} p_1(m-k)p_2(k). \tag{5.41}$$

It is evident from (5.37) that the quadrature squeezing is expressed as

$$\begin{aligned} \left\langle\left(\Delta \begin{matrix}\hat{Q}\\ \hat{P}\end{matrix}\right)^2\right\rangle &= 1+2[B-1\pm \mathrm{Re}(C)] = \exp(\mp 2r), \\ \lambda &= 1+2[B-1-|C|] = \exp(-2r), \end{aligned} \tag{5.42}$$

provided that $\varphi = \pi/2$ for simplicity.

First choose the initial phases such as $2\vartheta - \varphi = \pi/2$. As follows from (5.37) $A_1 = 0$ and $A_2 = |\xi|^2 \exp(2r)$. Therefore the dominating factor in (5.36) involves $A_2 \geq 0$ and $F - 1 \geq 0$ in correspondence to $p_2(k)$ and photon statistics are super-Poissonian for all $r$ (for all times $t$ from the beginning) because we have an amplified signal superimposed on positive noise. The opposite case appears if $2\vartheta - \varphi = -\pi/2$. In this case $A_1 = |\xi|^2 \exp(-2r)$ and $A_2 = 0$. The dominating factor in the generating function includes $A_1$ and $E - 1$, i.e. we have an attenuated signal superimposed on negative noise and therefore for small squeeze parameters (for short times after the switching on of the interaction) the photon statistics are sub-Poissonian and nonclassical radiation is generated compared to the Poisson statistics of coherent field ($E - 1 = F - 1 = 0$) at $r = 0$ ($gt = 0$). This is also seen from (5.38b) since the first term $2A_1(E-1) < 0$ is dominant in this case. To see the situation for higher $r$ (later times) we use (5.41). Adopting the phase condition $2\vartheta - \varphi = -\pi/2$ we have

$$p_1(k) = \frac{\Gamma(k+1/2)}{\sqrt{\pi F}\, k!} \left(1 - \frac{1}{F}\right)^k , \qquad (5.43a)$$

$$\begin{aligned} p_2(k) &= \frac{1}{\Gamma(k+1/2)\sqrt{E}} \left(1 - \frac{1}{E}\right)^k \exp\left(-\frac{|\xi|^2 \exp(-2r)}{E}\right) \\ &\times L_k^{-1/2}\left(-\frac{|\xi|^2 \exp(-2r)}{E(E-1)}\right) , \end{aligned} \qquad (5.43b)$$

where we have used that $L_k^{-1/2}(0) = \Gamma^2(k+1/2)/(\sqrt{\pi}\, k!)$ (the normalization of the Laguerre polynomials of Morse and Feshbach (1953) is used). We see that $p_1(k)$ is the geometrical distribution with the number of degrees of freedom $M = 1/2$ and $(1 - 1/F) = \tanh(r) \geq 0$ and $p_1(k)$ is non-negative for $r = gt \geq 0$. For the partial distribution $p_2(k)$ we obtain

$$\begin{aligned} p_2(k) &= \frac{(-1)^k \sqrt{2}}{\Gamma(k+1/2)\,[1+\exp(-2r)]^{1/2}} [\tanh(r)]^k \\ &\times \exp\left(-\frac{2|\xi|^2}{1+\exp(2r)}\right) L_k^{-1/2}\left(\frac{2|\xi|^2}{\sinh(2r)}\right) . \end{aligned} \qquad (5.44)$$

Now we can use the following approximate expressions for the Laguerre polynomials

$$L_j^M(x) \underset{x\to 0}{\approx} \frac{\Gamma^2(j+M+1)}{j!\,\Gamma(M+1)} \left(1 - \frac{jx}{M+1}\right) , \qquad (5.45a)$$

$$L_j^M(x) \underset{x\to\infty}{\simeq} \frac{\Gamma(j+M+1)}{j!} (-x)^j . \qquad (5.45b)$$

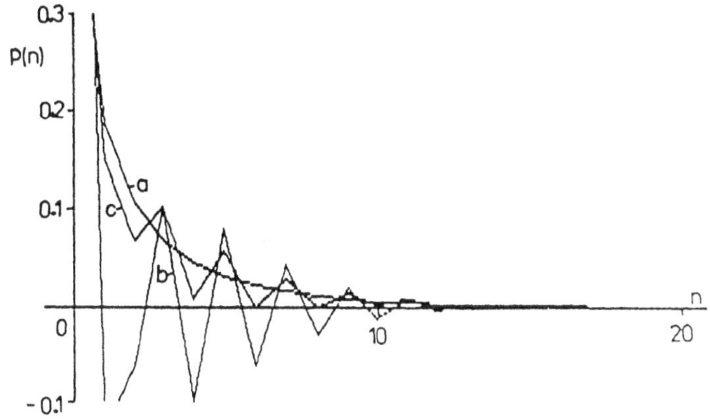


Figure 5.2: Photon distribution $p_1(n)$ (curve a), $p_2(n)$ (curve b), and their discrete convolution giving the actual photon distribution $p(n)$ (curve c); oscillations result from negative values of the distribution $p_2(n)$, which is a consequence of negative "mean number of chaotic photons"; $|\xi| = 0.8$, $r = 1$, $2\vartheta - \varphi = -\pi/2$ [after J. Peřina and J. Bajer, *Phys. Rev.* **A 41** (1990) 516].

For small and large values of the argument $2|\xi|^2/\sinh(2r)$ in (5.44) we obtain

$$p_2(k) \approx \frac{(-1)^k \sqrt{2}\,\Gamma(k+1/2)}{\sqrt{\pi}\,k!\,[1+\exp(-2r)]^{1/2}} [\tanh(r)]^k \times \exp\left(-\frac{2|\xi|^2}{1+\exp(2r)}\right)\left(1-\frac{4k|\xi|^2}{\sinh(2r)}\right), \quad (5.46a)$$

$$p_2(k) \simeq \frac{\sqrt{2}}{k!\,[1+\exp(-2r)]^{1/2}} \left(\frac{|\xi|^2}{\cosh^2(r)}\right)^k \times \exp\left(-\frac{2|\xi|^2}{1+\exp(2r)}\right), \quad (5.46b)$$

respectively. We see from (5.46a) that for small $|\xi|^2$ and/or large $r$ the partial distribution $p_2(k)$ takes on successively positive and negative values for $k$ even and odd, respectively ($k|\xi|^2/\sinh(2r) \ll 1$). Therefore the convolution (5.41) representing the resulting photon distribution is oscillating and super-Poissonian [Mišta, Peřinová, Peřina and Braunerová (1977), Peřina and Bajer (1990)]. For large $|\xi|^2$ and/or small $r$, the oscillations are missing as is seen from the asymptotic expression (5.46b), which always provides positive values. This is demonstrated in Fig. 5.2. A time development of the photon-number distribution in the process under discussion is shown in Fig. 5.3.

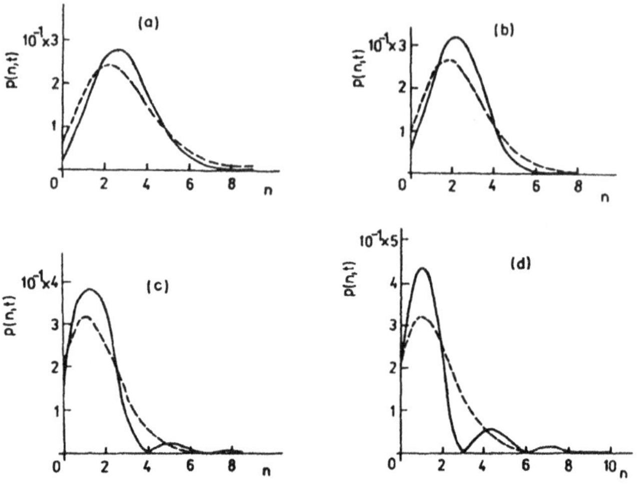


Figure 5.3: The photocount distribution for a degenerate parametric amplifier, $|\xi| = 2$, $2\vartheta - \varphi = -\pi/2$, $g = 10^4$ s$^{-1}$ ($\gamma = \langle n_R \rangle = 0$) (a) $t = 2 \times 10^{-5}$ s, (b) $t = 3 \times 10^{-5}$ s, (c) $t = 6 \times 10^{-5}$ s, and (d) $t = 8 \times 10^{-5}$ s. The full curves represent the actual distributions, the dashed curves are the corresponding Poisson distributions with the same $\langle n \rangle$ [after L. Mišta, V. Peřinová, J. Peřina and Z. Braunerová, *Acta Phys. Pol.* **A 51** (1977) 739].

For the spontaneous process $|\xi|^2 = 0$ and

$$p(2n) = \frac{(2n)!}{(2^n n!)^2 \cosh(r)} [\tanh(r)]^n ,$$

$$p(2n+1) = 0, \qquad (5.47)$$

which shows that in this case $p(n)$ is zero for $n$ odd as a reflection of the fact that the two-photon process is discussed.

For other values of $|\xi|^2$ and $r$, various behaviour of $p(n)$ is possible, e.g. maxima situated at $n$ even can change to minima of $p(n)$ and minima for $n$ odd can change to maxima for larger $n$, further larger period of oscillations or a double modulation can occur.

Thus the negative values of the quasiprobability $p_2(k)$ are just a consequence of the quantum properties of the squeezed state and they represent a source of oscillations in the photon-number distribution as well as a source of sub-Poissonian statistics if the initial phases are appropriate.

If dissipation is introduced with the mean number of dissipative particles $\langle n_R \rangle$, then we have

$$\begin{aligned} E - 1 &= \frac{1}{2}\exp(-\gamma t)[\exp(-2gt) - 1] + \langle n_R \rangle [1 - \exp(-\gamma t)] , \\ F - 1 &= \frac{1}{2}\exp(-\gamma t)[\exp(2gt) - 1] + \langle n_R \rangle [1 - \exp(-\gamma t)] . \end{aligned} \qquad (5.48)$$

Hence with increasing $\langle n_R \rangle$ the reduced negative values of quantum noise $(E-1)\exp(\gamma t)$ are further compensated by the positive dissipative term $\langle n_R \rangle [1-\exp(-\gamma t)]$ and oscillations as well as other nonclassical behaviour are degraded.

Some undirect explanation of oscillations in photon-number distribution can be based on interferences in phase space, as mentioned above. The corresponding calculations are rather sophisticated [Schleich and Wheeler (1987, 1987a)], however the principle can be understood easily. In the phase space a squeeze state is represented by the noise ellipse (cut-area of the $\Phi_{\mathcal{A}}(\alpha)$-quasidistribution), whereas the Fock-state $|n\rangle$ is represented by a circle as having uncertain phase. Considering an area between circles for $|n\rangle$ and $|n+1\rangle$ we have no overlapping with the noise ellipse for $n < |\xi(t)|^2$, whereas for $n > |\xi(t)|^2$ two symmetrically located areas of overlap are arising giving the same contribution to the resulting probability but with opposite phases, which leads to the $\cos^2$-modulation as in the Young two-slit experiment, however not in the ordinary space but in phase space.

A nice review of interferences in phase space was presented by Vogel and Schleich (1992).

## 5.5 Quantum phase

The phase of optical fields plays the decisive role in many optical phenomena, particularly in interference and diffraction of light. However, when the intensity of light is decreased on a quantum level the classical notion of phase loses its sense and complications occur to define it on a quantum level.

A phase operator $\hat{\varphi}$ has been intoduced by Dirac [see e.g. Heitler (1954) and Akhiezer and Berestetsky (1965)] in analogy to a classical procedure of introducing the phase,

$$\hat{a} = \exp(i\hat{\varphi})\,\hat{n}^{1/2}\,, \tag{5.49}$$

$\hat{n}$ being the number operator $\hat{a}^\dagger \hat{a}$. Assuming that $\hat{\varphi}$ may be a Hermitian operator (then $\exp(i\hat{\varphi})$ is unitary) and using the commutation rule

$$[\exp(i\hat{\varphi}), \hat{n}] = \exp(i\hat{\varphi})\,, \tag{5.50a}$$

and

$$[\hat{\varphi}, \hat{n}] = -i\hat{1}\,, \tag{5.50b}$$

it folows that

$$\langle(\Delta\hat{n})^2\rangle\,\langle(\Delta\hat{\varphi})^2\rangle \geq \frac{1}{4}\,. \tag{5.51}$$

But the difficulty arises from the fact that the operator $\hat{U} = \exp(i\hat{\varphi})$ is not unitary, since $\hat{U}\hat{n}\hat{U}^\dagger = \hat{n} + \hat{1}$. To avoid this difficulty Susskind and Glogower

(1964) (see Carruthers and Nieto (1968)) defined the Hermitian operators

$$\hat{C} = \frac{1}{2}\left(\hat{U} + \hat{U}^\dagger\right), \qquad \hat{S} = \frac{1}{2i}\left(\hat{U} - \hat{U}^\dagger\right) \tag{5.52}$$

fulfilling the commutation rules

$$[\hat{C}, \hat{n}] = i\hat{S}, \qquad [\hat{S}, \hat{n}] = -i\hat{C} \tag{5.53}$$

and consequently

$$\begin{aligned} \langle(\Delta\hat{n})^2\rangle\,\langle(\Delta\hat{C})^2\rangle &\geq \frac{\langle\hat{S}\rangle^2}{4}, \\ \langle(\Delta\hat{n})^2\rangle\,\langle(\Delta\hat{S})^2\rangle &\geq \frac{\langle\hat{C}\rangle^2}{4}. \end{aligned} \tag{5.54}$$

It is more convenient to define

$$\hat{U} = (\hat{n}+1)^{-1/2}\hat{a}, \qquad \hat{U}^\dagger = \hat{a}^\dagger(\hat{n}+1)^{-1/2}, \tag{5.55}$$

for which it holds that

$$\hat{U}\hat{U}^\dagger = \hat{1}, \qquad \hat{U}^\dagger\hat{U} = \hat{1} - |0\rangle\langle 0|, \tag{5.56}$$

which means that the unitary properties can be satisfied by antinormal ordering of operators $\hat{U}$ and $\hat{U}^\dagger$.

From (5.54) we obtain more general uncertainty relation [see review by Lukš and Peřinová (1994) and references therein]

$$\left[\langle(\Delta\hat{n})^2\rangle + \frac{1}{4}\right] V \geq \frac{1}{4}, \tag{5.57}$$

where the phase variance equals $V = D^2 = 1 - \langle\hat{C}\rangle^2 - \langle\hat{S}\rangle^2 = 1 - |\langle\exp(i\hat{\varphi})\rangle|^2$ as given in (3.51) (an unimportant vacuum term is neglected). It should be noted at this moment that the inequality (5.57) is more fundamental one replacing the inequality (5.51) valid in most practicaly cases, but not without exception. The inequality (5.57) has strong physical consequences. The operator representation of physical quantities in quantum theory necessary for discussion of spontaneous effects leads to occurrence of i) statistics, ii) such inequalities leading to the complementarity of observation of ideal wave or particle phenomena as exhibitions of wave-particle duality, experimentally well verified for photons, electrons, neutrons, atoms and ions. Consider an interferometer with single photons provided that one cannot distinguish to which arm of the interferometer a photon belongs. Then we have uncertainty in the photon number $n$ and we may observe ideal interference fringes (with the maximum visibility) if the phase is quite certain, i.e. $V = 0$, which means that the phase has a certain value $\bar{\varphi}$ with the phase probability distribution $P(\varphi) = \delta(\varphi - \bar{\varphi})$. Therefore (5.57) demands $\langle(\Delta\hat{n})^2\rangle \to \infty$. Now try to determine during the measurement to which arm the photon belongs. If we are successful, we specify the Fock state $|1\rangle$ and $\langle(\Delta\hat{n})^2\rangle = 0$. In this case (5.57) leads to $V \geq 1$ and hence $V = 1$, $P(\varphi) = 1/2\pi$, i.e. the phase is quite uncertain over the interval $(0, 2\pi)$ and $\langle\exp(i\hat{\varphi})\rangle = 0$. The interference fringes have to be missing necessarily. In the case of uncertainty in $n$ and the occurrence of interference fringes with one photon the quantized Maxwellian wave fields $\hat{E}$ or $\hat{H}$ represent probability waves propagating through the arms of the interferometer and forming the interference pattern they give the probabilities of registration of photons during many expositions. The propagation of a photon through any arm is quite stochastic. Therefore the questions such as "through which arm a photon propagates" or "how the photon knows that both arms are open" have no sense. Only in the classical limit of strong fields when the density of photons is so high that amounts of the energy $\hbar\omega$ are negligible we obtain the Maxwellian classical wave fields $E$ and $H$ identical to the original probability waves. From this point of view there is no EPR paradox because a measurement on an EPR-particle leads to collapsed wave function as a consequence of the increased information of the observer and it does not implicate new physical properties of the particles, which are determined in the interaction during their creation. Such a quantum statistical description is necessarily obtained when the operator representation is adopted. The non-existence of the real de Broglie guided waves (these must be probability waves) was experimentally supported by Wang, Zou and Mandel (1991). Thus as ideal phenomena we can observe either particle or wave effects, as is stated by the complementarity principle. However if non-ideal effects are observed both can occur simultaneously [Mizobuchi and Ohtaké (1992), Ghose, Horne and Agarwal (1992)].

Similar conclusions, although not so clear, are following from the fluctuation formula (2.123). The ideal interference appears if the generalized wave fluctuations $\langle(\Delta W)^2\rangle_{\mathcal{N}} = 0$ and we have uncertainty in $n$, $\langle(\Delta n)^2\rangle = \langle n\rangle$. On the other hand a which-path information means that $\langle(\Delta n)^2\rangle = 0$ and the fourth-order wave uncertainty is $\langle(\Delta W)^2\rangle_{\mathcal{N}} = -\langle n\rangle$ leading to the washing out of the interference fringes.

The normalized minimum uncertainty states for the photon numbers and phase were constructed by Jackiw (1968) as a solution of the eigenvalue problem

$$\left(\Delta\hat{n} + i\xi\Delta\hat{S}\right)|\psi\rangle = 0\,, \tag{5.58}$$

$\xi$ being a complex parameter. Then the phase operator can be defined with the help of the relation

$$\widehat{\exp}(i\varphi) = \exp(i\bar{\varphi}) \sum_{n=0}^{\infty} |n\rangle\langle n+1| , \tag{5.59}$$

where $\bar{\varphi}$ is obtained as a dominant phase $\arg\langle\widehat{\exp}(i\varphi)\rangle$. The Jackiw states are also called the crescent states as having the uncertainty (noise) area in the form of a crescent.

Pegg and Barnett (1989) have constructed the Hermitian phase operator

$$|\theta_m, s\rangle = \frac{1}{\sqrt{s+1}} \sum_{n=0}^{s} \exp(in\theta_m)\, |n\rangle , \tag{5.60}$$

where

$$\theta_m = \theta + 2\pi\frac{m}{s+1} , \qquad m = 0, 1, \ldots, s , \tag{5.61}$$

$\theta$ being a chosen value, which uses the Loudon's definition of phase states [Loudon (1973)]. These states form an orthogonal system and a phase operator is defined as

$$\hat{\Phi}_{\theta,s} = \sum_{m=0}^{s} \theta_m |\theta_m, s\rangle\langle\theta_m, s| , \tag{5.62}$$

and any phase function can be represented by the operator

$$\hat{M}_{\theta,s} = \sum_{m=0}^{s} M(\theta_m)|\theta_m, s\rangle\langle\theta_m, s| . \tag{5.63}$$

The values of phase available by measurements are then obtained by putting $s \rightarrow \infty$.

A phase distribution $P(\varphi)$ is obtained as

$$\begin{aligned} P(\varphi) &= \lim_{s\rightarrow\infty} \frac{s+1}{2\pi} \langle\theta_m, s|\hat{\varrho}|\theta_m, s\rangle \\ &= \frac{1}{2\pi} \sum_{m,n=0}^{\infty} \exp\left[-i(m-n)\varphi\right] \varrho_{nm} . \end{aligned} \tag{5.64}$$

With $P(\varphi)$ we can define the phase variance $\langle(\Delta\hat{\Phi}_\theta)^2\rangle = \langle\hat{\Phi}_\theta^2\rangle - \langle\hat{\Phi}_\theta\rangle^2$, $\langle\Phi_\theta^k\rangle = \int_\theta^{\theta+2\pi} \varphi^k P(\varphi)\, d\varphi$, fulfilling the uncertainty relation

$$\langle(\Delta\hat{n})^2\rangle\langle(\Delta\hat{\Phi}_\theta)^2\rangle \geq \frac{1}{4} . \tag{5.65}$$

The use of the definition (5.55) makes it possible to represent any function $M(\varphi)$ by the operator

$$\hat{M} = \frac{1}{2\pi} \int_\theta^{\theta+2\pi} M(\varphi)|\varphi\rangle\langle\varphi|\, d^2\varphi , \tag{5.66}$$

where the phase states

$$|\varphi\rangle = \sum_{n=0}^{\infty} \exp(in\varphi)\,|n\rangle \tag{5.67}$$

fulfil an approximate orthogonal relation

$$\langle\varphi|\varphi'\rangle = \pi\delta(\varphi-\varphi') + \frac{1}{1-\exp\left[-i(\varphi-\varphi')\right]}\,. \tag{5.68}$$

Particularly one can define

$$\begin{aligned}\hat{C} &= \frac{1}{2\pi}\int_{\theta}^{\theta+2\pi} \cos(\varphi)\,|\varphi\rangle\langle\varphi|\,d\varphi\,,\\ \hat{S} &= \frac{1}{2\pi}\int_{\theta}^{\theta+2\pi} \sin(\varphi)\,|\varphi\rangle\langle\varphi|\,d\varphi\,.\end{aligned} \tag{5.69}$$

The phase distribution equals $P(\varphi) = \langle\varphi|\hat{\varrho}|\varphi\rangle/2\pi$.

There are other definitions of the phase operator, e.g. Paul (1974) defined

$$\hat{\Phi} = \frac{1}{2\pi i}\int \left[\ln(\alpha) - \ln(\alpha^*)\right]|\alpha\rangle\langle\alpha|\,d^2\alpha\,, \tag{5.70}$$

and also a set of operators can be established as follows

$$\begin{aligned}\hat{\Phi}_k &= \frac{1}{\pi}\int \frac{\alpha^k}{|\alpha|^k}\,|\alpha\rangle\langle\alpha|\,d^2\alpha\,,\\ \hat{\Phi}_{-k} &= \frac{1}{\pi}\int \frac{\alpha^{*k}}{|\alpha|^k}\,|\alpha\rangle\langle\alpha|\,d^2\alpha\,, \qquad k = 1,2,\ldots\end{aligned} \tag{5.71}$$

describing higher powers of the phase.

It seems that various quantum phase concepts are to be related to various experimental arrangements for phase measurements.

Recently Noh, Fougères and Mandel (1992, 1992a) suggested operational definition of the quantum phase. If we have two inputs on a beam splitter, two photodetectors placed beyond the splitter (Fig. 5.4) will provide a photocounts with the instantaneous integrated intensities $W$,

$$W_{3,4} = \frac{1}{2}\left[W_1 + W_2 \mp 2W_{12}\sin(\phi)\right]\,, \tag{5.72}$$

when $\phi = \phi_1 - \phi_2$ is the difference of phases of single beams if a symmetric beam splitter is adopted. Performing the second measurement using the $\lambda/4$-plate as indicated in Fig. 5.4, we can similarly determine

$$W_{5,6} = \frac{1}{2}\left[W_1 + W_2 \pm 2W_{12}\cos(\phi)\right]\,; \tag{5.73}$$

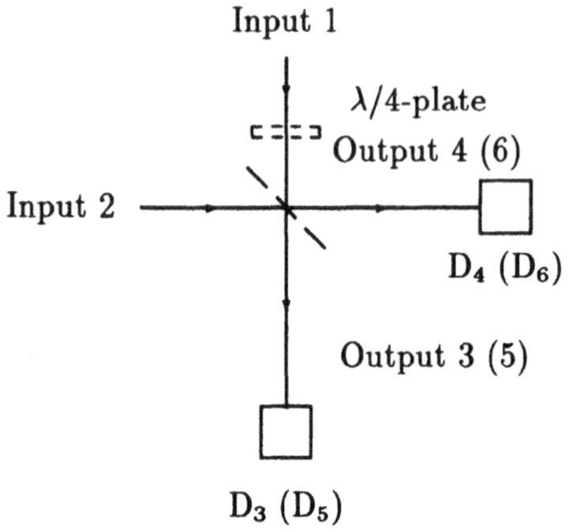


Figure 5.4: A scheme of measurements of sine and cosine of the phase difference of two beams.

here

$$
\begin{aligned}
W_j &= \eta \int_t^{t+T} I_j(t')\, dt' , \\
W_{12} &= \eta \int_t^{t+T} [I_1(t') I_2(t')]^{1/2}\, dt' .
\end{aligned}
\tag{5.74}
$$

The mutual quantity $W_{12}$ need not be directly measured. Squaring and adding (5.72) and (5.73), we obtain for $W_{12}^2$

$$4W_{12}^2 = (W_4 - W_3)^2 + (W_6 - W_5)^2 \tag{5.75}$$

and the sine and cosine of the phase difference of both the beams can be operationally defined and measured as follows

$$\sin(\phi) = \frac{W_4 - W_3}{\left[(W_4 - W_3)^2 + (W_6 - W_5)^2\right]^{1/2}} , \tag{5.76a}$$

$$\cos(\phi) = \frac{W_5 - W_6}{\left[(W_4 - W_3)^2 + (W_6 - W_5)^2\right]^{1/2}} . \tag{5.76b}$$

Combining two systems in Fig. 5.4 one can measure $S = \sin(\phi)$ and $C = \cos(\phi)$ simultaneously. Noh et al. also analyzed higher-order moments $\langle S^r \rangle$ and $\langle C^r \rangle$ and demonstrated that the $\hat{S}$- and $\hat{C}$-operators can be written in terms of the corresponding annihilation and creation operators in the form

$$
\begin{aligned}
\hat{S} &= iK_2 \left[\exp(i\varphi)\, \hat{a}_2^\dagger \hat{a}_1 - \exp(-i\varphi)\, \hat{a}_1^\dagger \hat{a}_2\right] , \\
\hat{C} &= K_1 \left[\exp(i\varphi)\, \hat{a}_2^\dagger \hat{a}_1 + \exp(-i\varphi)\, \hat{a}_1^\dagger \hat{a}_2\right] ,
\end{aligned}
\tag{5.77}
$$

where $K_1$ and $K_2$ are constants and $\varphi$ is a phase. It holds that $[\hat{C}, \hat{S}] = 2iK_1K_2(\hat{n}_2 - \hat{n}_1)$ and $[\hat{S}, \hat{n}_1 + \hat{n}_2] = [\hat{C}, \hat{n}_1 + \hat{n}_2] = \hat{0}$. If one of the beams were strong to be described classically, then (5.77) would be reduced to a definition of quadratures, i.e. (5.77) are their operator generalizations. The authors applied this procedure to weak fields, phase state, coherent state, ensemble of coherent states, Fock state and to single-photon state. Noh, Fougères and Mandel (1993) succeeded to measure phase distributions. Fougères, Noh, Grayson and Mandel (1994) and Fougères, Torgerson and Mandel (1994) applied the operational method of determining the quantum phase to partially coherent fields with the degree of coherence under control.

The theoretical as well as experimental analysis of Noh et al. demonstrated that there is agreement with the conclusion that the phase difference is well defined only when there is large uncertainty in photon numbers. Conversely, the moments of the sine and cosine operators correspond to completely uncertain phase difference when the photon numbers are certain. Further the authors found that when fluctuations of phases and amplitudes are correlated there are difficulties to measure the phase difference of weak quantum as well as classical fields. It seems that a phase operator has to be introduced with respect to measurement schemes which can lead to various phase operators and a correspondence with classical phase concept should be followed because in classical optics phase arises from a natural definition. It is therefore difficult or impossible to find a universal quantum phase operator representing the phase difference between two quantum fields.

An alternative definition of quantum phase was presented by Beck, Smithey, Cooper and Raymer (1993) based on the determination of quasidistributions by means of the optical homodyne tomography (Sec. 5.6). The phase problem can also be related to quantum Stokes parameters [Sánchez-Soto and Luis (1994)].

In the number squeezed states fulfilling equality in (5.51) [Kitagawa and Yamamoto (1986)] the squeezing occurs if $\langle(\Delta\hat{n})^2\rangle < \langle\hat{n}\rangle$ and the uncertainty contour takes on the form of a crescent (crescent states). For perfect squeezing the mean photon number $\langle\hat{n}\rangle$ remains finite and the state approaches a photon-number state (Fock state) with the uncertainty contour in the form of a ring.

Finally it should be noted that from the physical point of view the Susskind-Glogower or operational definitions seem to be most useful because the phase of the field is available through the functions sine or cosine in real field measurements whereas direct phase definitions of the phase, such as the Pegg-Barnett definition, are rather mathematically sophisticated. Similar arguments are appropriate for phase distributions.

A contemporary state of this research was reviewed in a special issue of Physica Scripta edited by W. Schleich and S. M. Barnett (1993).

## 5.6 Indirect measurements of quasidistributions

It is impossible to measure directly the quasidistributions such as the Glauber-Sudarshan quasidistribution $\Phi_{\mathcal{N}}(\alpha)$, the quasidistribution $\Phi_{\mathcal{A}}(\alpha)$ related to antinormal ordering of field operators or the Wigner quasidistribution $\Phi_{\mathcal{S}}(\alpha)$ because $\mathrm{Re}(\alpha)$ and $\mathrm{Im}(\alpha)$ are the eigenvalues of the operators $(\hat{a}+\hat{a}^\dagger)/2$ and $(\hat{a}-\hat{a}^\dagger)/2i$, respectively which do not commute and consequently $\mathrm{Re}(\alpha)$ and $\mathrm{Im}(\alpha)$ cannot be measured simultaneously with arbitrary accuracy. Nevertheless Vogel and Risken (1989) suggested to derive these quasidistributions by using measurements of the appropriate probability distributions of certain quadrature components. The method was realized for the first time by Smithey, Beck, Raymer and Faridani (1993).

Define the characteristic function (cf. (4.56))

$$C(\beta, s) = \mathrm{Tr}\left\{\hat{\varrho}\exp\left(\frac{s}{2}|\beta|^2 + \beta\hat{a}^\dagger - \beta^*\hat{a}\right)\right\}, \tag{5.78}$$

from which the quasidistribution is obtained by the Fourier transformation

$$\Phi(\alpha, s) = \frac{1}{\pi^2}\int C(\beta, s)\exp(\alpha\beta^* - \alpha^*\beta)\, d^2\beta\,. \tag{5.79}$$

We easily obtain $\Phi_{\mathcal{N}}(\alpha) = \Phi(\alpha, 1)$, $\Phi_{\mathcal{A}} = \Phi(\alpha, -1)$ and $\Phi_{\mathcal{S}}(\alpha) = \Phi(\alpha, 0)$.

Adopting the homodyne detection providing

$$\hat{x}(\vartheta) = \hat{x}^\dagger(\vartheta) = \frac{1}{2}\left[\hat{a}^\dagger\exp(i\vartheta) + \hat{a}\exp(-i\vartheta)\right] = \hat{a}_r\cos(\vartheta) + \hat{a}_i\sin(\vartheta)\,, \tag{5.80}$$

where $\hat{a}_r = (\hat{a}+\hat{a}^\dagger)/2 = \hat{x}(0)$ and $\hat{a}_i = i(\hat{a}^\dagger - \hat{a})/2 = \hat{x}(\pi/2)$ are related to the quantities describing the squeezing of vacuum fluctuations. The operator $\hat{x}$ can be called, with respect to (5.80), the rotated quadrature phase. The complete information for calculating any time expectation value of the rotated quadrature phase $\hat{x}(\vartheta)$ is given by the probability distribution $p(x,\vartheta)$, which can be defined as the Fourier transform of the characteristic function

$$c(\eta, \vartheta) = \mathrm{Tr}\left\{\hat{\varrho}\exp(i\eta\hat{x})\right\} \tag{5.81}$$

in the forn

$$p(x, \vartheta) = \frac{1}{2\pi}\int_0^{2\pi} c(\eta, \vartheta)\exp(-i\eta x)\, d\eta\,. \tag{5.82}$$

Here $\eta$ is a real parameter. Because $\hat{x}(\vartheta)$ is a Hermitian operator, $p(x,\vartheta)$ is well behaved and positive.

It is possible to establish one-to-one correspondence between quasidistribution $\Phi(\alpha, s)$ and probability distribution $p(x, \vartheta)$. From (5.78) and (5.81) we observe that

$$c(\eta, \vartheta) = C\left(\frac{i\eta \exp(i\vartheta)}{2}, s\right) \exp\left(-\frac{s\eta^2}{8}\right), \tag{5.83}$$

or using the real variables $\beta = \beta_r + i\beta_i$, $C(\beta, s) = C(\beta_r, \beta_i, s)$, we obtain

$$c(\eta, \vartheta) = C\left(-\frac{1}{2}\eta \sin(\vartheta), \frac{1}{2}\eta \cos(\vartheta), s\right) \exp\left(-\frac{s\eta^2}{8}\right). \tag{5.84}$$

Knowing $c(\eta, \vartheta)$ for all values $-\infty < \eta < \infty$ and for all $0 \leq \vartheta < \pi$, the characteristic function $C(\beta_r, \beta_i, s)$ is known in the whole complex $\beta$-plane. This one-to-one correspondence between the characteristic functions determines the one-to-one correspondence between the quasidistribution $\Phi(\alpha, s)$ and the probability distribution $p(x, \vartheta)$. Performing the Fourier transformation of (5.83) we obtain

$$\begin{aligned} p(x, \vartheta) &= \frac{1}{2\pi} \int\int\int \Phi\left[u \cos(\vartheta) - v \sin(\vartheta), u \sin(\vartheta) + v \cos(\vartheta), s\right] \\ &\quad \times \exp\left[-\frac{s\eta^2}{8} + i(u - x)\eta\right] du\, dv\, d\eta \end{aligned} \tag{5.85}$$

or

$$\begin{aligned} p(x, \vartheta) &= \sqrt{\frac{2}{\pi s}} \int\int \Phi\left[u \cos(\vartheta) - v \sin(\vartheta), u \sin(\vartheta) + v \cos(\vartheta), s\right] \\ &\quad \times \exp\left[-\frac{(u - x)^2}{s}\right] du\, dv\,. \end{aligned} \tag{5.86}$$

Particularly for $s = 0$

$$p(x, \vartheta) = \int \Phi_S\left[x \cos(\vartheta) - v \sin(\vartheta), x \sin(\vartheta) + v \cos(\vartheta)\right] dv\,. \tag{5.87}$$

The inversion provides in general

$$\begin{aligned} \Phi(\alpha_r, \alpha_i, s) &= \frac{1}{4\pi^2} \int_{-\infty}^{\infty} \int_{-\infty}^{\infty} \int_0^{2\pi} p(x, \vartheta) \exp\Big(\frac{s\eta^2}{8} \\ &\quad + i\eta\left[x - \alpha_r \cos(\vartheta) - \alpha_i \sin(\vartheta)\right]\Big) |\eta|\, dx\, d\eta\, d\vartheta\,. \end{aligned} \tag{5.88}$$

Consequently the quasidistributions $\Phi(\alpha, s)$ can uniquely be determined from the probability distributions and vice versa.

Smithey et al. (1993) determined the Wigner function

$$\Phi_{\mathcal{S}}(x,p) = \frac{1}{\pi}\int_{-\infty}^{\infty} \langle x + x'|\hat{\varrho}|x - x'\rangle \exp(-2ipx')\, dx' \tag{5.89}$$

using the above expressions on the basis of measurements of the quadrature phase distribution $p(x,\vartheta)$ for both vacuum and quadrature-squeezed states produced in an optical parametric down-conversion process combined with the balanced homodyne detection. The inversion of (5.89) then provides the elements of the statistical operator

$$\langle x + x'|\hat{\varrho}|x - x'\rangle = \int_{-\infty}^{\infty} \Phi_{\mathcal{S}}(x,p) \exp(2ipx')\, dp\,. \tag{5.90}$$

The method was called the optical homodyne tomography. Quite interesting curves of $p(x,\vartheta)$ were obtained in relation to the changes of variance with respect to the phase of the local oscillator including regimes of squeezed fluctuations under the vacuum fluctuations. The Wigner function was clearly reconstructed including its topology in the form of ellipses. The method of optical homodyne tomography is suitable to be applied to any states including macroscopic superposition states and squeezed states which have Wigner quasidistribution with negative values. It is able to extract irregular mathematical properties (negative values, singularities) of quasidistributions reflecting quantum properties of light beams [Smithey, Beck, Raymer and Faridani (1993)].

Beck, Smithey, Cooper and Raymer (1993) and Smithey, Beck, Cooper and Raymer (1993) applied this method of optical homodyne tomography to the experimental determination of the number-phase uncertainty relations.

## 5.7 Relations of quantum features of light

From the above treatment it is obvious that negative values of Glauber-Sudarshan quasidistribution $\Phi_{\mathcal{N}}$ and its singular mathematical behaviour reflecting the quantum properties of optical fields are necessary conditions to have optical beams exhibiting quadrature squeezing, photon antibunching, sub-Poissonian behaviour of photon statistics and related phenomena of oscillations in photon-number distribution, as explained in Sec. 5.4. These are various quantum features of optical fields related to various order correlation effects, e.g. the quadrature squeezing is the second-order effect, photon antibunching is the fourth-order effect, sub-Poissonian behaviour includes all-order moments and photon-number oscillations as well. Therefore a general conclusion is that all these quantum features are generally independent. However this does not exclude that in many cases some of them may occur together. In this section we restrict ourselves to considerations in the quantization volume of the field. Thus photon antibunching and sub-Poissonian statistics will be synonyms (Sec. 5.3).

The simplest case is to discuss the two-photon squeezed states. For them $B-|C|-1=\langle\Delta\hat{a}^\dagger\Delta\hat{a}\rangle-|\langle(\Delta\hat{a})^2\rangle|<0$ is always the necessary and sufficient condition for squeezing of vacuum fluctuations but it is only necessary for sub-Poissonian behaviour, as we have seen in Sec. 5.4. However, all antibunched (sub-Poissonian) fields are squeezed. More generally we can derive for photon-number fluctuations ($\varphi$ is squeeze phase and $\vartheta=\arg(\xi)$)

$$\begin{aligned}\langle(\Delta\hat{n})^2\rangle &= |\xi|^2\left[\exp(-2r)\cos^2\left(\vartheta-\frac{\varphi}{2}\right)+\exp(2r)\sin^2\left(\vartheta-\frac{\varphi}{2}\right)\right]\\ &\quad+2\sinh^2(r)\cosh^2(r)\,,\end{aligned} \tag{5.91}$$

which again shows that the squeezed vacuum ($|\xi|=0$) is super-Gaussian since $\langle(\Delta\hat{n})^2\rangle=2\langle\hat{n}\rangle(1+\langle\hat{n}\rangle)$, $\langle\hat{n}\rangle=\sinh^2(r)$. Assume that $|\xi|^2\gg\exp(2r)$ and $\langle\hat{n}\rangle=|\xi|^2+\sinh^2(r)\approx|\xi|^2$ so that the squeezing is only a small effect on a coherent background. Then the last quantum term in (5.91) can be neglected and we have

$$\langle(\Delta\hat{n})^2\rangle=|\xi|^2\left[\exp(-2r)\cos^2\left(\vartheta-\frac{\varphi}{2}\right)+\exp(2r)\sin^2\left(\vartheta-\frac{\varphi}{2}\right)\right]. \tag{5.92}$$

Now we see that if the squeezing is along the coherent amplitude, the state is sub-Poissonian. If the squeezing is perpendicular to the coherent amplitude, the photon statistics are super-Poissonian.

This particular case can be generalized to adopt more general inequalities and eigenvalue problems. Quantum optics prefer to use moments of the operator representing wave field instead of using the complete information contained in the statistical operator $\hat{\varrho}$. As this operator is normalizable, $\mathrm{Tr}\{\hat{\varrho}\}=1$, Hermitian, $\hat{\varrho}=\hat{\varrho}^\dagger$, and positive-definite, the necessarily $\mathrm{Tr}\{\hat{\varrho}^2\}\leq 1$ and the equality holds for pure states, whereas $\mathrm{Tr}\{\hat{\varrho}^2\}<1$ for mixed states. Consequently, some intrinsic restrictions appear, which apply to all possible moments of the field operators. These can be called generalized Schwarz inequalities [Klauder and Sudarshan (1968)]. Their origin is similar as the origin of violation of Bell inequalities (Chap. 7), i.e. they are predictions of quantum theory. Let us restrict ourselves to relations between moments of the operators $\hat{a}$ and $\hat{a}^\dagger$ up to the fourth order. In such a way we can obtain some general consequences for the relation between squeezing of vacuum fluctuations and photon antibunching, related to the second-order and fourth-order effects, respectively. We have mentioned some special relations of these effects and some other may be shown. We have also seen that squeezed vacuum is always super-Gaussian and bunched. Bondurant and Shapiro (1984) found maximum antibunching for ideal squeezed states given by $\langle(\Delta\hat{n})^2\rangle = \langle\hat{n}\rangle^{2/3}$ for large mean photon numbers $\langle\hat{n}\rangle$. Still stronger antibunching given by $\langle(\Delta\hat{n})^2\rangle = \langle\hat{n}\rangle^{1/3}$ was reported by Kitagawa and Yamamoto (1986) for states generated with the help of Kerr nonlinearity. More general treatments need to introduce the so-called noise minimum states [Hradil (1989, 1991, 1992)]. They are defined as states minimizing the root-mean-square of operator $\hat{n}$ under the constraints of lower-order moments (up to the second order). They can be specified as extremum states of generalized Schwarz inequalities and as solutions of variational equations.

The generalized Schwarz inequality for two bounded operators $\hat{A}$ and $\hat{B}$ can be written as

$$|\langle\hat{A},\hat{B}\rangle|^2 \le \|\hat{A}\|^2\,\|\hat{B}\|^2\,, \tag{5.93}$$

where the scalar product is defined as $\langle\hat{A},\hat{B}\rangle = \mathrm{Tr}\{\hat{\varrho}\hat{A}^\dagger\hat{B}\}$ and the norm is defined as $\|\hat{A}\|^2 = \mathrm{Tr}\{\hat{A}^\dagger\hat{A}\}$. The scalar product admits the possibility $\|\hat{A}\| = 0$ for $\hat{A} \neq \hat{0}$. The equality in (5.93) occurs only for the operators fulfilling $\|\hat{A}\| = \|\hat{B}\| = 0$ or $\|\hat{A}\|\,\|\hat{B}\| \neq 0$ and $(\hat{A} - \lambda\hat{B})\hat{\varrho} = \hat{0}$ for some parameter $\lambda$. We point out that inequality (5.93) is always valid in quantum theory, whereas the classical Schwarz inequality can be violated by quantum phenomena.

For lower-order moments we obtain simple consequences in the form of the following inequalities:

$$|\langle\hat{a}\rangle|^2 \le \langle\hat{n}\rangle\,, \tag{5.94a}$$

$$|\langle\hat{a}^2\rangle - \langle\hat{a}\rangle^2|^2 \le \left(\langle\hat{n}\rangle - |\langle\hat{a}\rangle|^2\right)^2 + \langle\hat{n}\rangle - |\langle\hat{a}\rangle|^2\,. \tag{5.94b}$$

The equality in (5.94a) occurs for coherent states and in (5.94b) for squeezed coherent states, because (5.94b) is just the Schrödinger-Robertson inequality (5.10). More complicated situation is met if the relation is to be found between the second-order moments and the fourth-order quantities such as $\gamma_N^{(2)} = \langle\hat{a}^{\dagger 2}\hat{a}^2\rangle/\langle\hat{n}\rangle^2$ and $\langle(\Delta\hat{n})^2\rangle = \langle\hat{n}^2\rangle - \langle\hat{n}\rangle^2 = \langle\hat{n}\rangle + \langle\hat{a}^{\dagger 2}\hat{a}^2\rangle - \langle\hat{n}\rangle^2$. Substituting $\hat{A} = \hat{a}^2$, $\hat{B} = \hat{1}$ to (5.94) we obtain

$$\gamma_N^{(2)} \ge \frac{|\langle\hat{a}^2\rangle|^2}{\langle\hat{n}\rangle^2}\,. \tag{5.95}$$

As the extremum states we find the SU(1,1) coherent states or even (odd) coherent states (Sec. 4.3). This condition forbids some fields to be antibunched. All squeezed vacuum states ($\langle\hat{a}\rangle = 0$) are a simple example as a consequence of the condition (5.11) for the principal squeezing. Mixed states described by the generalized superposition of signal and noise (Sec. 4.8) also exhibit a simple relation between both the nonclassical effects (Secs. 5.1–5.4) since all antibunched (sub-Poissonian) single-mode states are squeezed. Greater details have been presented by Hradil (1989, 1990).

Another approach can be based on an estimation of photon-number variance following from the Heisenberg uncertainty relations for non-commuting quadrature operators

$$\begin{aligned} \hat{Q}(\vartheta) &= \hat{a}\exp(-i\vartheta) + \hat{a}^\dagger \exp(i\vartheta), \\ \hat{P}(\vartheta) &= -i\left[\hat{a}\exp(-i\vartheta) - \hat{a}^\dagger \exp(i\vartheta)\right] \end{aligned} \tag{5.96}$$

and the photon number operator $\hat{n}$,

$$[\hat{n}, \hat{Q}(\vartheta)] = -i\hat{P}(\vartheta). \tag{5.97}$$

From here it follows that [Hradil (1991)]

$$\sqrt{\langle(\Delta\hat{n})^2\rangle} \geq |\langle\hat{a}\rangle| \frac{\lambda}{\lambda_1\lambda_2}, \tag{5.98}$$

where $\lambda_{1,2}$ are the half-axes (5.9) of the noise ellipse and $\lambda^2 = \lambda_1^2 \sin^2(\phi) + \lambda_2^2 \cos^2(\phi)$, $\phi = \arg\left([\langle\hat{a}^2\rangle - \langle\hat{a}\rangle^2]/2\right) - \arg(\langle\hat{a}\rangle)$. The extremum states associated with the inequality (5.98) may be found among the eigenvectors of the non-Hermitian operator $\hat{n} - i|\xi|\hat{Q}(\vartheta)$,

$$\left[\hat{n} - i|\xi|\hat{Q}(\vartheta)\right] |\psi\rangle = \Omega|\psi\rangle, \tag{5.99}$$

where $\xi = i|\xi|\exp(i\vartheta)$ is a complex parameter. The solution can be expressed as the state

$$|\psi\rangle = N \exp\left[|\xi|\hat{P}(\vartheta)\right] |M\rangle, \tag{5.100}$$

where the normalization factor $N$ is obtained as a function of the continuous and discrete parameters $\xi$ and $M$, respectively, $N(\xi, M) = \exp(-|\xi|^2)$ $[M!/L_M^0(-4|\xi|^2)]^{1/2}$, $L_M^0$ being the Laguerre polynomial. Such states are crescent states, as discussed in Sec. 5.5, because cuts of the quasidistribution $\Phi_{\mathcal{A}}(\alpha)$ have a crescent shape. Similar states were discussed as near-photon-number eigenstates generated by the state reduction during down conversion, as displaced Fock states, or stationary states in medium with the Kerr nonlinearity. Such crescent states are obviously non-Gaussian and may also be associated with the number-phase minimum uncertainty states in the Shapiro-Wagner phase concept, as discussed by Hradil (1992).

The generalized Schwarz inequalities enable us to find the minimum value of the photon-number fluctuations $\langle(\Delta\hat{n})^2\rangle$ and the fourth-order degree of coherence $\gamma_N^{(2)}$ as functions of values of $\langle\hat{a}\rangle$, $\langle\hat{a}^\dagger\rangle$, $\langle\hat{a}^2\rangle$, $\langle\hat{a}^{\dagger 2}\rangle$ and $\langle\hat{a}^\dagger\hat{a}\rangle$ which are corresponding to the extremum states. More general solutions require a variational approach has to be used. All states minimizing the functional $\mathrm{Tr}\{\hat{\varrho}\hat{n}^2\}$ under the restrictions $\mathrm{Tr}\{\hat{\varrho}\} = 1$, $\mathrm{Tr}\{\hat{\varrho}\hat{a}\} = \langle\hat{a}\rangle$ and $\mathrm{Tr}\{\hat{\varrho}\hat{n}\} = \langle\hat{n}\rangle$

are noise minimum states whose statistical operator $\hat{\varrho}_{\min}$ obeys the Lagrange-Euler equation

$$\left(\hat{n}^2 + \gamma\hat{n} - \xi^{*2}\hat{a}^2 - \xi^2\hat{a}^{\dagger 2} + \alpha\hat{a}^\dagger + \alpha^*\hat{a}\right)\hat{\varrho}_{\min} = \mu\hat{\varrho}_{\min}\,, \tag{5.101}$$

$\mu$, $\gamma$ being the real Lagrange multipliers and $\xi$ and $\alpha$ are complex multipliers. Equation (5.101) represents the steady state solution for the nonlinear hamiltonian, containing terms appropriate for the Kerr nonlinearity together with the terms for the generation of squeezed states. If we knew the solution for the ground state of the physical system specified by $\hat{\varrho}_{\min}$ with determined $\mu_{\min}$ as functions of parameters $\xi$, $\alpha$, $\gamma$ and if we used them together with the parameters $\langle\hat{a}\rangle$, $\langle\hat{a}^2\rangle$ and $\langle\hat{n}\rangle$, we could obtain the minimum for the photon-number variance. Nevertheless, such a problem represents rather complicated procedure because physically such a hamiltonian involves a variety of quantum optical effects, such as antibunching, squeezing, bistability, etc. However, some particular solutions can be obtained [Hradil (1989, 1990)].

It should be noted that from the point of view of detection it is possible to detect squeezed light as sub-Poissonian light if the registered squeezed light is superimposed, before the detection, on a coherent component $\gamma$ [Mandel (1982)]. For the variance of the integrated intensity we have for the short detection times $T$

$$\begin{aligned}
\eta^2\langle(\Delta W)^2\rangle_{\mathcal{N}} &= \langle(\Delta\hat{n})^2\rangle - \langle\hat{n}\rangle = (\eta T)^2 \\
&\times\left[\left\langle\left(\hat{A}^{(-)}+\gamma^*\right)^2\left(\hat{A}^{(+)}+\gamma\right)^2\right\rangle - \left\langle\left(\hat{A}^{(-)}+\gamma^*\right)\left(\hat{A}^{(+)}+\gamma\right)\right\rangle^2\right] \\
&\approx (\eta T)^2|\gamma|^2\Big[\langle:(\Delta\hat{Q})^2:\rangle\cos^2(\theta) + \langle:(\Delta\hat{P})^2:\rangle\sin^2(\theta) \\
&\quad + \frac{1}{2}\langle\{\Delta\hat{Q},\Delta\hat{P}\}\rangle\sin(2\theta)\Big]\,,
\end{aligned} \tag{5.102}$$

provided that the reference coherent field is sufficiently intense; here $\theta$ is again the phase of $\gamma$ and the normally ordered quantities $\langle:(\Delta\hat{Q})^2:\rangle = \langle(\Delta\hat{Q})^2\rangle - 1$ and $\langle:(\Delta\hat{P})^2:\rangle = \langle(\Delta\hat{P})^2\rangle - 1$. For particular values of the phase $\theta$ we arrive at

$$\langle(\Delta\hat{n})^2\rangle - \langle\hat{n}\rangle = \begin{cases} (\eta T)^2|\gamma|^2\langle:(\Delta\hat{Q})^2:\rangle, & \theta = 0\,, \\ (\eta T)^2|\gamma|^2\langle:(\Delta\hat{P})^2:\rangle, & \theta = \pi/2\,. \end{cases} \tag{5.103}$$

Therefore if in-phase component or out-of-phase component are squeezed ( $\langle:(\Delta\hat{Q})^2:\rangle = \langle(\Delta\hat{Q})^2\rangle - 1 < 0$ or $\langle:(\Delta\hat{P})^2:\rangle = \langle(\Delta\hat{P})^2\rangle - 1 < 0$ ), then the corresponding photocount statistics are sub-Poissonian. Of course, this is always the exhibition of quantum properties of optical fields. If classical field fluctuations are squeezed (not below the quantum limit), their quadratures may be measured with arbitrary accuracies and such a field always behaves like sub-Poissonian. In this way classical and quantum squeezed states can be distinguished.

Some other relations of nonclassical properties can be presented for nonlinear optical processes, such as second harmonic generation, three-field interaction and Raman scattering in a short time approximation [Peřina, Peřinová and Kodousek (1984)], described by the effective hamiltonians

$$\hat{H} = \sum_{j=1}^{2} \hbar\omega_j \hat{a}_j^\dagger \hat{a}_j - \hbar g \left( \hat{a}_1^2 \hat{a}_2^\dagger + \hat{a}_1^{\dagger 2} \hat{a}_2 \right) , \tag{5.104a}$$

$$\hat{H} = \sum_{j=1}^{3} \hbar\omega_j \hat{a}_j^\dagger \hat{a}_j - \hbar g \left( \hat{a}_1 \hat{a}_2 \hat{a}_3^\dagger + \hat{a}_1^\dagger \hat{a}_2^\dagger \hat{a}_3 \right) , \tag{5.104b}$$

$$\hat{H} = \sum_{j=L,S,A,V} \hbar\omega_j \hat{a}_j^\dagger \hat{a}_j - \hbar \left( g \hat{a}_L \hat{a}_S^\dagger \hat{a}_V^\dagger + \kappa^* \hat{a}_L \hat{a}_V \hat{a}_A^\dagger + \text{h.c.} \right) , \tag{5.104c}$$

respectively, where $L$, $S$, $A$, $V$ denote successively the laser, Stokes, anti-Stokes and vibrational modes. The corresponding frequency resonance conditions read

$$\omega_2 = 2\omega_1 , \tag{5.105a}$$

$$\omega_1 + \omega_2 = \omega_3 , \tag{5.105b}$$

$$\omega_S = \omega_L - \omega_V , \qquad \omega_A = \omega_L + \omega_V \tag{5.105c}$$

and the phase matching is assumed in (5.104a,b) for simplicity.

In the fundamental mode of the second harmonic generation ($\xi_2 = 0$) principal squeezing of the amount $\lambda_1 = 1 - 2g^2t^2|\xi_1|^2$ is exhibited, assuming the initial field in the coherent state $|\xi_1\rangle|\xi_2\rangle$. This just corresponds to the sub-Poissonian statistics in the fundamental mode of the amount $\langle(\Delta W_1)^2\rangle_\mathcal{N}/\langle W_1\rangle_\mathcal{N} = -2g^2t^2|\xi_1|^2$. Also the intermodal principal squeezing is possible since $\lambda_{12} = 1 - g^2t^2|\xi_1|^2$ in relation to the sub-Poissonian compound statistics $\langle(\Delta W_{12})^2\rangle_\mathcal{N} = \langle(\Delta W_1)^2\rangle_\mathcal{N} + \langle(\Delta W_2)^2\rangle_\mathcal{N} + 2\langle\Delta W_1 \Delta W_2\rangle_\mathcal{N} = -2g^2t^2|\xi_1|^4$.

Quite analogous results are available for the above non-degenerate process (5.104b). The compound sub-frequency principal squeezing $\lambda_{12} = 1 - g^2t^2|\xi_1||\xi_2|$ may occur, assuming the initial coherent state $|\xi_1\rangle|\xi_2\rangle|\xi_3\rangle$. This nonclassical effect corresponds to the sub-Poissonian statistics $\langle(\Delta W_{12})^2\rangle_\mathcal{N}/\langle W_1\rangle_\mathcal{N}\langle W_2\rangle_\mathcal{N} = -2g^2t^2|\xi_1||\xi_2|$. In this approximation the second-harmonic mode 2 and the sum-frequency mode 3 have tendency to be coherent. Nevertheless, in higher order they also can exhibit nonclassical behaviour. For example in the degenerate case and $\xi_2 = 0$ we obtain for principal squeezing of vacuum fluctuations $\lambda_2 = 1 - (4/3)g^4t^4|\xi_1|^4$ in relation to sub-Poissonian effect $\langle(\Delta W_2)^2\rangle_\mathcal{N} = -(4/3)g^6t^6|\xi_1|^8$. Also in the stimulated process ($\xi_1$, $\xi_2 \neq 0$) principal squeezing is possible as being $\lambda_2 = 1 - (16/3)g^3t^3|\xi_1|^2|\xi_2|$ and in the case of sub-harmonic generation ($\xi_1 = 0$) it holds that $\lambda_2 = 1 - (8/3)g^4t^4|\xi_2|^2$.

In the non-degenerate case and for the process of the sum-frequency generation ($\xi_3 = 0$) it holds that $\lambda_3 = 1 - (1/3)g^4t^4|\xi_1|^2|\xi_2|^2$ in correspondence to the sub-Poissonian effect $\langle(\Delta W_3)^2\rangle_{\mathcal{N}} = -(1/3)g^6t^6|\xi_1|^4|\xi_2|^4$. Further we can derive that $\lambda_3 = 1 - (4/3)g^3t^3|\xi_1||\xi_2||\xi_3|$ for the spontaneous sub-frequency generation ($\xi_1 = \xi_2 = 0$). Thus in all these cases squeezing effects are observable in correspondence to antibunching of photons and their sub-Poisson behaviour. In general, signal-signal quantum correlations are always stronger than signal-pump correlations [Mišta and Peřina (1977), see also Peřina (1991), Sec. 10.2].

Also in the case of Raman or hyper-Raman scattering including the coupling of Stokes and anti-Stokes modes, interesting nonclassical effects are available. In general convenient conditions for observing nonclassical behaviour are given if $|g||\xi_S| < |\kappa||\xi_A|$. For example laser photon-phonon antibunching $\langle(\Delta W_{LV})^2\rangle_{\mathcal{N}} = -(|g|^2 + |\kappa|^2)t^2|\xi_L|^2|\xi_V|^2$ in the course of the spontaneous scattering ($\xi_S = \xi_A = 0$) is derived assuming the initial coherent state $|\xi_L\rangle|\xi_S\rangle|\xi_A\rangle|\xi_V\rangle$ and also $\langle(\Delta W_{LA})^2\rangle_{\mathcal{N}} = -2|g||\kappa|t^2|\xi_L|^2|\xi_S||\xi_A|\cos[2\vartheta_L - \vartheta_S - \vartheta_A + \arg(g) - \arg(\kappa)]$. Stokes photon-phonon squeezing can be obtained as $\lambda_{SV} = 1 - |g|t|\xi_L|$ including the spontaneous scattering.

Particular effects are available in hyper-Raman scattering described by the effective hamiltonian in (5.104c) with $\hat{a}_L$ substituted by $\hat{a}_L^2$ in the degenerate case and by $\hat{a}_{L_1}\hat{a}_{L_2}$ in the nondegerate case ($\omega_S = 2\omega_L - \omega_V$, $\omega_A = 2\omega_L + \omega_V$ or $\omega_S = \omega_{L_1} + \omega_{L_2} - \omega_V$, $\omega_A = \omega_{L_1} + \omega_{L_2} + \omega_V$). This means that the self-interaction in the laser modes takes place similarly as in the fundamental mode in the course of the second-harmonic generation. In these cases we can derive that $\lambda_L = 1 - 2[(|g|^2 + |\kappa|^2)|\xi_V|^2 + |g|^2]t^2|\xi_L|^2$ for the degenerate case or $\lambda_{L_1L_2} = 1 - [(|g|^2 + |\kappa|^2)|\xi_V|^2 + |g|^2]t^2|\xi_{L_1}||\xi_{L_2}|$ for the non-degenerate case in complete analogy to photon antibunching specified by $\langle(\Delta W_L)^2\rangle_{\mathcal{N}}/\langle W_L\rangle_{\mathcal{N}} = -2[(|g|^2 + |\kappa|^2)|\xi_V|^2 + |g|^2]t^2|\xi_L|^2$ and $\langle\Delta W_{L_1}\Delta W_{L_2}\rangle_{\mathcal{N}}/\langle W_{L_1}\rangle_{\mathcal{N}}\langle W_{L_2}\rangle_{\mathcal{N}} = -[(|g|^2 + |\kappa|^2)|\xi_V|^2 + |g|^2]t^2|\xi_{L_1}||\xi_{L_2}|$.

Thus in many cases squeezing effects correspond to photon antibunching or sub-Poissonian photon statistics in nonlinear optical phenomena. In some cases, hovewer, antibunching can occur and squeezing connot and vice versa, which is a consequence of the different orders of these phenomena. For instance in the parametric down-conversion process there is strong quantum correlation of sub-frequency photons, $\langle\Delta W_1\Delta W_2\rangle_{\mathcal{N}} = g^2t^2|\xi_3|^2 > 0$, even if squeezing of vacuum fluctuations can occur, $\lambda_{12} \approx 1 - 2gt|\xi_3|\sin(\vartheta_3)$ ($\xi_1 = \xi_2 = 0$). We can note that antibunching and sub-Poissonian behaviour are independent of fundamental optical oscillations, whereas squeezing effects always include these oscillations.

More details can be found in the book by Peřina (1991) (Chap. 10) and in reviews by Raymer and Walmsley (1990) and by Miranowicz and Kielich (1993). Nonclassical properties of light generated in nonlinear optical processes were considered in a series of books, book chapters, reviews and proceedings [Kielich (1981), Paul (1982), Mandel and Wolf (1978, 1984, 1990), Schubert and Wilhelmi (1986), Eberly, Mandel and Wolf (1989), Peřina (1991a) and Evans and Kielich (1993)].

## 5.8 Interaction of a single atom with light — cavity quantum electrodynamics and cavity quantum optics

In this section we provide elements of interaction of a single atom with light giving a basis for cavity quantum electrodynamics and cavity quantum optics. The corresponding experimental results will be treated in Sec. 8.5. Greater details are available in recent reviews [Zaheer and Zubairy (1991), Milonni and Singh (1991), Meystre and Sargent III (1991), Chap. 13, Meystre (1992)].

As a basic description of a radiation mode with a single atom the Jaynes-Cummings model [Jaynes and Cummings (1963)] can serve based on the hamiltonian

$$\hat{H} = \hat{H}_0 + \hat{H}_{\mathrm{int}}, \tag{5.106}$$

where the free hamiltonian equals

$$\hat{H}_0 = \hbar\omega\hat{a}^\dagger\hat{a} + \frac{1}{2}\hbar\Omega\hat{\sigma}_z, \tag{5.107a}$$

representing the free energy of the radiation mode (neglecting zero-point energy $\hbar\omega/2$ in the field playing no role in this problem) and of a two level atom with the transition frequency $\Omega$, $\hat{\sigma}_z$ being the Pauli matrix, and the interaction hamiltonian is expressed as

$$\hat{H}_{\mathrm{int}} = -e\hat{\mathbf{r}}.\hat{\mathbf{E}} = \hbar g\left(\hat{a}^\dagger\hat{\sigma}_- + \hat{a}\hat{\sigma}_+\right), \tag{5.107b}$$

where $\hat{\sigma}_-$ and $\hat{\sigma}_+$ are Pauli lowering and raising atomic operators, fulfilling the commutation rule

$$[\hat{\sigma}_+, \hat{\sigma}_-] = \hat{\sigma}_z, \tag{5.108}$$

$\hat{\mathbf{E}}$ is the electric strength operator, $e\hat{\mathbf{r}}$ the atomic dipole momentum operator, $g = \mathcal{P}\mathcal{E}\sin(\mathbf{k}.\mathbf{r})/2\hbar$ is the coupling constant between the electromagnetic field and the atom, $\mathcal{P}$ is the electric-dipole matrix element, $\mathcal{E}$ is the electric field per photon if the field is in a volume and the atom is situated at $\mathbf{r}$ [for greater details, see Meystre and Sargent III (1991), Chap. 13]. We have also used the rotating wave approximation neglecting the rotating terms such as $\hat{a}^\dagger\hat{\sigma}_+$ and $\hat{a}\hat{\sigma}_-$, which are virtual and do not conserve the energy. The Jaynes-Cummings model represents the important description in quantum optics because it provides a simple model for emission and it can be solved in a closed form providing various dynamical effects really observed, such as collapses and revivals of atomic population inversion when the two-level atom interacts with a field, for instance initially in a coherent state [Eberly, Narozhny and Sanchez-Mondragon (1980)]. This model can be realized in micromasers where many effects of atom–field dynamics predicted by such a model can be observed [see Sec. 8.5].

The energy eigenvalues of the hamiltonian (5.106) can be written in the form

$$E_{1n} = \hbar\left(n+\frac{1}{2}\right)\omega + \frac{\hbar}{2}\Omega_n = \hbar\left[-\frac{\Omega}{2} + (n+1)\omega + \frac{1}{2}(\Omega_n+\delta)\right], \quad (5.109a)$$

$$E_{2n} = \hbar\left(n+\frac{1}{2}\right)\omega - \frac{\hbar}{2}\Omega_n = \hbar\left[+\frac{\Omega}{2} + n\omega - \frac{1}{2}(\Omega_n+\delta)\right], \quad (5.109b)$$

where $\delta = \Omega - \omega$ is the atom–field frequency detuning and the n-photon Rabi flopping frequency

$$\Omega_n = \sqrt{\delta^2 + 4g^2(n+1)} \quad (5.110)$$

has been introduced. The corresponding eigenvectors can be written in the form

$$\begin{aligned} |1,n\rangle &= \sin(\vartheta_n)\,|a,n\rangle + \cos(\vartheta_n)\,|b,n+1\rangle, \\ |2,n\rangle &= \cos(\vartheta_n)\,|a,n\rangle - \sin(\vartheta_n)\,|b,n+1\rangle, \end{aligned} \quad (5.111)$$

where the states $|a\rangle$ and $|b\rangle$ are the upper and lower atomic states, respectively and $|n\rangle$ are the Fock number states of the field. The angle $\vartheta$ is defined as

$$\begin{aligned} \cos(\vartheta_n) &= \frac{\Omega_n - \delta}{[(\Omega_n-\delta)^2 + 4g^2(n+1)]^{1/2}}, \\ \sin(\vartheta_n) &= \frac{2g\sqrt{n+1}}{[(\Omega_n-\delta)^2 + 4g^2(n+1)]^{1/2}}. \end{aligned} \quad (5.112)$$

Particularly it holds that

$$\cos(2\vartheta_n) = -\frac{\delta}{\Omega_n}, \qquad \sin(2\vartheta_n) = \frac{2g\sqrt{n+1}}{\Omega_n}, \quad (5.113)$$

and for $n = 0$ and on frequency resonance $\omega = \Omega$, the levels $|1,0\rangle$ and $|2,0\rangle$ are separated by the frequency $\Omega_0 = 2g$, giving the vacuum Rabi splitting.

The above results are sufficient to deduce the dynamical properties of the Jaynes-Cummings model. The most interesting feature of this model is the occurrence of collapses and revivals of the population inversion and related mean photon number, photon statistics and quadrature fluctuations. One can deduce for resonance $\omega = \Omega$ that, assuming the two-level atom initially in its upper state $|a\rangle$ and the field in the Fock state $|n\rangle$, the probability $|c_{an}(t)|^2$ for the atom to be in the upper state at time $t$ is given by

$$|c_{an}(t)|^2 = \cos^2\left(g\sqrt{n+1}\,t\right) , \tag{5.114a}$$

and if the atom is initially in its lower state $|b\rangle$, we have

$$|c_{bn}(t)|^2 = \sin^2\left(g\sqrt{n}\,t\right) . \tag{5.114b}$$

This shows that the population oscillates periodically at the Rabi frequency, similarly as in the case of classical fields. The difference is that for the quantum field the Rabi frequency is different if the atom starts in the upper level or in the lower level, because the creation and annihilation operators $\hat{a}^\dagger$ and $\hat{a}$ do not commute.

If the initial field is an arbitrary initial state described by the diagonal matrix elements $\varrho(n,n)$ of the statistical operator $\hat{\varrho}$, we obtain for the corresponding probabilities

$$|c_a(t)|^2 = \sum_{n=0}^{\infty} \varrho(n,n)\cos^2\left(g\sqrt{n+1}\,t\right) , \tag{5.115a}$$

$$|c_b(t)|^2 = \sum_{n=0}^{\infty} \varrho(n,n)\sin^2\left(g\sqrt{n}\,t\right) . \tag{5.115b}$$

If the initial field is in a coherent state $|\alpha\rangle$, then $\varrho(n,n) = |\alpha|^{2n}\exp(-|\alpha|^2)\,/\,n!$. From (5.115a) we then obtain for strong coherent field ($|\alpha|^2 \gg 1$) and $gt \ll |\alpha|$ that [Meystre, Quattropani and Baltes (1974)]

$$p_a = |c_a(t)|^2 \approx \frac{1}{2} + \frac{1}{2}\cos\left(2gt|\alpha|\right)\exp\left(-g^2t^2\right) , \tag{5.116}$$

leading to collapses and revivals of the probability $p_a$, as shown in Fig. 5.5. The revivals are a consequence of the quantal nature of the field as the atomic evolution is determined by the individual field quanta. The revivals are successively broader, they overlap and give a quasi-random time evolution. These quantum effects can be really observed [Rempe, Walther and Klein (1987)]. If a continuous spectrum of vacuum fluctuations occurs, the resulting lower-level probability amplitudes interfere with each other, giving the macroscopic exponential decay. For initial chaotic field the collapses and revivals are much less pronounced. The collapses are created by destructive interference of quantum Rabi floppings at different frequencies and may occur also under the influence of a classical field with intensity fluctuations, whereas the

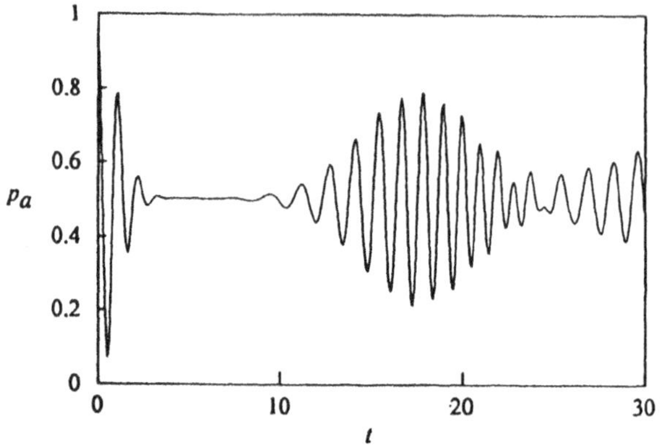


**Figure 5.5:** Collapses and revivals in the interaction of a quantum light field with an atom.

revivals are of purely quantum origin [Knight and Radmore (1982), Barnett, Filipowicz, Javanainen, Knight and Meystre (1986), Eberly, Narozhny and Sanchez-Mondragon (1980)].

It may be noted that the revivals and collapses can also be interpreted in terms of interferences of quasidistributions $\Phi_{\mathcal{A}}(\alpha)$ arising by splitting a single-peaked distribution for the initial coherent state and the atom in its upper state [Eiselt and Risken (1990)].

For $n = 0$ in (5.114a,b) we describe the spontaneous effect since for an atom initially in the upper state we have $|c_{a0}(t)|^2 = \cos^2(gt)$, whereas for the atom initially in its lower state we have $|c_{b0}(t)|^2 = 0$. The above quantum effect is related to the quantum result for the vacuum expectation value of the square of the electric strength operator $\langle \hat{E}^2 \rangle \sim \langle 0|(\hat{a}+\hat{a}^\dagger)^2|0\rangle = 1$. This means that vacuum fluctuations stimulate the excited atom to emit a spontaneous photon. Of course, no spontaneous emission is possible if the atom is in the ground state.

It should be noted that if the virtual terms $\hat{a}^\dagger\hat{\sigma}_+$ and $\hat{a}\hat{\sigma}_-$ are not neglected in the hamiltonian, chaotic behaviour can arise in the atom-field system [Milonni, Shih and Ackerhalt (1987)].

In free space the atom interacts with a continuum of modes of the electromagnetic field so that the probability amplitudes corresponding to the interaction of the atom with single modes can interfere to provide the exponential decay for the upper population state. In this case the hamiltonians (5.107a,b) are replaced by the multimode hamiltonians

$$\hat{H}_0 = \hbar \sum_\lambda \omega_\lambda \hat{a}_\lambda^\dagger \hat{a}_\lambda + \frac{1}{2}\hbar\Omega\hat{\sigma}_z\,, \tag{5.117a}$$

$$\hat{H}_{\rm int} = \hbar \sum_\lambda \left( g_\lambda \hat{a}_\lambda^\dagger \hat{\sigma}_- + {\rm h.c.}\right)\,, \tag{5.117b}$$

where $\omega_\lambda$ is the frequency of the $\lambda$-th radiation mode, and $\hat{a}_\lambda$ and $\hat{a}_\lambda^\dagger$ are the corresponding annihilation and creation operators, respectively, $g_\lambda$ are the dipole coupling constants between the atom and the mode $\lambda$ of the field. The standard Wigner-Weisskopf analysis [Louisell (1973), Meystre and Sargent III (1991), Peřina (1991)] leads to the irreversible exponential decay of the upper population state without any revivals and collapses or a periodic exchange of energy between the atom and the field. This means that the probability amplitudes for single events interfere destructively when summing over the continuum of modes.

If the atom is situated inside a one-dimensional high $Q$-cavity ($Q = \Omega/\Delta\omega_c$, $\Omega$ being the atomic resonance frequency to which the cavity is tuned, $\Delta\omega_c$ is the cavity line width of a Lorentzian spectrum) and the cavity is tuned near the atomic resonance frequency $\Omega$, then the decay rate of the cavity can be modified such as

$$\gamma_{\text{cav}} \approx \gamma_{\text{free}} \frac{\lambda^3}{V} Q\,, \tag{5.118}$$

$V$ being the volume of the field. For the detuned cavity from the atomic transition we obtain approximately

$$\gamma_{\text{cav}} \approx \gamma_{\text{free}} \frac{\lambda^3}{VQ}\,. \tag{5.119}$$

Consequently for sufficiently hight cavity $Q$ factors the spontaneous emission rate can be enhanced or inhibited compared to its free-space value. More detailed discussion of these questions can be found in a review by Meystre (1992) and concerning the corresponding experiments in a review by Haroche and Kleppner (1989).

Experiments with a single atom in interaction with radiation can be made in the microwave region, where high $Q$ and single-mode cavities are available if highly excited Rydberg atoms are used (Sec. 8.5), for which the dipole matrix element $g$ between neighbouring levels scales as $n^2$, $n$ being the principal quantum number. Then for sufficiently high $n$ stimulated effects can overcome spontaneous emission for small photon numbers. In the visible region the dipole matrix elements are small and linear size of cavities available in laboratories is much larger than a transition wavelength $\lambda$. Therefore cooperative effects are to be used to obtain large effective coupling constants and to realize cavity quantum optics experiments in this region.

A collection of $N$ two-level atoms in a single-mode cavity can be described by the hamiltonian

$$\hat{H}_{\text{coop}} = \hbar\omega\hat{a}^\dagger\hat{a} + \frac{1}{2}\hbar\Omega\sum_{j=1}^{N}\hat{\sigma}_{zj} + \hbar\sum_{j=1}^{N}\left(g\hat{a}^\dagger\hat{\sigma}_{-j} + \text{h.c.}\right)\,, \tag{5.120}$$

assuming that the atoms are in a volume that is small compared with the volume $\lambda^3$. Introducing the collective operators

$$\hat{S}_z = \sum_j \hat{\sigma}_{zj}, \qquad \hat{S}_\pm = \sum_j \hat{\sigma}_{\pm j} \tag{5.121}$$

fulfilling the standard commutation rules for the angular momentum operators, the hamiltonian (5.120) can be written as

$$\hat{H}_{\rm coop} = \hbar\omega\hat{a}^\dagger\hat{a} + \frac{1}{2}\hbar\Omega\hat{S}_z + \hbar\left(g\hat{a}^\dagger\hat{S}_- + {\rm h.c.}\right). \tag{5.122}$$

This hamiltonian permits to introduce the angular momentum states analogous to the Fock (number) states, the so-called Dicke states and the states analogous to the coherent states, the so-called atomic coherent states — Bloch states, appropriate for the quantum description of cooperative effects such as superradiance, etc. [Dicke (1954), Allen and Eberly (1975), Haroche and Raimond (1985), Peřina (1991), Sec. 4.10.2 and references therein].

More complicated situation occurs if strong coupling between the atoms and radiation takes place, which needs to involve decay rates for the field and the atomic system and a mutual atom-field decay rate. For a review we refer the reader to the paper by Meystre (1992).

For the modification of the spontaneous emission rate of the atom a phase-conjugating mirror can be adopted. The atom can be many wavelengths away from the phase-conjugating mirror to achieve a considerable modification [Milonni, Bochove and Cook (1989, 1989a) and references therein]. In this case the decay rate can be increased by $(1 + \mathcal{R})$, where $\mathcal{R}$ is the intensity reflection of the phase-conjugating mirror. The effect of the phase conjugation on the atom arises from the quantum noise radiation produced by the pumped phase conjugating mirror, which leads to the excitation of the ground state of the atom near the mirror. A quantum noise treatment can be presented along the following line [Gaeta and Boyd (1988, 1989), Bajer and Peřina (1991), Bochove (1992), Arnoldus and George (1991, 1992)].

The dynamics of the phase-conjugation process can be described by the operator relation

$$\hat{b} = r\hat{a}^\dagger + \sqrt{1+|r|^2}\,\hat{L}, \tag{5.123}$$

where $\hat{b}$ is the annihilation operator of the phase-conjugated beam (with removed wave-front aberrations), $r$ is reflectivity coefficient ($|r|$ can become larger than unity due to the pumping of the phase-conjugating mirror), $\hat{a}^\dagger$ is the creation operator of an incident signal beam, and a quantum noise operator $\hat{L}$ fulfils the boson commutation rule $[\hat{L}, \hat{L}^\dagger] = \hat{1}$. It also holds that $[\hat{a}, \hat{a}^\dagger] = [\hat{b}, \hat{b}^\dagger] = \hat{1}$. As the first and second terms in (5.123) commute, quantum noise is generated in the process and the generation of nonclassical light from a coherent light due to dynamics of the phase conjugation process cannot be expected. This can be seen calculating the normal quantum characteristic function (Sec. 4.3.3) for the phase-conjugated beam

$$\begin{aligned} C_{\mathcal{N}}^{(b)}(\eta) &= \mathrm{Tr}\left\{\hat{\varrho}\exp\left(\eta\hat{b}^{\dagger}\right)\exp\left(-\eta^{*}\hat{b}\right)\right\} \\ &= \exp\left(-|r|^{2}|\eta|^{2}\right)\int\int d^{2}\alpha\, d^{2}\lambda\, \Phi_{\mathcal{N}}^{(a)}(\alpha)\, \Phi_{\mathcal{N}}^{(L)}(\lambda) \\ &\quad \times \exp\left[-\eta^{*}\left(r\alpha^{*}+\sqrt{1+|r|^{2}}\lambda\right)+\eta\left(r^{*}\alpha+\sqrt{1+|r|^{2}}\lambda^{*}\right)\right], \end{aligned} \tag{5.124}$$

where $\eta$ is the parameter of the characteristic function, $\alpha$ and $\lambda$ are the complex amplitudes corresponding to the operators $\hat{a}$ and $\hat{L}$ in their coherent states, and $\Phi_{\mathcal{N}}^{(a)}(\alpha)$ and $\Phi_{\mathcal{N}}^{(L)}(\lambda)$ are the corresponding Glauber-Sudarshan quasidistributions, respectively (Sec. 4.3). In deriving (5.124) we have used the Baker-Hausdorff identity (4.18) (Sec. 4.3.1). Performing the Fourier transformation, we arrive at the resulting Glauber-Sudarshan quasidistribution for the phase conjugated beam

$$\begin{aligned} \Phi_{\mathcal{N}}^{(b)}(\beta) &= \frac{1}{\pi|r|^{2}}\int\int \Phi_{\mathcal{N}}^{(a)}(\alpha)\, \Phi_{\mathcal{N}}^{(L)}(\lambda)\exp\left[-\frac{\left|\beta-\left(r\alpha^{*}+\sqrt{1+|r|^{2}}\lambda\right)\right|^{2}}{|r|^{2}}\right] \\ &\quad \times d^{2}\alpha\, d^{2}\lambda\,, \end{aligned} \tag{5.125}$$

where $\beta$ is the resulting complex amplitude of the phase conjugated mode b. This demonstrates that for initially coherent modes with the complex amplitudes $\alpha_0$, $\lambda_0$ [i.e. $\Phi_{\mathcal{N}}^{(a)}(\alpha)=\delta(\alpha-\alpha_0)$, $\Phi_{\mathcal{N}}^{(L)}(\lambda)=\delta(\lambda-\lambda_0)$] the phase conjugated field is the superpositon of the signal $r\alpha_0^{*}+\sqrt{1+r|^{2}}\lambda_0$ and the noise $|r|^2$ determined by the reflection coefficient. In the case that the additive quantum noise represents vacuum fluctuations, we put $\lambda_0=0$. Thus, no quantum effects can arise in this case. If quantum statistical state of the incident radiation is arbitrary and $L$ is noisy or at least coherent, we obtain the similar conclusion from (5.125) considering only the vacuum state of $L$,

$$\Phi_{\mathcal{N}}^{(b)}(\beta)=\frac{1}{\pi|r|^{2}}\int \Phi_{\mathcal{N}}^{(a)}(\alpha)\exp\left(-\frac{|\beta-r\alpha^{*}|^{2}}{|r|^{2}}\right)d^{2}\alpha=\frac{1}{|r|^{2}}\Phi_{\mathcal{A}}^{(b)}\left(\frac{\beta^{*}}{r^{*}}\right), \tag{5.126}$$

where $\Phi_{\mathcal{A}}$ is the quasidistribution related to the antinormal ordering, which is always positively semidefinite (Sec. 4.3). Only when noise mode L is nonclassical, can the conjugated beam occur in a nonclassical state.

The principal squeeze variance $\lambda^{(b)}$, as defined in Sec. 5.1, is obtained in the form [Bajer and Peřina (1991)]

$$\lambda^{(b)}=1+2|r|^{2}(1+B_a)+2(1+|r|^{2})B_L-2\left|r^{2}C_a^{*}+(1+|r|^{2})C_L\right|, \tag{5.127}$$

where $B_a$, $B_L$, $C_a$ and $C_L$ are the corresponding noise functions as defined in Sec. 4.8 [equation (4.180)] given for normal ordering and for modes a and L. Analyzing this expression we can conclude that in the case of a coherent incident light and squeezed noise L, it holds that

$$\lambda^{(b)} = |r|^2 + (1 + |r|^2)(1 + 2B_L - 2|C_L|) \geq |r|^2 . \tag{5.128}$$

Hence only in the case that the phase-conjugated mode is attenuated ($|r|^2 < 1$) squeezing of vacuum fluctuations ($\lambda^{(b)} < 1$) can occur. If the value of the squeeze parameter of L is denoted as $s$, then squeezing of vacuum fluctuations in the conjugated beam occurs under the condition that

$$\exp(-2s) < \frac{1 - |r|^2}{1 + |r|^2} . \tag{5.129}$$

If modes a and L have arbitrary statistical behaviour and assuming an optimum phase condition, squeezing of vacuum fluctuations takes place provided that

$$|r|^2 < \frac{1 - \langle (\Delta \hat{Q}_L)^2 \rangle}{\langle (\Delta \hat{Q}_a)^2 \rangle + \langle (\Delta \hat{Q}_L)^2 \rangle} , \tag{5.130}$$

where the quadrature fluctuations are defined in Sec. 5.1.

For integrated intensity fluctuations of the phase-conjugated beam we obtain

$$\langle (\Delta W_b)^2 \rangle_{\mathcal{N}} = B_b^2 + |C_b|^2 + 2B_b|\beta|^2 + \mathrm{Re}(C_b^* \beta^2) , \tag{5.131}$$

where

$$\begin{aligned} \beta &= r\alpha^* + \sqrt{1 + |r|^2}\lambda , \\ B_b &= |r|^2(1 + B_a) + (1 + |r|^2)B_L , \\ C_b &= r^2 C_a^* + (1 + |r|^2)C_L \end{aligned} \tag{5.132}$$

are expressed in terms of quantities for the incident light. The photon-number fluctuations of the phase-conjugated beam are given by the standard expression

$$\langle (\Delta \hat{n}_b)^2 \rangle = \langle \hat{n}_b \rangle + \langle (\Delta W_b)^2 \rangle_{\mathcal{N}} . \tag{5.133}$$

We can again find [Bajer and Peřina (1991)] that sub-Poisson behaviour is possible only if the L-mode is squeezed and the phase-conjugated beam attenuated ($|r| < 1$), in agreement with the condition (5.129) for squeezing of vacuum fluctuations. In Fig. 5.6 we illustrate the influence of the reflectivity coefficient $r$ on the photon-number distribution $p(n)$. For small $|r|$ the photon statistics are sub-Poissonian or they oscillate (Sec. 5.4). Both these quantum effects rapidly disappear when the reflectivity of the phase conjugating mirror grows.

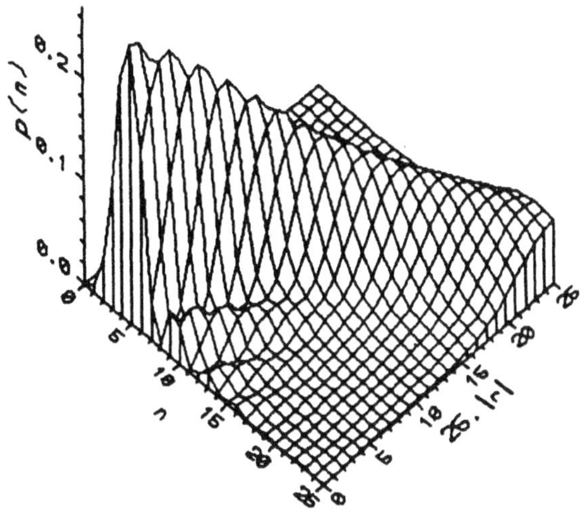


Figure 5.6: Dependence of the photon-number distribution on the reflectivity $|r|$ for $s = 1$, $\alpha = \lambda = 2$ [after J. Bajer and J. Peřina *Opt. Commun.* **85** (1991) 261].

To control the statistics of the mode L we can use the standard four-wave mixing with counterpropagating beams in a nonlinear medium with the third-order susceptibility $\chi^{(3)}$ and of the length $l$, as illustrated in Fig. 5.7. Denoting the annihilation operator for the signal beam as $\hat{a}_1(0)$ and that for the conjugated beam by $\hat{a}_2(0)$, it holds that [e.g. Yariv and Yeh (1984)]

$$\hat{a}_2(0) = \frac{\hat{a}_2(l)}{\cos(|\kappa| l)} - i\frac{\kappa^*}{|\kappa|}\hat{a}_1^\dagger(0)\tan(|\kappa| l)\,, \tag{5.134}$$

where $\kappa = g\alpha_3^*\alpha_4^*$ ($g$ being the coupling constant), $\alpha_3$ and $\alpha_4$ are strong (classical) pumping complex amplitudes. Comparing (5.134) with (5.123) we see that

$$\hat{a} = \hat{a}_1(0)\,, \quad \hat{b} = \hat{a}_2(0)\,, \quad r = -i\frac{\kappa^*}{|\kappa|}\tan(|\kappa| l)\,, \quad \hat{L} = \hat{a}_2(l)\,. \tag{5.135}$$

Thus we can control the quantum statistics of the additive noise $\hat{L} = \hat{a}_2(l)$ by injecting a signal $\hat{a}_2(l)$ of requested statistical properties, including the non-classical ones, at the free entrance of the phase conjugating mirror (Fig. 5.7).

An interesting conclusions can be obtained taking into account the quantum correlations of the atom and the field, i.e. taking into account their entanglement, which means that the properties of one subsystem can be deduced from the properties of the other subsystem. In particular if the partial entropy, defined in terms of the statistical operator as

$$S = -\mathrm{Tr}\,\{\hat{\varrho}\ln(\hat{\varrho})\}\,, \tag{5.136}$$

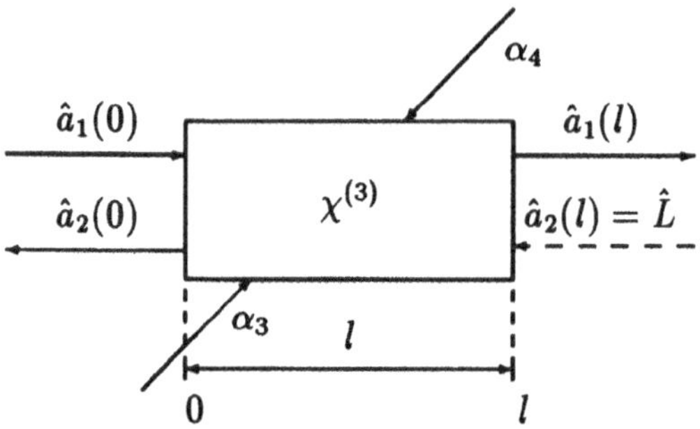


Figure 5.7: Stimulated backward four-wave mixing for phase conjugation realizing the control of the noise $L$ in terms of the stimulating injected signal $\hat{a}_2(l)$; $\hat{a}_1(0)$ is the incident signal, $\hat{a}_2(0)$ the conjugated signal, $\alpha_3$ and $\alpha_4$ are strong pumping amplitudes.

for one of the subsystems decreases, then the subsystem evolves to a pure state in which $S = 0$ and $\mathrm{Tr}\{\hat{\varrho}^2\} = \mathrm{Tr}\{\hat{\varrho}\} = 1$, whereas an increase of the entropy of the subsystem indicates that the two subsystems are losing their individuality and are getting to be quantum correlated or entangled.

There are various extensions of the Jaynes-Cummings model. In general all quantum features of the model are very sensitive to dissipations, e.g. the revival effect is associated with the constructive interference of the probability amplitudes, which is smoothed out by dissipative phenomena. Multi-level atoms can be assumed in the generalized Jaynes-Cummings model and generalized interactions can be considered where the operator $\hat{a}$ in (5.107b) is replaced by the number operator $\hat{a}^\dagger\hat{a}$, by $\hat{a}(\hat{a}^\dagger\hat{a})$ representing the intensity dependence of the coupling constant, by $\hat{a}^k$ representing $k$-th photon effect, by $\hat{a}^{\dagger k}\hat{a}^k$ involving multiphoton annihilations and creations, or they can involve Raman scattering process, or the Kerr effect described by the interaction hamiltonian $\hbar g\hat{a}^{\dagger 2}\hat{a}^2$. There are many papers devoted to these generalizations, we mention papers by Bužek and Jex (1989, 1990), Bužek (1990) and Drobný and Jex (1992) and references therein. A topical review of the Jaynes-Cummings model including various generalizations and including also an extended reference source on this topic was published by Shore and Knight (1993).

A nice review of the Jaynes-Cummings model was presented by Barnett and Phoenix (1993). Quantum effects in atomic optics were further discussed by Walls, Collett, Storey and Tan (1993), Storey, Collett and Walls (1993), Walls (1993), Haroche (1992) and Knight and Bužek (1995) among others.

# Chapter 6

# Photon interferences and correlations

Interference is one of the most important phenomena in optical fields representing their wave properties. Such two optical fields can in principle be in constructive or destructive interference according to their mutual phase (value of difference of their simultaneous phases) giving rise to an increase in the intensity (white interference fringes) or to a decrease in the intensity (dark interference fringes). Such correlations of two wave amplitudes are quite natural to be understood in a wave picture whereas it is more difficult to understand them in terms of particles-photons. Nevertheless making use of principles of quantum theory, which assumes probability waves, also particle-like theory of interference can be constructed. It is a basis for explanation of photon interferences given by Dirac (1958) stating that "each photon interferes only with itself" (for a discussion, see Mandel (1976)). Such point of view is related to the use of standard two beam interferometers detecting single photons. Moreover Dirac assumed an ideal case of maximum visibility interference fringes. The observation of interference pattern is related, from the point of view of quantum theory, to uncertainties in observables determined by commutation relations for the corresponding operators (energy–time, coordinate–momentum, photon number–phase, etc.). The wave-particle duality and complementarity principle play the principal role here. If a particle cannot be distinguished with respect to these uncertainties, the interference pattern may be observed because the number of particles is uncertain in the measurement and the phase can be certain. If particles in the interferometer can be distinguished, a Fock state is determined, the number of particles is quite certain, the corresponding phase is uncertain and the interference pattern is missing. Broader varieties of interesting interference and correlation phenomena are arising if the Hanbury Brown-Twiss intensity correlation measurements are adopted and multiphoton detections are considered.

## 6.1 Second-order interferences

The Young two slit experiment is the best arrangement to present classical as well as quantum treatments of the second-order interferences.

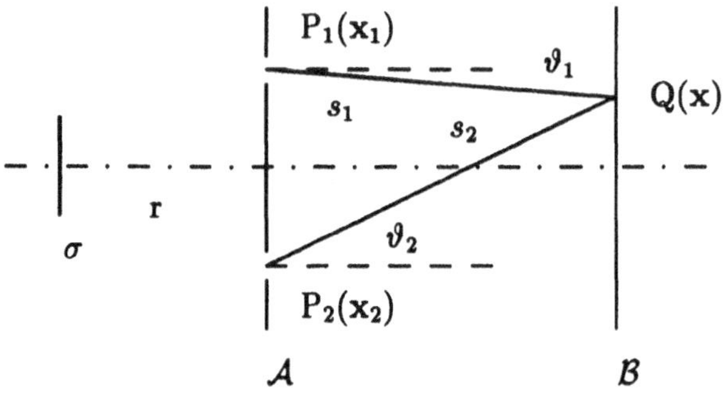


Figure 6.1: Second-order interference; $\sigma$ is an extended quasimonochromatic source, $\mathcal{A}$ is a screen with two small pinholes $P_1(\mathbf{x}_1)$ and $P_2(\mathbf{x}_2)$ and $\mathcal{B}$ is the screen of observation with a typical point $Q(\mathbf{x})$.

Consider such an arrangement as shown in Fig. 6.1. For a detailed treatment we refer the reader to the texts by Born and Wolf (1965), Beran and Parrent (1964) and Peřina (1985). The quantum mean intensity at the point Q at time $t$ of the screen $\mathcal{B}$ is equal to

$$\begin{aligned} I(\mathbf{x},t) &\equiv \Gamma_{\mathcal{N}}^{(1,1)}(x,x) = \mathrm{Tr}\left\{\hat{\varrho}\left[a_1^*\hat{A}^{(-)}(x_1) + a_2^*\hat{A}^{(-)}(x_2)\right]\right. \\ &\quad \left.\times\left[a_1\hat{A}^{(+)}(x_1) + a_2\hat{A}^{(+)}(x_2)\right]\right\} \\ &= |a_1|^2\,\Gamma_{\mathcal{N}}^{(1,1)}(x_1,x_1) + |a_2|^2\,\Gamma_{\mathcal{N}}^{(1,1)}(x_2,x_2) \\ &\quad + 2|a_1a_2|\,\mathrm{Re}\left[\Gamma_{\mathcal{N}}^{(1,1)}(x_1,x_2)\right], \end{aligned} \tag{6.1}$$

where $x \equiv (\mathbf{x},t)$, $a_j = i\bar{k}\cos(\vartheta_j)/2\pi s_j$ are a propagation factors considered at the mean frequency ($\bar{k} = 2\pi/\bar{\lambda}$) and respecting angle deviations of propagating polarized radiation (inclination angles $\vartheta_j$) and a decrease of the intensity during propagation (all distances are assumed to be much larger than the mean wavelength $\bar{\lambda}$). The occurrence of the interference fringes depends on non-vanishing interference term $2|a_1a_2|\,\mathrm{Re}\left[\Gamma_{\mathcal{N}}^{(1,1)}(x_1,x_2)\right]$, which expresses impossibility to distinguish from which pinhole a photon came to Q (the interference can be observed only if photons are within one cell of the quantum phase space or within the coherence volume defined as $V_c = c(r\bar{\lambda})^2/\Delta v S$, $c$ being the light velocity, $\Delta v$ the spectral half-width and $S$ the area of the source). Introducing the intensities $I_j(x) = |a_j|^2\,\Gamma_{\mathcal{N}}^{(1,1)}(x_j,x_j)$ at $x$ arising from light at $x_j$ and the degree of coherence (2.101) fulfilling the inequality (2.102), the interference law (6.1) reads

$$I(\mathbf{x},t) = I_1 + I_2 + 2\sqrt{I_1 I_2}\,|\gamma(x_1,x_2)|\cos\left[\phi(x_1,x_2)\right] , \tag{6.2}$$

where $\phi(x_1,x_2)$ is the phase of the second-order degree of coherence $\gamma_{\mathcal{N}}^{(1,1)}(x_1,x_2) \equiv \gamma(x_1,x_2)$. The visibility of interference fringes is defined as

$$C = \frac{I_{\max} - I_{\min}}{I_{\max} + I_{\min}} = \frac{2}{\sqrt{I_1/I_2} + \sqrt{I_2/I_1}}\,|\gamma(x_1,x_2)| , \tag{6.3}$$

where $I_{\max}$ and $I_{\min}$ are the maximum and minimum values of the intensity (envelops of white and dark fringes), respectively. If $I_1 = I_2$, then $C = |\gamma|$. It is obvious that the maximum visibility of the interference pattern is reached with coherent light when $|\gamma(x_1,x_2)| = 1$ for all $x_1$, $x_2$. In this case the real amplitudes $\sqrt{I_j}$ are superimposed with the mutual angle $\phi$. If both the beams are incoherent, then $|\gamma(x_1,x_2)| = 0$ and the resulting intensity is the sum of independent partial intensities. In cases $0 < |\gamma(x_1,x_2)| < 1$ both the beams are partially coherent providing the interference fringes with less visibility. Define the probabilities $p_j = I_j/(I_1+I_2)$, $j = 1,2$ of coherent light ($|\gamma| = 1$) that a given photon belongs to the $j$-th beam. We see from (6.3) that

$$(p_1 - p_2)^2 = 1 - C^2 , \qquad p_1 p_2 = \frac{C^2}{4} . \tag{6.4}$$

Hence, if there is certain that a photon belongs to the first or the second beam ($p_1 = 1$ or $p_2 = 1$), the visibility of interference fringes $C = 0$ and the interference pattern is not observable. However, if it is impossible to distinguish to which beam the photon belongs, then $p_1 = p_2 = 1/2$ and $C = 1$, and the interference fringes with the maximum visibility are observed. We see that the modulus of the degree of coherence $\gamma(x_1,x_2)$ ( the correlation function $\Gamma_{\mathcal{N}}^{(1,1)}(x_1,x_2)$) can directly be measured in such an interference experiment. The phase $\phi$ of the correlation function can in principle be determined from positional measurements. Note that partially coherent optical fields can be considered as a sum of a coherent field with an incoherent field, since

$$\begin{aligned} I(\mathbf{x},t) &= |\gamma(x_1,x_2)|\left[I_1 + I_2 + 2\sqrt{I_1 I_2}\cos(\phi)\right] \\ &\quad + \left[1 - |\gamma(x_1,x_2)|\right]\left[I_1 + I_2\right] . \end{aligned} \tag{6.5}$$

Further note that for stationary fields, $I(\mathbf{x},t)$ is independent of $t$ and $\Gamma_{\mathcal{N}}^{(1,1)}(x_1,x_2) \equiv \Gamma_{\mathcal{N}}^{(1,1)}(\mathbf{x}_1,\mathbf{x}_2,t_1-t_2)$.

The maximum visibility of the interference pattern requires the equality

$$|\Gamma_{\mathcal{N}}^{(1,1)}(x_1,x_2)|^2 = \Gamma_{\mathcal{N}}^{(1,1)}(x_1,x_1)\,\Gamma_{\mathcal{N}}^{(1,1)}(x_2,x_2) . \tag{6.6}$$

Writing

$$\Gamma_{\mathcal{N}}^{(1,1)}(x_1,x_2) = A(x_1)B(x_2) = \Gamma_{\mathcal{N}}^{(1,1)*}(x_2,x_1) = A^*(x_2)B^*(x_1), \qquad (6.7)$$

a real constant $K = A(x_1)/B^*(x_1) = A^*(x_2)/B(x_2)$ exists. Therefore $A(x) = KB^*(x)$ and introducing a field $V(x) = \sqrt{K}B(x)$, we see that a necessary and sufficient condition for the second-order coherence as the stable maximum visibility interference pattern is the factorization of the correlation function

$$\Gamma_{\mathcal{N}}^{(1,1)}(x_1,x_2) = V^*(x_1)V(x_2)\,. \qquad (6.8)$$

The consequences for higher-order correlation functions are expressed also by factorization conditions (see (2.78) for the fully coherent field) and they will be discussed in Sec. 6.3.

The interference pattern (6.2) is observable if some effective time delay $\tau = t_2 - t_1$ of the beams is small compared to the coherence time $1/\Delta v$. In this case photons belonging to two beams cannot be distinguished due to the Heisenberg uncertainty relation $\Delta t \Delta E \geq h$. In this case the spectral properties of beams are conserved and beams may be called to be cross-spectrally pure, which means that temporal coherence $(\gamma_{\mathcal{N}}(\mathbf{x},\mathbf{x},\tau))$ and spatial coherence $(\gamma_{\mathcal{N}}(\mathbf{x}_1,\mathbf{x}_2,0))$ are independent $(\gamma_{\mathcal{N}}(\mathbf{x}_1,\mathbf{x}_2,\tau) = \gamma_{\mathcal{N}}(\mathbf{x}_1,\mathbf{x}_1,\tau)\gamma_{\mathcal{N}}(\mathbf{x}_1,\mathbf{x}_2,0))$. On the other hand, if $\tau \gg 1/\Delta v$, the interference fringes in the space-time are missing, but the spectrum of radiation is modulated [Mandel (1962), Mandel and Wolf (1965)],

$$G_{\mathcal{N}}(\mathbf{x},v) \equiv G_{\mathcal{N}}^{(1,1)}(\mathbf{x},v) = G_{\mathcal{N}}(\mathbf{x}_1,\mathbf{x}_1,v) + G_{\mathcal{N}}(\mathbf{x}_2,\mathbf{x}_2,v)$$
$$+\, 2\sqrt{G_{\mathcal{N}}(\mathbf{x}_1,\mathbf{x}_1,v)G_{\mathcal{N}}(\mathbf{x}_2,\mathbf{x}_2,v)}\,|g_{\mathcal{N}}(\mathbf{x}_1,\mathbf{x}_2,v)|\cos(\psi), \qquad (6.9)$$

where

$$g_{\mathcal{N}}(\mathbf{x}_1,\mathbf{x}_2,v) = \frac{G_{\mathcal{N}}(\mathbf{x}_1,\mathbf{x}_2,v)}{\sqrt{G_{\mathcal{N}}(\mathbf{x}_1,\mathbf{x}_1,v)G_{\mathcal{N}}(\mathbf{x}_2,\mathbf{x}_2,v)}} \qquad (6.10)$$

is the spectral degree of coherence also fulfilling

$$0 \leq |g_{\mathcal{N}}(\mathbf{x}_1,\mathbf{x}_2,v)| \leq 1\,, \qquad (6.11)$$

and $\psi$ is the phase of $g_{\mathcal{N}}$. The half-widths of $\gamma_{\mathcal{N}}(\tau)$ and $g_{\mathcal{N}}(v)$ are inversely proportional, like those of the Fourier pairs and consequently the coherence properties in $(\mathbf{x},t)$ for $\tau < 1/\Delta v$ and in $(\mathbf{k},v)$ for $\tau > 1/\Delta v$ are complementary. Such quantum measurements have been performed by Zou, Grayson and Mandel (1992) using the frequency down conversion and they are discussed in Chap. 8. These complementary interference properties may be suggested to play a role in the explanation of the Einstein-Podolsky-Rosen "paradox" [Rauch (1993)] and "nonlocal" quantum effects.

Consider now particular states. Assume first an interference of two fields described by the coherent states $|\{\beta_\lambda^{(1)}\}\rangle$, $|\{\beta_\lambda^{(2)}\}\rangle$. This does not mean that both the fields are generated by a common source. Their origin can be in two independent light sources, e.g. ideal lasers. The Glauber-Sudarshan quasidistributions of such coherent beams are

$$\Phi_{\mathcal{N}}^{(j)}(\{\alpha_{\lambda_j}\}) = \prod_\lambda \delta\left(\alpha_{\lambda_j} - \beta_{\lambda_j}\right), \quad j = 1,2, \tag{6.12}$$

and

$$\gamma_{\mathcal{N}}(x_1, x_2) = \frac{V^*(x_1)V(x_2)}{|V(x_1)|\,|V(x_2)|}, \tag{6.13}$$

$V(x_j) \equiv V\left(x_j, \{\beta_{\lambda_j}\}\right)$, i.e. the interference pattern has the maximum visibility since $|\gamma_{\mathcal{N}}(x_1, x_2)| = 1$. Thus we have the ideal interference fringes including independent sources. This requests also the knowledge of phases of the coherent states. Such an information is in the optical range almost always lost. Then the distributions (6.12) have to be averaged over the phases of $\{\alpha_{\lambda_j}\}$, which gives the distributions independent of phases,

$$\Phi_{\mathcal{N}}^{(j)}(\{\alpha_{\lambda_j}\}) = \prod_\lambda \frac{1}{\pi}\delta\left(|\alpha_{\lambda_j}|^2 - |\beta_{\lambda_j}|^2\right), \quad j = 1,2. \tag{6.14}$$

Now the interference term vanishes and no interference effect can be observed in the intensity pattern. This is in agreement with the fact that such an interference effect can occur only when there is at least partial coherence between the beams.

Consider now single photon two-mode field described by the state

$$|\psi_1\rangle = \hat{b}^\dagger|0\rangle = \cos(\vartheta)|1,0\rangle + \sin(\vartheta)|0,1\rangle, \tag{6.15}$$

where

$$\hat{b}^\dagger = \hat{a}_1^\dagger\cos(\vartheta) + \hat{a}_2^\dagger\sin(\vartheta), \tag{6.16}$$

$\vartheta$ is the angle between interfering beams described by annihilation operators $\hat{a}_1$ and $\hat{a}_2$, respectively. It is obviously $[\hat{b}, \hat{b}^\dagger] = \hat{1}$. The interference law reads now

$$\begin{aligned} I(\mathbf{x},t) &= \text{const}\left\{\cos^2(\vartheta)\langle 1,0|\hat{a}_1^\dagger\hat{a}_1|1,0\rangle + \sin^2(\vartheta)\langle 0,1|\hat{a}_2^\dagger\hat{a}_2|0,1\rangle\right. \\ &\quad \left. + \sin(2\vartheta)\cos(\phi)\langle 1,0|\hat{a}_1^\dagger\hat{a}_2|0,1\rangle\right\} \\ &= \text{const}\,[1 + \sin(2\vartheta)\cos(\phi)], \end{aligned} \tag{6.17}$$

using the definition of annihilation and creation operators and the orthogonality of the states. For $\vartheta = \pi/4$ we clearly have maximum fringe visibility and it is apparent that an interference pattern may be built up from a succession of single-photon interference events. This also supports the interpretation that the interference arises from a single photon interfering with itself. In the case of two independent sources the interference pattern is observable if one cannot distinguish, as a consequence of the Heisenberg uncertainty relations, from which source a given photon came. Localization of a photon at $(\mathbf{x}, t)$ with a resolution better than one fringe width undetermines the photon momentum to such an extent that it is impossible to ascribe the photon to either of the two sources separately.

Let us consider the $n$-photon state

$$|\psi_n\rangle = \frac{\hat{b}^{\dagger n}}{\sqrt{n!}}|0\rangle = \sum_{j=0}^{n}[\cos(\vartheta)]^j[\sin(\vartheta)]^{n-j}\left[\frac{n!}{j!(n-j)!}\right]^{1/2}|j, n-j\rangle . \quad (6.18a)$$

Putting again $\vartheta = \pi/4$ for simplicity we obtain for the two-photon state

$$|\psi_2\rangle = \frac{1}{2}\left[|0,2\rangle + \sqrt{2}|1,1\rangle + |2,0\rangle\right] \quad (6.18b)$$

and for the quantum expectation value of the intensity it holds that

$$\begin{aligned} I(\mathbf{x}, t) &= \frac{\text{const}}{4}\Big[\langle 2,0|\hat{a}_1^\dagger \hat{a}_1|2,0\rangle + \langle 0,2|\hat{a}_2^\dagger \hat{a}_2|0,2\rangle \\ &\quad + 2\langle 1,1|\hat{a}_1^\dagger \hat{a}_1|1,1\rangle + 2\langle 1,1|\hat{a}_2^\dagger \hat{a}_2|1,1\rangle \\ &\quad + 4\sqrt{2}\cos(\phi)\langle 0,2|\hat{a}_2^\dagger \hat{a}_1|1,1\rangle\Big] \\ &= 2\,\text{const}\,[1+\cos(\phi)] , \end{aligned} \quad (6.19)$$

providing again the maximum visibility pattern. To obtain a non-zero value of the second-order correlation function and observable interference, it is necessary to have states differing by a single photon. Such a state can be written as

$$|\psi\rangle = \frac{1}{\sqrt{2}}\left[|n, n-1\rangle + |n-1, n\rangle\right] . \quad (6.20)$$

The interference law for this case is in (6.2), where $I_j = |\alpha_j|^2(n-1/2)$, $j = 1, 2$ and $\gamma = n\alpha_1^*\alpha_2/[(2n-1)|\alpha_1||\alpha_2|]$. Thus such an n-photon field produces interference with the modulus of the degree of coherence $|\gamma| = 1/(2-1/n)$, i.e. such a field is partially coherent. A single-photon field is coherent because $|\gamma| = 1$ providing the maximum visibility (interferences of a photon with itself) and the result is the same as with the coherent state involving cooperation of photons in an infinite superposition of weighted Fock occupation number states. For $n \to \infty$ the modulus of the degree of coherence $|\gamma|$ tends to 1/2 and the resulting optical field is partially coherent.

Although the most important experiments will be discussed in Chapter 8, we would like to point out here that all interference experiments including those with single photons of nonclassical (antibunched) light are in full agreement with the quantum interpretation. Probability waves arising from operator representations are forming an interference pattern governing registrations of single photons on the interference screen. In a strong field limit (strong field or many single-photon realizations) a classical pattern is obtained with classical waves identical to the original probability waves governing microobjects.

It should be noted that Jaeger, Horne and Shimony (1993) derived an interesting complementarity of one-particle visibilities $C_j$, $j = 1, 2$ and two-particle visibility $C_{12}$ of interference fringes (Sec. 6.2) expressed by the condition $C_j^2 + C_{12}^2 \leq 1$, or $C_j C_{12} \leq 1/2$.

## 6.2 Fourth-order interferences — Hanbury Brown-Twiss correlations

The first experiment involving the higher-order correlations of the complex amlitudes than the second-order correlations of standard interferometric measurements was realized by Hanbury Brown and Twiss (1956) with the goal to measure the angular diameters of stars with higher resolving power and to eliminate atmospheric distortions. The scheme of their apparatus is in Fig. 6.2. Light from a natural source (mercury lamp, star) S was split by a half-silvered mirror M and fell on two photocells $P_1$ and $P_2$, whose outputs were sent through band-limited amplifiers $A_1$ and $A_2$ to a correlator or a coincidence device C (classical strong fields or low intensity quantum fields are registered), where the outputs of the detectors are multiplied and averaged. This experiment was repeated many times with strong classical as well as low intensity quantum fields. The most comprehensive review of the intensity correlation effects was presented by Mandel (1963).

To explain this experiment we must look more deeply into a physical model involving the field intensity $I$ rather than the integrated intensity $W$ (cf. eq. (2.115) of Sec. 2.9.6). Assume first linearly polarized light and calculate fluctuations of the number of photoelectrons emitted by a photocathode exposed to light intensity $I(t)$. Using the photodetection equation (2.113), we arrive at

$$\begin{aligned} \langle n^2 \rangle &= \langle n \rangle + \eta^2 \int_t^{t+T} \int_t^{t+T} \langle I(t') I(t'') \rangle_{\mathcal{N}} \, dt' dt'' \\ &= \langle n \rangle + \eta^2 \int_0^T \int_0^T \langle I(t + t' - t'') I(t) \rangle_{\mathcal{N}} \, dt' dt'' , \end{aligned} \qquad (6.21)$$

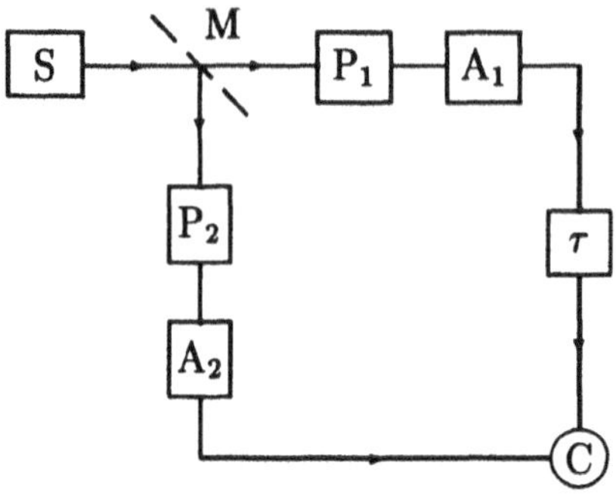


Figure 6.2: An outline of the apparatus for demonstrating the Hanbury Brown-Twiss correlations between intensity fluctuations; S is natural source, M is half-silvered mirror, $P_1$ and $P_2$ are photomultipliers, $A_1$ and $A_2$ are amplifiers, C is an electric correlator or a coincidence device, $\tau$ is a time delay element.

where the mean number of photoelectrons $\langle n \rangle = \eta \langle I \rangle_{\mathcal{N}} T$ and we have assumed a stationary field. The correlation of intensities of natural light is expressed by the factorization formula (5.23) and introducing the quantity

$$\xi(T) = \frac{1}{T} \int_0^T \int_0^T |\gamma_{\mathcal{N}}(t' - t'')|^2 \, dt' dt'' = \frac{2}{T} \int_0^T (T - \tau) |\gamma_{\mathcal{N}}(\tau)|^2 \, d\tau \,, \tag{6.22}$$

we obtain from (6.21)

$$\langle (\Delta n)^2 \rangle = \langle n^2 \rangle - \langle n \rangle^2 = \langle n \rangle \left( 1 + \langle n \rangle \frac{\xi(T)}{T} \right) , \tag{6.23}$$

i.e. we have got the multifold Bose-Einstein (Gaussian) statistics with $T/\xi(T) = M$ degrees of freedom. As $|\gamma_{\mathcal{N}}(\tau)| \leq 1$ it follows from (6.22) that $\xi(T) \leq T$ and $M \geq 1$. If the detection time $T \ll \tau_c = 1/\Delta v$, then $|\gamma_{\mathcal{N}}(\tau)|$ is non-vanishing only within an interval of the order of $1/\Delta v$ and we can write

$$\xi(T) \approx \xi(\infty) = \int_{-\infty}^{+\infty} |\gamma_{\mathcal{N}}(\tau)|^2 \, d\tau \,, \tag{6.24}$$

which can be considered as the coherence time $\tau_c = \xi(\infty)$. The number of degrees of freedom $M = T/\tau_c$ and the photocount distribution is given by the Mandel-Rice formula (2.130). This formula can be considered as good approximation for natural radiation for any detection time intervals.

Departures from the Poisson counting statistics depend on the degenaracy parameter

$$\delta = \frac{\langle n \rangle}{T} \tau_c = \eta \langle I \rangle_{\mathcal{N}} \tau_c \,, \tag{6.25}$$

giving the mean number of counts (photons, for $\eta = 1$) per cell of the phase space (more generally $\delta = \langle n \rangle V_c/V$, $V_c$ being the coherence volume and $V$ the detection volume). For chaotic (thermal) light $\delta \ll 1$ and it is difficult to observe directly the excess photon noise. The so-called pseudothermal light, obtained by randomization of laser light using rotating ground glass disk or nematic liquid crystals in the electric field, can remove this problem. In the laser light $\delta \gg 1$, which expresses many photon cooperation in the coherent state.

Of course in (6.23) we can also include the spatial and polarization degrees of freedom substituting $T/\xi(T) \to M = 2TS/\left[\xi(T)A_c(1+P^2)\right]$, $S$ being the detection area, $A_c$ the coherence area and $P$ the degree of polarization.

The correlation between counts of the two photodetectors may be obtained in the similar way,

$$\begin{aligned} \langle n_1 n_2 \rangle &= \eta_1\eta_2 \int_t^{t+T}\int_t^{t+T} \langle I_1(t')I_2(t'')\rangle_{\mathcal{N}}\, dt'dt'' \\ &= \eta_1\eta_2 \int_0^T\int_0^T \langle I_1(t+t'-t'')I_2(t)\rangle_{\mathcal{N}}\, dt'dt''\,, \end{aligned} \tag{6.26}$$

where $\eta_1$, $I_1$ and $\eta_2$, $I_2$ are related to the first and the second photodetectors, respectively. Assuming cross-spectral purity condition $\gamma_{\mathcal{N},12}(\tau) = \gamma_{\mathcal{N},12}(0)\gamma_{\mathcal{N},11}(\tau)$ expressing the independence of spatial and temporal coherence, we obtain for correlation of fluctuations of photoelectrons from the detectors

$$\langle \Delta n_1 \Delta n_2 \rangle = \langle (n_1 - \langle n_1 \rangle)(n_2 - \langle n_2 \rangle)\rangle = \langle n_1 \rangle \langle n_2 \rangle |\gamma_{\mathcal{N},12}(0)|^2 \frac{\tau_c}{T}, \tag{6.27}$$

respectively involving the finite detection area $S$ and partial polarization $P$,

$$\langle \Delta n_1 \Delta n_2 \rangle = \frac{1}{2}\left(1+P^2\right)\langle n_1 \rangle \langle n_2 \rangle |\gamma_{\mathcal{N},12}(0)|^2 \frac{\tau_c A_c}{TS}. \tag{6.28}$$

Thus, in general, the correlation fluctuations measurements provide information on spectral properties (temporal coherence), the spatial coherence, and the polarization properties of light beams.

It is interesting to note that similar calculations can be performed for non-Gaussian (non-chaotic) light. For the intensity correlation we can write

$$\begin{aligned} \langle I_1(t)I_2(t+\tau)\rangle_{\mathcal{N}} &= \langle I_1\rangle_{\mathcal{N}}\langle I_2\rangle_{\mathcal{N}} + \langle \Delta I_1(t)\Delta I_2(t+\tau)\rangle_{\mathcal{N}} \\ &= \langle I_1\rangle_{\mathcal{N}}\langle I_2\rangle_{\mathcal{N}}\left(1+|\xi_{12}(\tau)|^2\right), \end{aligned} \tag{6.29}$$

where

$$|\xi_{12}(\tau)|^2 = \frac{\langle \Delta I_1(t)\Delta I_2(t+\tau)\rangle_{\mathcal{N}}}{\langle I_1(t)\rangle_{\mathcal{N}}\langle I_2(t)\rangle_{\mathcal{N}}}. \tag{6.30}$$

Obviously for chaotic light $|\xi_{12}(\tau)| \equiv |\gamma_{\mathcal{N},12}(\tau)|$. The photocount correlations for polarized non-Gaussian light are equal to

$$\langle \Delta n_1 \Delta n_2 \rangle = \langle n_1 \rangle \langle n_2 \rangle \frac{\bar{\xi}(T)}{T}, \tag{6.31}$$

where

$$\begin{aligned} \bar{\xi}(T) &= \frac{1}{T\langle I_1\rangle_{\mathcal{N}}\langle I_2\rangle_{\mathcal{N}}} \int_0^T \int_0^T \langle \Delta I_1(t')\Delta I_2(t'')\rangle_{\mathcal{N}}\, dt'dt'' \\ &= \frac{2}{T\langle I_1\rangle_{\mathcal{N}}\langle I_2\rangle_{\mathcal{N}}} \int_0^T (T-\tau)\langle \Delta I_1(\tau)\Delta I_2(0)\rangle_{\mathcal{N}}\, d\tau, \end{aligned} \tag{6.32}$$

if the stationary condition is again assumed. Of course, for ideal lasers, $\Delta I_j \approx 0$, $\langle \Delta n_1 \Delta n_2 \rangle \approx 0$ and no Hanbury Brown-Twiss correlations occur. There are no departures from the Poisson statistics in this case.

The above correlations of fluctuations can be adopted to define terms of coherence time, coherence area and coherence volume for states of any statistical properties,

$$\tau_c = \int_{-\infty}^{+\infty} \left| \frac{\zeta(\mathbf{x},\mathbf{x},\tau)}{\zeta(\mathbf{x},\mathbf{x},0)} \right|^2 d\tau, \tag{6.33a}$$

$$A_c = \int_{-\infty}^{+\infty} \int_{-\infty}^{+\infty} \left| \frac{\zeta(\mathbf{x}_1,\mathbf{x}_2,0)}{\zeta(\mathbf{x},\mathbf{x},0)} \right|^2 d\mathbf{x}, \tag{6.33b}$$

$$V_c = c \int_{-\infty}^{+\infty} \int_{-\infty}^{+\infty} \int_{-\infty}^{+\infty} \left| \frac{\zeta(\mathbf{x}_1,\mathbf{x}_2,\tau)}{\zeta(\mathbf{x},\mathbf{x},0)} \right|^2 d\mathbf{x}\, d\tau, \tag{6.33c}$$

where $\zeta(\mathbf{x}_1,\mathbf{x}_2,\tau) \equiv \zeta(\mathbf{x}_1 - \mathbf{x}_2,\tau)$ is assumed and $\mathbf{x}_1 - \mathbf{x}_2 = \mathbf{x}$.

The main advantage of such correlation measurements is that they involve the intensity $I = |V|^2$ which varies slowly in comparison with the rapidly varying complex amplitude $V$, and the phase of $V$ is not involved at all, i.e. the degree of coherence appears only in the modulus $|\gamma_{12}(\tau)|$. Consequently phase distortions due to turbulence in the atmosphere, local changes of its index of refraction, various mechanical vibrations, etc. do not affect the correlation measurements, although they may be intolerable in the standard interferometers. Consequently much larger basis of the correlation interferometer may be used making it possible to increase its resolving power. On the other hand, the phase information about the field (the degree of coherence) is lost in such correlation measurements and other methods may have to be adopted to recover the phase [for a review see e.g. Peřina (1985)]. Such methods are necessary to be applied for the determination of the spectrum from the correlation measurements based on the Wiener-Khintchin theorem (2.87).

An interesting example is to apply the intensity correlation technique to the interference of two independent beams. We have seen that two stabilized laser sources with random phases described by the quasidistributions (6.14) provide no interference pattern. However, if we calculate the correlation of intensities, we obtain

$$\Gamma_{\mathcal{N}}^{(2,2)}(x_1, x_2, x_2, x_1) = \langle I(x_1)I(x_2)\rangle_{\mathcal{N}} = \langle [I_1(x_1) + I_2(x_1)]^2\rangle$$
$$+ 2\langle I_1(x_1)\rangle\langle I_2(x_1)\rangle \cos\left[(\omega_1 - \omega_2)(t_1 - t_2) + (\mathbf{k}_1 - \mathbf{k}_2)(\mathbf{x}_2 - \mathbf{x}_1)\right], \quad (6.34)$$

because the multiplication of the interference terms in $I(x_1)$ and $I(x_2)$ compensates random phases and creates the interference term in (6.34) providing the interference fringes (in space) and beats (in time) in the fourth-order correlation pattern. In (6.34) we have also assumed that $|t_1 - t_2| \ll 1/\Delta v$ and $|\mathbf{x}_1 - \mathbf{x}_2| \ll 1/\Delta k$. Hence, although the independent phase averaged coherent states cannot produce the interferences in the second-order pattern, they can in the fourth-order enabling thus to observe the interference of optical fields from independent sources. From (6.34) we obtain for coherent and chaotic fields

$$\Gamma_{\mathcal{N}}^{(2,2)}(x_1, x_2, x_2, x_1) = \langle I\rangle^2$$
$$\times \left\{1 + \frac{1}{2}\cos\left[(\omega_1 - \omega_2)(t_1 - t_2) + (\mathbf{k}_1 - \mathbf{k}_2)(\mathbf{x}_2 - \mathbf{x}_1)\right]\right\} \quad (6.35a)$$

and

$$\Gamma_{\mathcal{N}}^{(2,2)}(x_1, x_2, x_2, x_1) = \frac{3}{2}\langle I\rangle^2$$
$$\times \left\{1 + \frac{1}{3}\cos\left[(\omega_1 - \omega_2)(t_1 - t_2) + (\mathbf{k}_1 - \mathbf{k}_2)(\mathbf{x}_2 - \mathbf{x}_1)\right]\right\} \quad (6.35b)$$

respectively, where $\langle I\rangle = \langle I_1\rangle + \langle I_2\rangle$, $\langle I_1\rangle = \langle I_2\rangle$.

It is interesting to compare these results with the expressions for the intensity correlation functions for two different shining atoms [Mandel (1983), Teich, Saleh and Peřina (1984), Paul (1986)]. Considering two-level atoms described by Fermi-Dirac annihilation and creation operators fulfilling the anticommutation relation, some fourth-order terms representing the probability of detection of two photons disappear because the atom can radiate only one photon. On the other hand, such terms contribute to the classical correlation function. Different results for quantum and classical approaches reflect the impossibility of the classical theory to describe correctly the particle properties. Compared to (6.35a) we obtain from this fully quantum description

$$\Gamma_{\mathcal{N}}^{(2,2)}(x_1, x_2, x_2, x_1) \sim 1 + \cos\left[(\omega_1 - \omega_2)(t_1 - t_2) + (\mathbf{k}_1 - \mathbf{k}_2)(\mathbf{x}_2 - \mathbf{x}_1)\right], \quad (6.36)$$

which means that the quantum theory requires twice of the classical visibility. This is important for experimental verification of such quantum effects [Sec. 8.6, Ghosh and Mandel (1987)]. Comparing (6.35a,b) we observe that the visibility for coherent fields is higher than for chaotic light. Equation (6.36) shows that the quantum character of light further increases the degree of order of light and the visibility of the fourth-order interference pattern. The interference law (6.36) exhibits the possibility to place both the detectors in such a way that both can simultaneously register the emission of a photon or to avoid any coincidence.

Generalizing these results to two sources containing $M$ and $N$ atoms, respectively, the quantum correlation functions are

$$\Gamma_{\mathcal{N}}^{(2,2)}(x_1, x_2, x_2, x_1) \sim N(N-1) + M(M-1) \\ + MN\{1 + \cos[(\omega_1 - \omega_2)(t_1 - t_2) + (\mathbf{k}_1 - \mathbf{k}_2)(\mathbf{x}_2 - \mathbf{x}_1)]\} \quad (6.37a)$$

in the quantum case, whereas

$$\Gamma_{\mathcal{N}}^{(2,2)}(x_1, x_2, x_2, x_1) \sim (M+N)^2 + M(M-1) + N(N-1) \\ + 2MN\cos[(\omega_1 - \omega_2)(t_1 - t_2) + (\mathbf{k}_1 - \mathbf{k}_2)(\mathbf{x}_2 - \mathbf{x}_1)] \quad (6.37b)$$

for classical correlation functions. In the limiting case $M$, $N \to \infty$ both the correlation functions tend to the classical value for chaotic radiation. The same result is reached when averaging with the Poisson distribution over both the sources. This means that to measure quantum properties one has to control the number of emitting atoms.

A discussion of interferences between different photons from two incoherent sources was presented and experimental demonstration was provided by Endo and Toyoshima (1992).

## 6.3 Higher-order phenomena

The fourth-order Hanbury Brown-Twiss correlations can be generalized to any order making use of $N$ photodetectors instead of two photodetectors, as outlined in Fig. 6.3. Taking into account real point detectors and polarized light we obtain for the probability of detecting a photon in the time interval $(0, T_1)$ at the point $\mathbf{x}_1$, a photon in the time interval $(0, T_2)$ at the point $\mathbf{x}_2$, etc., and a photon at the time interval $(0, T_N)$ at the point $\mathbf{x}_N$

$$p^{(N)}(T_1, \ldots, T_N) = \int_0^{T_1} \cdots \int_0^{T_N} \Gamma_{\mathcal{N}}^{(N,N)}(\mathbf{x}_1, \ldots, \mathbf{x}_N, \mathbf{x}_N, \ldots, \mathbf{x}_1, \\ t'_1, \ldots, t'_N, t''_N, \ldots, t''_1) \prod_{j=1}^{N} S_j(t'_j - t''_j)\, dt'_j\, dt''_j, \quad (6.38a)$$

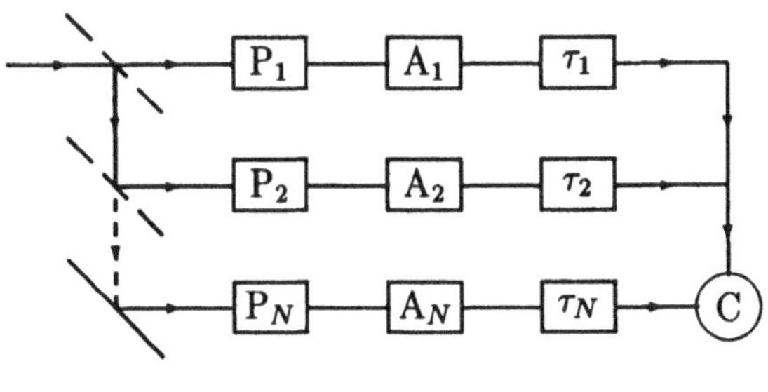


Figure 6.3: An outline for the measurement of the higher-order correlation function $\Gamma_{\mathcal{N}}^{(N,N)}(x_1,\ldots,x_N,x_N,\ldots,x_1)$; $P_j$ are photodetectors, $A_j$ amplifiers, $\tau_j$ delay elements, and C is the correlator or a coincidence counter.

where $\mathcal{S}_j(t'_j - t''_j)$ is the temporal response function of the $j$-th detector (the Fourier transformation of the spectral characteristic of the detector). For broadband photodetectors $\mathcal{S}(t'-t'')$ is proportional to $\delta(t'-t'')$. If extended (volume) detectors are used

$$p^{(N)} = \int_0^{T_1}\cdots\int_0^{T_N}\int_{V_1}\cdots\int_{V_N}\Gamma_{\mathcal{N}}^{(N,N)}(x'_1,\ldots,x'_N,x''_N,\ldots,x''_1) \times\prod_{j=1}^{N}\mathcal{S}_j(x'_j - x''_j)\,d^4x'_j\,d^4x''_j\,. \tag{6.38b}$$

For point broadband detectors situated at $\mathbf{x}_j$, $\mathcal{S}_j(x'_j - x''_j) = \eta_j\delta(\mathbf{x}'_j - \mathbf{x}''_j)\,\delta(\mathbf{x}'_j - \mathbf{x}_j)\delta(t'_j - t''_j)$, $\eta_j$ being the photoefficiency.

The multifold photodetection equation reads

$$p(n_1,\ldots,n_N,T_1,\ldots,T_N,t_1,\ldots,t_N) = \left\langle\prod_{j=1}^{N}\frac{[\eta_j W_j(T_j,t_j)]^{n_j}}{n_j!}\exp\left[-\eta_j W_j(T_j,t_j)\right]\right\rangle_{\mathcal{N}}, \tag{6.39}$$

where the integrated intensity equals

$$W_j(T_j,t_j) = \int_{t_j}^{t_j+T_j} I_j(\mathbf{x}_j,t'_j)\,dt'_j\,, \tag{6.40}$$

giving the probability of detecting $n_1$ photons at the time interval $(t_1,t_1+T_1)$ and at the point $\mathbf{x}_1$, etc., and $n_N$ photons at $(t_N,t_N+T_N)$ and $\mathbf{x}_N$. The average in (6.39) is taken with the multifold quasidistribution

$$P_{\mathcal{N}}(W_1,\ldots,W_N) = \int\Phi_{\mathcal{N}}(\{\alpha_\lambda\})\,\delta\left(|\alpha_\lambda|^2 - W_1\right)\cdots\delta\left(|\alpha_\lambda|^2 - W_N\right)\,d^2\{\alpha_\lambda\}\,. \tag{6.41}$$

From (6.39) it follows the fluctuation formula

$$\langle \Delta n_1 \ldots \Delta n_N \rangle = \eta_1 \ldots \eta_N \langle \Delta W_1 \ldots \Delta W_N \rangle_{\mathcal{N}} , \tag{6.42}$$

indicating that for ideal laser light source the $2N$th order photon number correlations $\langle \Delta n_1 \ldots \Delta n_N \rangle \approx 0$ because $\langle \Delta W_1 \ldots \Delta W_N \rangle_{\mathcal{N}} \approx 0$.

The complete coherence in the sense of the second-order factorization (6.8) [Titulaer and Glauber (1965, 1966)] leads to

$$\Gamma_{\mathcal{N}}^{(m,n)}(x_1, \ldots, x_{m+n}) = \gamma^{(m,n)} \prod_{j=1}^{m} V^*(x_j) \prod_{k=m+1}^{m+n} V(x_k) , \tag{6.43}$$

where the factorization coefficient is

$$\gamma^{(m,n)} = \frac{\Gamma_{\mathcal{N}}^{(m,n)}(x_0, \ldots, x_0)}{\left[\Gamma_{\mathcal{N}}^{(1,1)}(x_0, x_0)\right]^{(m+n)/2}} ; \tag{6.44}$$

as $x_0$ is an arbitrary point, (6.44) must be independent of $x_0$.

By analogy with the factorization (6.8) for the second-order coherence we may define the complete coherence by the factorization (6.43) with $\gamma^{(m,n)} = 1$ for all $m, n$ giving (2.78). We can speak of $2N$-th-order coherence if this holds for all $m, n \leq N$. If such factorization conditions hold for all $m, n$ (in practice mostly for all $m = n$ since the phase of such an optical field is not usually under control), then the field possesses complete coherence. This means from the classical point of view that a fully coherent field is a noiseless field the distribution function of which is the Dirac $\delta$-function. In the sense of the even-order factorization, such fields may possess uniform phase fluctuations. This illustrates a close relation between noiselessness of optical fields and full coherence. For quantum fields there also exist quantum states in which the factorization (2.78) holds. These are the coherent states as eigenstates of the photon annihilation operator. They are noiseless as the quantum states may be, i.e. they involve only fluctuations of the physical vacuum. However, its contribution to photodetection process is zero because the vacuum expectation values of the normal products of annihilation and creation operators vanish. Further we can state that the field $V(\mathbf{x}, t)$ may have any spectral composition so that full coherence in general does not require monochromaticity of the field. Only for stationary fields, for vhich $\Gamma_{\mathcal{N}}^{(1,1)}(t_1, t_2) = \Gamma_{\mathcal{N}}^{(1,1)}(t_1 - t_2) = V^*(t_1)V(t_2)$, which is satisfied only by an exponential function $V(x) \sim \exp(-i2\pi v_0 t)$ involving a constant $v_0$, is the monochromaticity the necessary and sufficient condition for the field to be coherent. A connection between fluctuations of the field and its coherence can be demonstrated as follows. Let the field be uniformly fluctuating, $V(x) = cV_{\mathrm{det}}(x)$, $c$ being a complex random variable and $V_{\mathrm{det}}$ is a deterministic field. The second-order coherence demands

$\langle|c|^2\rangle = \int P(c)|c|^2\, d^2c = 1$ ($P(c)$ being a probability distribution of $c$, and the integration is taken over the whole complex plane). Hence, some phase and amplitude fluctuations are admissible by the second-order coherence requesting the maximum visibility interference pattern. From the fourth-order coherence we obtain $\langle|c|^4\rangle = 1 = \langle|c|^2\rangle^2$ giving $|c|^2 = 1$, i.e. only uniform phase fluctuations are admissible, etc.

The physical significance of the factorization conditions for the coherent field can be written in terms of counting rates of the system of $N$ photodetectors in the form

$$\Gamma_{\mathcal{N}}^{(N,N)}(x_1,\ldots,x_N,x_N,\ldots,x_1) = \prod_{j=1}^{N} \Gamma_{\mathcal{N}}^{(1,1)}(x_j,x_j)\,, \tag{6.45}$$

which means that the $N$-fold joint counting rate of the system of $N$ photodetectors is equal to the product of the counting rates of single photodetectors, i.e. the responses of the counters are statistically independent of each other. If the average intensity of the field is independent of time (the field is stationary), the counters detect no tendency towards any correlations in the arrival times of photons. In this case there is no Hanbury Brown-Twiss correlation of intensity fluctuations. Thus fields for which this effect occurs cannot be coherent in the fourth and higher orders. Obviously the photon statistics of the coherent field are Poissonian. For instance for a single mode $\langle\hat{a}^{\dagger n}\hat{a}^n\rangle = \langle\hat{a}^\dagger\hat{a}\rangle^n$ and for the normal characteristic function we have

$$\begin{aligned}\langle\exp(is\hat{n})\rangle_{\mathcal{N}} &= \sum_{n=0}^{\infty}\frac{(is)^n}{n!}\langle\hat{a}^{\dagger n}\hat{a}^n\rangle \\ &= \sum_{n=0}^{\infty}\frac{(is)^n}{n!}\langle\hat{a}^\dagger\hat{a}\rangle^n = \exp(is\langle\hat{n}\rangle)\,,\end{aligned} \tag{6.46}$$

which is the generating function of the Poisson distribution

$$p(n) = \langle n|\hat{\varrho}|n\rangle = \frac{\langle\hat{n}\rangle^n}{n!}\exp(-\langle\hat{n}\rangle)\,. \tag{6.47}$$

The coefficient of factorization $\gamma^{(m,n)}$ appearing in (6.43) can simply be calculated for the Fock state and the chaotic state. For the Fock state $|n\rangle$ we obtain

$$\gamma^{(k,k)} = \frac{\langle\hat{a}^{\dagger k}\hat{a}^k\rangle}{\langle\hat{a}^\dagger\hat{a}\rangle^k} = \frac{n!}{(n-k)!\,n^k}\,, \tag{6.48}$$

being $\gamma^{(k,l)} = 0$ for $k \neq l$ and $\gamma^{(k,k)} = 0$, $k > n$. The sub-Poissonian behaviour of the Fock state is characterized by the negative normal variance $\langle(\Delta W)^2\rangle_{\mathcal{N}} = \langle\hat{a}^{\dagger 2}\hat{a}^2\rangle - \langle\hat{a}^\dagger\hat{a}\rangle^2 = n(n-1) - n^2 = -n < 0$.

For chaotic (thermal) radiation we clearly have

$$\gamma^{(k,l)} = \delta_{kl}\,k!\,. \tag{6.49}$$

## 6.4 Single photon interferences

Consider first a beam splitter from a point of view of single photons. A scheme is given in Fig. 6.4. Classically the beam splitter is described by a transmittance $\mathcal{T}$ and reflectance $\mathcal{R}$. The intensity of the tramsmitted wave $I_t$ and the intensity of reflected wave $I_r$ are obtained as $I_t = \mathcal{T}I$ and $I_r = \mathcal{R}I = (1-\mathcal{T})I$, $I$ being the incident light intensity. In terms of single photons probabilistic interpretation must be adopted. The probability that a photon is transtmitted is equal to $\mathcal{T}$ and the probability that it is reflected is equal to $\mathcal{R} = 1 - \mathcal{T}$.

Denoting the mean number of photons incident on the beam splitter during time $T$ as $\langle n \rangle$, then the mean number of transmitted photons is equal to $\langle n \rangle \mathcal{T}$, whereas the number of reflected photons in that time is equal to $\mathcal{R}\langle n \rangle = (1-\mathcal{T})\langle n \rangle$. We can calculate photon-number statistics under random partitioning by the beamsplitter. If the incident beam contains just $n$ photons, then the probability $p(m)$ that $m$ photons are transmitted is equal from elementary probability theory

$$p(m) = \binom{n}{m} \mathcal{T}^m (1-\mathcal{T})^{n-m}, \qquad m = 0, 1, \ldots, n. \tag{6.50}$$

The mean number of transmitted photons is $\langle m \rangle = \mathcal{T} n$. For the variance we obtain

$$\langle (\Delta m)^2 \rangle = \mathcal{T}(1-\mathcal{T})n = (1-\mathcal{T})\langle m \rangle. \tag{6.51}$$

For large intensities photons will be partitioned between the two transmitted and reflected beams with respect to classical optics of strong beams as $\mathcal{T}$ and $1-\mathcal{T}$, respectively. If the incident photon distribution equals $p_0(n)$, we obtain for the probability of finding $m$ photons transmitted through the beam

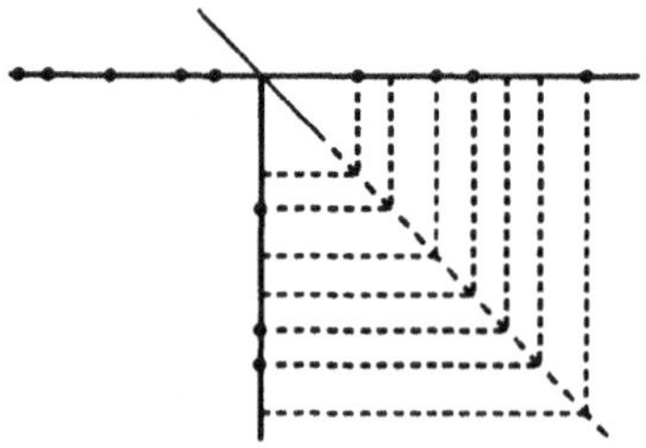

Figure 6.4: Random partitioning of photons by a beam splitter.

splitter

$$p(m) = \sum_{n=m}^{\infty} \binom{n}{m} \mathcal{T}^m (1-\mathcal{T})^{n-m} p_0(n), \qquad n = 0, 1, 2, \dots . \tag{6.52}$$

It is easily to show that for typical non-pathological distributions such as the Poisson distribution or the Bose-Einstein distribution $p(m)$ and $p_0(n)$ are of the same shape with the reduced mean number $\langle m \rangle = \mathcal{T}\langle n \rangle$. For the Fock state $|N\rangle$, $p_0(n) = \delta_{nN}$ and the form of the distribution is changed. Of course, equation (6.52) is also correct for the problem of the photon detection if $\mathcal{T}$ is interpreted as the photodetection efficiency $\eta$ (Sec. 2.9.6).

If a single photon is incident on the beam splitter, clearly anticorrelation of photons must be observed between outputs of photodetectors placed in both the beams. A general quantum theory of a lossless beamsplitter was developed by Campos, Saleh and Teich (1989). Assume two separate photons are in coincidence falling on the 50/50 beam splitter along two different entrance paths. Let $p(n, m)$ be the probability that $n$ and $m$ photons are in the one and the other outgoing beams, respectively. Then for bunched bosons (photons) $p(2, 0) = p(0, 2) = 1/2$, $p(1, 1) = 0$, whereas for antibunched fermions $p(2, 0) = p(0, 2) = 0$, $p(1, 1) = 1$ as a consequence of exclusion principle [see e.g. Loudon (1989a), Campos (1994)].

The quantum statistical properties of optical fields can be deduced from photoelectric measurements providing the photocount distribution, its factorial moments or correlation functions. As an effective tool for the detection of squeezed vacuum fluctuations homodyne detection can serve [Yuen and Shapiro (1978, 1980)]. Consider now this method in greater detail, as outlined in Fig. 6.5 (for physical basis, see Sec. 3.3.2). The input signal mode a represented by the annihilation operator $\hat{a}$ is superimposed on a local oscillator (reference) mode b having annihilation operator $\hat{b}$, using a lossless symmetric beam splitter M. Denoting its transmission and reflection amplitude coefficiens $t$ and $r$, respectively, then the output annihilation operators $\hat{c}$ and $\hat{d}$ are given by

$$\begin{bmatrix} \hat{c} \\ \hat{d} \end{bmatrix} = \begin{bmatrix} r & t \\ t & r \end{bmatrix} \begin{bmatrix} \hat{b} \\ \hat{a} \end{bmatrix} \tag{6.53}$$

and the unitarity of the transformation matrix (energy conservation law) requires that

$$|r|^2 + |t|^2 = 1 , \qquad rt^* + tr^* = 0 . \tag{6.54}$$

Two basic arrangements can be made to measure quantum effects in the input signal mode. In the ordinary homodyne detection $|r| \ll |t|$ and only the photodetector $P_1$ is useful. In the balanced homodyne detection $|r| = |t| = 1/\sqrt{2}$ and the output signal for processing is obtained as the difference

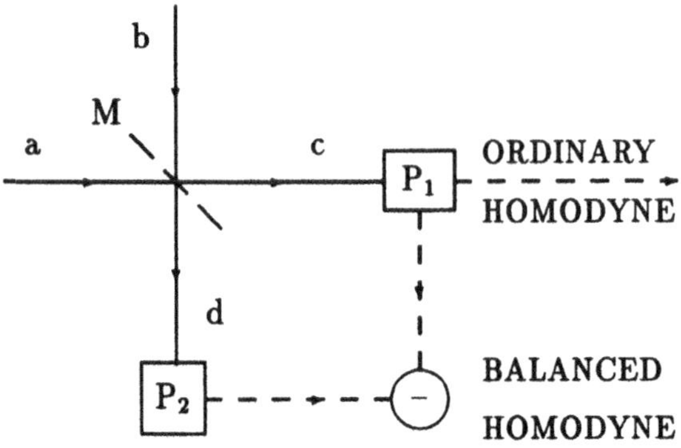


Figure 6.5: A scheme of homodyne detection; M is a beam splitter, $P_1$ and $P_2$ are photodetectors.

of the two photodetector currents. The second condition in (6.54) is fulfilled by $\arg(r) - \arg(t) = \pi/2$.

Considering ordinary homodyne detection we have for the mean number of detected photons

$$\langle \hat{c}^\dagger \hat{c} \rangle = |r|^2|\beta|^2 + |r||t||\beta|\langle \hat{Q}(\varphi)\rangle + |t|^2\langle \hat{a}^\dagger \hat{a}\rangle , \tag{6.55}$$

where the $\hat{Q}$-operator is defined as

$$\hat{Q}(\varphi) = \exp(i\varphi)\hat{a}^\dagger + \exp(-i\varphi)\hat{a} , \tag{6.56}$$

$\varphi = \arg(r) - \arg(t) + \arg(\beta) = \arg(\beta) + \pi/2$ and the local oscillator beam b has been assumed in the coherent state $|\beta\rangle$. Denoting $\alpha$ the complex amplitude of the signal mode a, we have for the ordinary homodyne detection $|r||\beta| \gg |t||\alpha|$. Under this assumption (6.55) provides the mean number of photons on the detector $P_1$,

$$\langle n_1 \rangle = |r|^2|\beta|^2 + |r||t||\beta|\langle \hat{Q}(\varphi)\rangle , \tag{6.57}$$

with the variance

$$\langle (\Delta n_1)^2 \rangle = |r|^2|\beta|^2 \left[ |r|^2 + |t|^2 \langle (\Delta \hat{Q})^2 \rangle \right] . \tag{6.58}$$

In this way the mean quadrature $\langle \hat{Q} \rangle$ and its variance $\langle (\Delta \hat{Q})^2 \rangle$ can be determined in the ordinary homodyne detection. Now with a real photodetector the quantum photoefficiency $\eta$ has to be taken into account. Using results of Sec. 2.9.6 and subtracting a constant level $|r|^2|\beta|^2$ in (6.57), we obtain for the mean number of counts relatively to this level

$$\langle m_1 \rangle = \eta|r||t||\beta|\langle \hat{Q}(\varphi) \rangle = 2\eta|r||t||\alpha||\beta|\cos(\varphi - \vartheta), \tag{6.59}$$

if the incident signal a is in the coherent state $|\alpha\rangle$ and $\vartheta = \arg(\alpha)$. The dominant expression for the correct variance $\langle (\Delta m_1)^2 \rangle$ is obtained on the basis of the Burgess variance expression (2.124) in Sec. 2.9.6

$$\langle (\Delta m_1)^2 \rangle = \eta|r|^2|\beta|^2 \left\{1 + \eta|t|^2 \left[\langle (\Delta \hat{Q})^2 \rangle - 1\right]\right\}. \tag{6.60}$$

In the coherent state $\langle (\Delta \hat{Q})^2 \rangle - 1 = 0$ and the first term in (6.60) represents the shot-noise corresponding to the Poisson statistics. For squeezed quantum fluctuations, $\langle (\Delta \hat{Q})^2 \rangle < 1$ and photocount statistics are sub-Poissonian, as discussed in greater detail at the end of Sec. 5.7. The deviation from Poisson statistics is dependent on the local oscillator phase through $\varphi$. This dependence is cancelled in the principal squeezing of vacuum fluctuations (Sec. 5.1). The measurement is performed in such a way that the input is first blocked to determine the shot-noise level. The variance of the signal is then determined in relation to this shot-noise level.

Consider now the balanced homodyne detection. The difference of the mean photon numbers equals from (6.55)

$$\langle n_{12} \rangle = \langle \hat{c}^\dagger \hat{c} - \hat{d}^\dagger \hat{d} \rangle = |\beta|\langle \hat{Q}(\varphi) \rangle. \tag{6.61}$$

The advantage is that balancing the two detectors, noise components of single detectors are cancelled and only the interference terms are conserved. The variance is also expressed in the simple form

$$\langle (\Delta n_{12})^2 \rangle = |\beta|^2 \langle (\Delta \hat{Q}(\varphi))^2 \rangle, \tag{6.62}$$

because $|\beta|^2 \gg |\alpha|^2$. We see that balancing also fully eliminates the local-oscillator noise. The corresponding mean number of counts and its variance are expressed as

$$\langle m_{12} \rangle = \eta|\beta|\langle \hat{Q}(\varphi) \rangle, \tag{6.63}$$

$$\langle (\Delta m_{12})^2 \rangle = \eta|\beta|^2 \left\{1 + \eta \left[\langle (\Delta \hat{Q}(\varphi))^2 \rangle - 1\right]\right\}. \tag{6.64}$$

The same arguments are correct as for the ordinary homodyne detection in (6.60).

For two-mode squeezed states heterodyne detection can be applied, when the frequency of the local oscillator is the central frequency between the frequencies of both the squeezed signals. However, in many cases the signal mode frequencies are sufficiently close or other simplifications can be used to

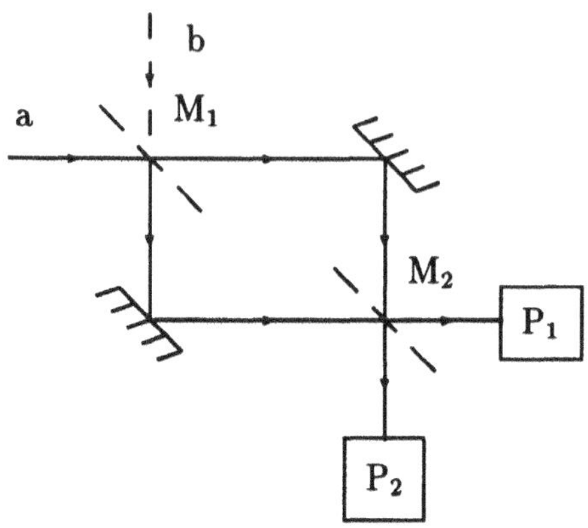


Figure 6.6: Mach-Zehnder interferometer; a (b) is a signal (reference) beam, $P_1$ and $P_2$ photodetectors, $M_1$ and $M_2$ are beam splitters.

obtain combined operators. Then heterodyne detection can be described in an analogous way as homodyne detection (Sec. 3.3.2).

Combining two beam splitters we have the Mach-Zehnder interferometer (Fig. 6.6) which is very important device for observations of quantum interferences. There are two entrance ports, if one (b) is empty, vacuum fluctuations are present there. If the splitter $M_2$ is missing, single photons are distinguished in branches of the interferometer and fringes at $P_1$ and $P_2$ cannot be observed because the corpuscular properties are observed (more precisely there is one maximum visibility, infinitely broad, white fringe on the detector representing the interference of the beam with itself, e.g. if the amplitude in one beam is $A = A/2 + A/2$, we obtain for the fringe intensity from the interference law $I = |A|^2 = |A|^2/4+|A|^2/4+2(|A|/2)(|A|/2)\cos[\arg(A)-\arg(A)]$). If the splitter $M_2$ is placed, interference fringes at $P_1$ and $P_2$ may be observed as a result of the superposition of both the beams 1 and 2 and wave properties are observed. Single photons will produce anticorrelated outputs from $P_1$ and $P_2$, i.e. the maxima and minima at $P_1$ and $P_2$ will be complementary. If two separate photons are in coincidence in each ports a and b in Fig. 6.6, two-photon interferometry is realized, whereas no interference can arise with fermions, as discussed above.

Raymer and Yang (1992) suggested to use optical parametric oscillators to realize a which-path interference experiment and examined the relation between the information gain and fringe visibility concluding that the information has an objective physical reality and, in relation to the complementarity principle, the information increase is related to the decrease of the fringe visibility and vice versa. An interesting analysis of the loss of coherence in the Young two-slit interferometer from the point of view of the which-path information, was presented by Tan and Walls (1993), showing that the loss of coherence may always be described in terms of a stochastic disturbance to the system due to coupling with the path detector. Decoherence effects in interferometric measurements were discussed by Bhandari (1992) and Namiki and Pascazio (1993). Hussain, Imoto and Loudon (1992) reexamined the single-photon nature of the second-order interference in the context of delayed-choice experiments.

## 6.5 Wave-particle duality

Wave-particle duality is well known fact experimentally verified for many sorts of particles. Traditional results refer to photons and electrons as basic particles of quantum electrodynamics. For broader reviews of wave-particle duality for electrons we refer the reader to works by Komrska (1971) and Matteucci (1990). Also classical experiments with neutrons provide nice confirmation of wave-particle duality, discussed by Badurek, Rauch and Tuppinger (1986) and Badurek, Rauch and Summhammer (1988), and reviewed by Gähler and Zeilinger (1991). In the last few years great attention has been devoted to atom and ion interferometers giving new frontiers in quantum optics. The extremely small wavelengths associated with atoms of about $10^4$ times shorter than those of visible light make it possible to increase substantionally the accuracy of measurements. The first two experiments [Carnal and Mlynek (1991), Keith, Ekstrom, Turchette and Pritchard (1991)] used nanofabricated structures to realize the Young's experiment for helium atoms and to build an atom interferometer for sodium atoms. In quite recent experiment Eichmann, Bergquist, Bollinger, Gilligan, Itano, Wineland and Raizen (1993) observed the interference effects in light scattered from two trapped atoms. The visibility of fringes can be explained in the framework of Bragg scattering and on the basis of which-path information for scattered photons. They demonstrated the increase of fringe spacing with decreasing ion separation varied from 5.4 to 3.7 $\mu$m. The fringe visibility was found highest for forward scattering directions and it decreases with increasing inclination angle. In the case that the initial and final states of ions after scattering are the same, one cannot distinguish from which ion a given photon came to the screen of observation and therefore the interference pattern is observable. On the other hand, if the final state differs from the initial one, one can, in principle, to distinguish the scattering event on the corresponding ion and to obtain the which-path information. Consequently, there is no interference in scattered light from both the ions. In this sense, the existence of interference fringes indicates wave-like behaviour, whereas the absence of fringes, consistent with a single photon trajectory beginning at the source, intersecting one of the ions, and going to the detector, indicates particle-like behaviour. These two ideally complementary pictures are essentially classical and they are contained in the quantum interpretation in probabilistic terms related to operator representations of observables, or in terms of Feynman path amplitudes the presence or absence of interference pattern depends on whether or not there are more or one possible paths from the initial to the final states.

The most interesting verification of the wave-particle duality for single photons has been presented by Aspect, Grangier and Roger (1989) (for greater details, see Sec. 8.1). In these papers a summary of low intensity interference experiments (the photon flux from $10^6$ to $10^2$ photons per second) is given in a table. These authors clearly experimentally demonstrated that classical interference patterns in the Mach-Zehnder interferometer on two photodetectors $P_1$ and $P_2$ (Fig. 6.6) are cumulative processes of single photon interferences with high visibility ($98.7 \pm 0.5$ %) and responses of $P_1$ and $P_2$ are anticorrelated, i.e. where there are maxima of fringes at $P_1$ minima of fringes must be at $P_2$ and vice versa, confirming the particle properties of photons. These experiments are direct illustration of the wave-particle duality applied to a single photon. The wave properties and particle properties are related as discussed in Sec. 5.5. Also the Einstein formula (2.123) represents the higher-order basis for the wave-particle duality (Sec. 5.5).

The wave-particle duality is also confirmed by the experiments with independent light beams, as mentioned in Secs. 6.1 and 6.2. For a review of original experiments we refer the reader to the review by Mandel and Wolf (1965) or to the book by Peřina (1991). However, the experiments by Pfleegor and Mandel (1967, 1968) and Radloff (1968, 1971) are to be particularly mentioned.

In the experiment of Pfleegor and Mandel interference fringes were observed on a cross-section of two beams from independent sources (lasers) under the conditions where light intensity was so low that the main time interval between photons was large compared with their transit time through the measuring device, i.e. it was a high probability that one photon was absorbed before the next one was emitted by one or another laser source. Since the intensities were very low, a photon correlation technique was required to observe the interference fringes. The interference pattern was received on a stack of thin glass plates, each of which had a thickness corresponding to about a half-fringe width. The plates were cut and arranged so that any light falling on the odd-numbered plates was fed to one photomultiplier, while light falling on the even-numbered plates was fed to the other photomultiplier. If the half-fringe spacing coincides with the plate thickness, and for example the fringe maxima fall on the odd-numbered plates, one photodetector will register nearly all the photons and the other will register no photons. The position of the fringe maxima is unpredictable and random, but if the number of photons registered by one photodetector increases, the number registered by the other must decrease, provided that the fringe spacing is correct for the plates. Thus there must be anticorrelation between the numbers of counts from the two photodetectors and such anticorrelation was really observed. Since the experimental conditions were arranged so that one photon was observed before the next one was emitted, the wave-particle duality for single photons is applied. Such experiments with independent light beams show that any localization of a photon in space-time automatically rules out the possibility of knowing its momentum, as a consequence of the uncertainty principle. Therefore one cannot say to which beam a given photon belongs — each photon is to be considered as being partly in both the beams and interfering with itself. However, also different photons can interfere producing partially coherent fields, as discussed in Sec. 6.1 and in great detail by Paul (1986).

The complementarity of wave and particle descriptions from the point of view of quantum optics was discussed by Garuccio and Lepore (1992). They have shown, in agreement with quantum theory and quantum optics description, that setting two Mach-Zehnder interferometers in the cascade, and adopting the frequency down conversion process for generation of two-photon correlated states, one can obtain the which-path information in the second interferometer without cancelling the correlation intensity interference pattern (Fig. 6.7). On the other hand, considering an arrangement as given in Fig. 6.8, the photon trajectories are displayed and the interference effect is smoothed out.

Scully and Walther (1989) suggested to use micromaser cavities as detectors for which-path information. Two neutron or atom beams are passing through the micromaser cavities. Let the state of the quantized field in the $j$-th micromaser be initially $|\phi_j^{(i)}\rangle$, $j = 1, 2$. If $\psi_j$, $j = 1, 2$ are wave functions of the corresponding neutron beams and denoting $|\phi_j^{(f)}\rangle$, $j = 1, 2$ the state of the maser field after the interaction, the resulting interferometric intensity equals

$$I = |\psi_1|^2 + |\psi_2|^2 + \left[\psi_1^* \psi_2 \langle \phi_1^{(f)}, \phi_2^{(i)} | \phi_1^{(i)}, \phi_2^{(f)} \rangle + \text{c.c.}\right] , \tag{6.65}$$

where the terms in the brackets describe the interference effects. If the fields in the maser cavities are prepared in the coherent states, i.e. $|\phi_j^{(i)}\rangle \equiv |\alpha_j\rangle$, then to a very good approximation we have

$$\langle \phi_1^{(f)}, \phi_2^{(i)} | \phi_1^{(i)}, \phi_2^{(f)} \rangle \approx \langle \alpha_1, \alpha_2 | \alpha_1, \alpha_2 \rangle = 1 \tag{6.66}$$

and the interaction in the maser cavities does not change the states of the wave functions and the interference pattern is observable. This is not surprising

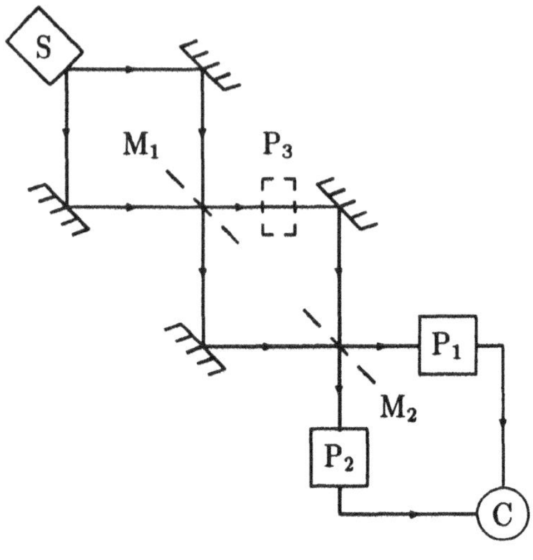


Figure 6.7: A cascade of two Mach-Zehnder interferometers; S is down converter, $M_1$ and $M_2$ are splitters, $P_1$–$P_3$ photodetectors, C is a coincidence device; the placing of the detector $P_3$ does not change the correlation intensity interference pattern.

because no which-path information is provided. On the other hand, if the masers are prepared in the Fock states $|n_j\rangle$, $j = 1, 2$, the interference terms are determined by the state vector product

$$\langle \phi_1^{(f)}, \phi_2^{(i)} | \phi_1^{(i)}, \phi_2^{(f)} \rangle = \langle n_1 + 1 | n_1 \rangle \langle n_2 | n_2 + 1 \rangle = 0 \tag{6.67}$$

and the interference terms are vanishing because now we have which-path information in the micromaser cavities. For instance, if we find that in the first maser cavity $(n_1+1)$ photons are present, we know that the corresponding path was followed.

We can say that we have here a new point of view on the wave-particle duality and observation in quantum theory. The loss of coherence is a result of the establishment and loss of quantum correlation between the system and the micromaser detectors and it is not caused by the random phase of both the beams, as is usually applied. This means that which-path information is enough to smooth out the interference pattern. This approach can be generalized [Kwiat, Steinberg and Chiao (1994)] and it leads to the conclusion that the loss of coherence and wave-like behaviour is not necessarily caused by the uncertainty relations, but it may be caused by a quantum entanglement of the interfering system with a measuring apparatus.

Instead of beams of neutrons one can use Rydberg atomic beams and quantum beats may be observed [Scully and Walther (1989)]. States of Rb-

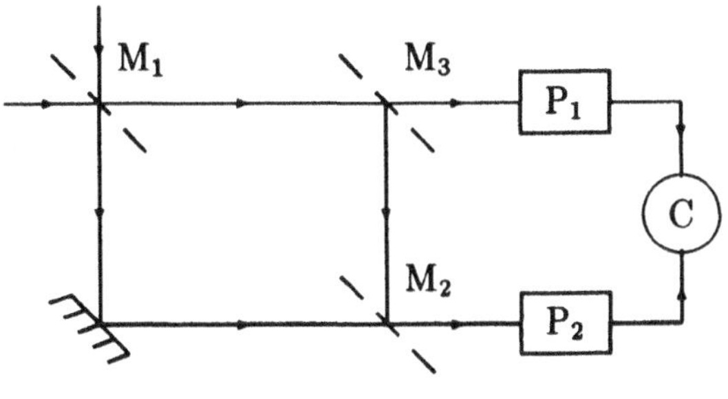


Figure 6.8: A detection arrangement providing which-path information and no coincidences; $M_1$–$M_3$ are splitters, $P_1$ and $P_2$ photodetectors and C is a coincidence device.

atoms with n=63 are suitable. The conclusions are similar as above and in this way new tests of wave-particle duality and complementarity are possible [Scully, Englert anf Walther (1991)].

A neutron interferometry realization of which-path experiment was suggested by Rauch and Vigier (1990). Diffraction and interference effects of atoms were discussed by Storey, Collett and Walls (1992).

Interesting considerations about the wave-particle duality can be performed based on spontaneous emissions of photons in nearly opposite directions from a molecule producing its recoil. Lai and Diels (1992) observed single-photon interference fringes created by spontaneous emissions in nearly opposite directions from a thin layer of Rhodamine 66. It was found that the probability amplitude of spontaneous emission from an atomic system is highly coherent in all directions, provided that no recoil of the atomic system can be measured. If the recoil is measurable, the interference pattern is ruled out. This points out the wave-particle duality of a photon in the emission.

New aspects about the wave-particle duality can be derived using an amplification process for beams [Swanson and Carlsten (1993)]. As we have seen as the evidence for the wave-like behaviour the fringe visibility can serve, whereas anticorrelation between outputs of two photodetectors exposed to both the branches of an interferometer is a measure of particle-like behaviour. Substituting amplifiers in the arms of the Mach-Zehnder interferometer in Fig. 6.6, Swanson and Carlsten (1993) demonstrated that, as a result of amplification, considering the average fringe visibility and the average anticorrelation between two amplifiers in the interferometer, the amplified signals from single-photon inputs provide evidence, in consistence with the input signals, for extremes of both the wave-like and particle-like characteristics. Degradation of the visibility and of the anticorrelation is a result of the quantum noise added during amplification and not of a trade-off between particle-like and wave-like behaviour. Thus the amplification provides simultaneous observation of non-ideal wave and particle behaviour.

Single-photon interferences were experimentally as well as theoretically analyzed in recent papers by Hariharan, Brown and Sanders (1993, 1993a), Hariharan, Roy, Robinson and O'Byrne (1993) and Hariharan, Fujima, Brown and Sanders (1993) including a discussion of the geometric Berry-Pancharatnam phase. They concluded that single-photon interferences are appropriate to explain the experimental results and they are to be attributed to the formation of quantum superposition states. The role of the vacuum state for interference of independent light beams was considered by Agarwal and Hariharan (1993).

## 6.6 Methods involving frequency down conversion and feedbacks

Frequency down conversion is one of the most important nonlinear optical processes because it provides the quantum correlated sub-frequency photons (entangled states) so useful for experimental verifications of quantum theory. Some elements of the degenerate sub-frequency process were presented in Sec. 5.4 based on the effective hamiltonian (5.31) describing generation of two correlated photons of the frequency $\omega$ if the pumping field has the frequency $2\omega$. For more detailed treatment including effects of dissipations we refer the reader to Sec. 10.1 of the book by Peřina (1991).

The main attention, from the point of view of verifications of quantum theory, is to be devoted to nondegenerate optical parametric process as specified by the effective hamiltonian

$$\begin{aligned} \hat{H} &= \hbar\omega_1\left(\hat{a}_1^\dagger\hat{a}_1+\frac{1}{2}\right)+\hbar\omega_2\left(\hat{a}_2^\dagger\hat{a}_2+\frac{1}{2}\right) \\ &\quad -\hbar\left[g\hat{a}_1\hat{a}_2\exp(i\omega t-i\varphi)+\text{h.c.}\right], \end{aligned} \tag{6.68}$$

considering two quantum modes with the annihilation (creation) operators $\hat{a}_j$ ($\hat{a}_j^\dagger$), $j=1,2$ coupled to the strong sum-frequency mode of the frequency $\omega=\omega_1+\omega_2$ with the coupling constant $g$ and the pumping phase $\varphi$. If the phase matching for wave vectors, $\mathbf{k}_3=\mathbf{k}_1+\mathbf{k}_2$, is assumed, the coupling constant $g$ is real and we can write down the corresponding Heisenberg-Langevin equations

$$\begin{aligned} \frac{d\hat{a}_1}{dt} &= -i\omega_1\hat{a}_1-\frac{\gamma_1}{2}\hat{a}_1+ig\hat{a}_2^\dagger\exp(-i\omega t+i\varphi)+\hat{L}_1, \\ \frac{d\hat{a}_2}{dt} &= -i\omega_2\hat{a}_2-\frac{\gamma_2}{2}\hat{a}_2+ig\hat{a}_1^\dagger\exp(-i\omega t+i\varphi)+\hat{L}_2, \end{aligned} \tag{6.69}$$

where the Langevin forces $\hat{L}_j$, $j = 1, 2$ are derived in Sec. 4.6 and $\gamma_j$ are the damping constants. Using the coherent-state technique we can derive the generalized Fokker-Planck equation by the standard way (Sec. 4.7) for the quasidistribution $\Phi_\mathcal{A}(\alpha_1, \alpha_2, t)$ related to antinormal ordering of field operators

$$\begin{aligned}\frac{\partial \Phi_\mathcal{A}}{\partial t} &= \left\{\sum_{j=1}^{2}\left[\left(i\omega_j + \frac{\gamma_j}{2}\right)\frac{\partial}{\partial \alpha_j}\alpha_j + \text{c.c.} + \gamma_j\left(\langle n_{Rj}\rangle + 1\right)\frac{\partial^2}{\partial \alpha_j \partial \alpha_j^*}\right]\right. \\ &\quad \left. + \left[ig\exp(i\omega t - i\varphi)\left(\alpha_1\frac{\partial}{\partial \alpha_2^*} + \alpha_2\frac{\partial}{\partial \alpha_1^*} + \frac{\partial^2}{\partial \alpha_1^* \partial \alpha_2^*}\right) + \text{c.c.}\right]\right\}\Phi_\mathcal{A}. \end{aligned} \tag{6.70}$$

Its solution can be obtained in the form of the generalized superposition of coherent fields and quantum noise (Sec. 4.8, for details, see Peřina (1991), Sec. 10.1.2).

Neglecting dissipations for simplicity, the equations of motion (6.69) can be solved in the form

$$\begin{aligned}\hat{a}_1(t) &= \exp(-i\omega_1 t)\left[\hat{a}_1(0)\cosh(gt) + i\hat{a}_2^\dagger(0)\exp(i\varphi)\sinh(gt)\right], \\ \hat{a}_2(t) &= \exp(-i\omega_2 t)\left[\hat{a}_2(0)\cosh(gt) + i\hat{a}_1^\dagger(0)\exp(i\varphi)\sinh(gt)\right]. \end{aligned} \tag{6.71}$$

In this case, in analogy to the degenerate case presented in Sec. 5.4, the noise functions $B_1$, $B_2$, $C_1$, $C_2$, $D_{12}$ and $\bar{D}_{12}$ as given in (4.180) of Sec. 4.8, are obtained as

$$\begin{aligned}B_1(t) &= B_2(t) = B_1(t)B_2(t) - |D_{12}(t)|^2 = \cosh^2(gt), \\ C_1(t) &= C_2(t) = \bar{D}_{12}(t) = 0, \\ D_{12}(t) &= \frac{i}{2}\sinh(2gt)\exp(-i\omega t + i\varphi). \end{aligned} \tag{6.72}$$

The variance of the integrated intensity for the combined mode (1,2) is obtained in the form [Mišta and Peřina (1977)]

$$\begin{aligned}\langle(\Delta W)^2\rangle_\mathcal{N} &= \langle(\Delta W_1)^2\rangle_\mathcal{N} + \langle(\Delta W_2)^2\rangle_\mathcal{N} + 2\langle\Delta W_1 \Delta W_2\rangle_\mathcal{N} \\ &= 2\sinh(gt)\left[\frac{\sinh(3gt) - \sinh(gt)}{2} + \sinh(3gt)\left(|\xi_1|^2 + |\xi_2|^2\right)\right. \\ &\quad \left. + 2\cosh(3gt)|\xi_1||\xi_2|\sin(\vartheta_1 + \vartheta_2 - \varphi)\right], \end{aligned} \tag{6.73}$$

where $\vartheta_j$ are the phases of the initial coherent amplitudes $\xi_j$, $j = 1, 2$, i.e. $\hat{a}_j(0)|\xi_j\rangle = \xi_j|\xi_j\rangle$. The two-mode variance is defined as $\langle(\Delta W)^2\rangle_\mathcal{N} =$

$\langle : \hat{n}^2(t) : \rangle - \langle \hat{n}(t) \rangle^2$, where $\hat{n}(t) = \hat{a}_1^\dagger(t)\hat{a}_1(t) + \hat{a}_2^\dagger(t)\hat{a}_2(t)$. Photon antibunching as a purely quantum phenomenon and equivalently sub-Poissonian photon statistics are obtained at the maximum level when $\vartheta_1 + \vartheta_2 - \varphi = -\pi/2$.

The two-mode principal squeezing defined in (5.12) is obtained as

$$\lambda_{12} = 2\left[1 + (B_1 - 1) + (B_2 - 1) - 2\mathrm{Re}\left(\bar{D}_{12}\right) - |C_1 + C_2 + 2D_{12}|\right]$$
$$= 2\exp(-2gt)\,, \tag{6.74a}$$

while

$$\left\langle \left( \Delta \begin{matrix} \hat{Q}_{12} \\ \hat{P}_{12} \end{matrix} \right)^2 \right\rangle = 2 \begin{pmatrix} \exp(-2gt) \\ \exp(2gt) \end{pmatrix}, \qquad \varphi = \pi/2\,. \tag{6.74b}$$

The two-mode quadrature operators are defined as

$$\begin{aligned} \hat{Q}_{12} &= \hat{a}_1 + \hat{a}_2 + \hat{a}_1^\dagger + \hat{a}_2^\dagger\,, \\ \hat{P}_{12} &= -i\left(\hat{a}_1 + \hat{a}_2 - \hat{a}_1^\dagger - \hat{a}_2^\dagger\right)\,. \end{aligned} \tag{6.75}$$

On the other hand, if one of the sub-frequency modes is the strong, classical pumping mode, then the effective hamiltonian has the form

$$\begin{aligned} \hat{H} &= \hbar\omega_1\left(\hat{a}_1^\dagger\hat{a}_1 + \frac{1}{2}\right) + \hbar\omega_2\left(\hat{a}_2^\dagger\hat{a}_2 + \frac{1}{2}\right) \\ &\quad + \hbar\left[g\hat{a}_1\hat{a}_2^\dagger\exp(i\omega t - i\varphi) + \text{h.c.}\right]\,, \end{aligned} \tag{6.76}$$

assuming the phase matching ($\mathbf{k} + \mathbf{k}_2 = \mathbf{k}_1$) so that $g$ is real, $\omega = \omega_1 - \omega_2$. In this case the solutions of the Heisenberg equations

$$\begin{aligned} \frac{d\hat{a}_1}{dt} &= -i\omega_1\hat{a}_1 - ig\hat{a}_2\exp(-i\omega t + i\varphi)\,, \\ \frac{d\hat{a}_2}{dt} &= -i\omega_2\hat{a}_2 - ig\hat{a}_1\exp(i\omega t - i\varphi) \end{aligned} \tag{6.77}$$

are oscillatory solutions

$$\begin{aligned} \hat{a}_1(t) &= \exp(-i\omega_1 t)\left[\hat{a}_1(0)\cos(gt) - i\hat{a}_2(0)\exp(i\varphi)\sin(gt)\right]\,, \\ \hat{a}_2(t) &= \exp(-i\omega_2 t)\left[\hat{a}_2(0)\cos(gt) - i\hat{a}_1(0)\exp(-i\varphi)\sin(gt)\right]\,. \end{aligned} \tag{6.78}$$

These equations do not include any creation operators and therefore the nonlinear dynamics of the process cannot change the initial photon statistics. This means that e.g. the signal of the frequency $\omega_1$ is down-converted to radiation of frequency $\omega_2 = \omega_1 - \omega$ (or the signal of the frequency $\omega_2$ is up-converted to radiation of the frequency $\omega_1 = \omega + \omega_2$) without any change of the photon statistics. This includes the frequency conversion and tuning of squeezed light [Huang and Kumar (1992), Kim and Kumar (1992)].

If all three modes are quantum dynamical and the sum-frequency mode is externally and classically pumped with the power $p_3$, we can describe this interaction by the effective hamiltonian for radiation

$$\hat{H} = \sum_{j=1}^{3} \hbar\omega_j \left(\hat{a}_j^\dagger \hat{a}_j + \frac{1}{2}\right) - \hbar \left(g\hat{a}_1\hat{a}_2\hat{a}_3^\dagger + \text{h.c.}\right) + i\hbar \left[p_3\hat{a}_3^\dagger \exp(-i\omega_3 t) - \text{h.c.}\right] \tag{6.79}$$

and the Heisenberg-Langevin equations are of the form

$$\begin{aligned} \frac{d\hat{a}_1}{dt} &= -\left(i\omega_1 + \frac{\gamma_1}{2}\right)\hat{a}_1 + ig^*\hat{a}_2^\dagger\hat{a}_3 + \hat{L}_1', \\ \frac{d\hat{a}_2}{dt} &= -\left(i\omega_2 + \frac{\gamma_2}{2}\right)\hat{a}_2 + ig^*\hat{a}_1^\dagger\hat{a}_3 + \hat{L}_2', \\ \frac{d\hat{a}_3}{dt} &= -\left(i\omega_3 + \frac{\gamma_3}{2}\right)\hat{a}_3 + ig\hat{a}_1\hat{a}_2 + \hat{L}_3' + p_3\exp(-i\omega_3 t), \end{aligned} \tag{6.80}$$

which can be rewritten in the interaction picture $[\hat{a}_j = \hat{A}_j \exp(-i\omega_j t)$, $\hat{L}_j = \hat{L}_j' \exp(i\omega_j t)]$ to the form

$$\begin{aligned} \frac{d\hat{A}_1}{dt} &= -\frac{\gamma_1}{2}\hat{A}_1 + ig^*\hat{A}_2^\dagger\hat{A}_3 + \hat{L}_1, \\ \frac{d\hat{A}_2}{dt} &= -\frac{\gamma_2}{2}\hat{A}_2 + ig^*\hat{A}_1^\dagger\hat{A}_3 + \hat{L}_2, \\ \frac{d\hat{A}_3}{dt} &= -\frac{\gamma_3}{2}\hat{A}_3 + ig\hat{A}_1\hat{A}_2 + \hat{L}_3 + p_3. \end{aligned} \tag{6.81}$$

The corresponding generalized Fokker-Planck equation can be derived in the form

$$\begin{aligned} \frac{\partial\Phi_{\mathcal{A}}}{\partial t} = \Bigg\{ \sum_{j=1}^{3} \left[ \left(i\omega_j + \frac{\gamma_j}{2}\right) \frac{\partial}{\partial\alpha_j}\alpha_j + \text{c.c.} + \gamma_j\left(\langle n_{Rj}\rangle + 1\right)\frac{\partial^2}{\partial\alpha_j\partial\alpha_j^*} \right] \\ + \left[ ig^* \left( \alpha_1^*\alpha_2^*\frac{\partial}{\partial\alpha_3^*} - \alpha_1^*\alpha_3\frac{\partial}{\partial\alpha_2} - \alpha_2^*\alpha_3\frac{\partial}{\partial\alpha_1} - \alpha_3\frac{\partial^2}{\partial\alpha_1\partial\alpha_2} \right) + \text{c.c.} \right] \Bigg\} \Phi_{\mathcal{A}}. \end{aligned} \tag{6.82}$$

Analytically it is possible to obtain all quantum statistical characteristics such as quantum characteristic function, quasidistribution, generating function, factorial moments, photocount distribution and quadrature variances (principal squeeze variance) up to $(|g|t)^2$, in special cases to higher orders in $|g|t$, or a symbolic computational method can be adopted to include any powers [Bajer and Lisoněk (1991), Bajer and Peřina (1992)]. Reviews can be found in the book by Peřina (1991) (Sec. 10.2) and in the chapter by

Peřina, Bajer, Peřinová and Hradil (1993). Particularly this permits to give tendencies for regimes with anticorrelations, antibunching of photons and sub-Poissonian behaviour of photons and to establish approximate expressions for quadrature squeezing of vacuum fluctuations [Peřina (1991), Sec. 10.2].

Consider for a moment the system of equations (6.81) as classical one with neglected Langevin noise and find a stationary solution. Denoting $\varrho_j = |A_j|$ and $\varphi_j = \arg(A_j)$, $j = 1,2,3$ we obtain from the stationary conditions $dA_j/dt = 0$ the following stationary solutions [Peřina, Jr. (1993a)]

i)

$$\varrho_{30} = \frac{2|p_3|}{\gamma_3}, \quad \varrho_{10} = \varrho_{20} = 0,$$
$$\varphi_{30} = \arg(p_3) + 2\pi l, \qquad l \text{ integer},$$
$$\varphi_{10} \text{ and } \varphi_{20} \text{ undetermined}, \tag{6.83a}$$

and

ii)

$$\varrho_{10} = \left[\frac{|p_3| - p_{\text{th}}}{|g|}\sqrt{\frac{\gamma_2}{\gamma_1}}\right]^{1/2}, \quad \varrho_{20} = \left[\frac{|p_3| - p_{\text{th}}}{|g|}\sqrt{\frac{\gamma_1}{\gamma_2}}\right]^{1/2},$$
$$\varrho_{30} = \frac{\sqrt{\gamma_1\gamma_2}}{2|g|}, \quad \varphi_{30} = \arg(p_3) + 2\pi l, \qquad l \text{ integer},$$
$$\varphi_{10} + \varphi_{20} - \arg(p_3) + \arg(g) = \frac{\pi}{2} + 2\pi l, \qquad l \text{ integer},$$
$$|p_3| > p_{\text{th}} = \frac{\gamma_3\sqrt{\gamma_1\gamma_2}}{2|g|}; \tag{6.83b}$$

$p_{\text{th}}$ denotes the threshold pumping amplitude.

The linear stability analysis leads to conclusions that the stationary point (6.83a) is stable provided that $|p_3| < p_{\text{th}}$. In the case of the stationary point (6.83b) we can substitute $A_j = A_{j0} + \delta A_j$ and we obtain in the polar coordinates neglecting Langevin forces and higher-order terms

$$\frac{d}{dt}\begin{bmatrix} \delta\varrho_1 \\ \delta\varrho_2 \\ \delta\varrho_3 \end{bmatrix} = \begin{bmatrix} -\gamma_1 & |g|\varrho_{30} & |g|\varrho_{20} \\ |g|\varrho_{30} & -\gamma_2 & |g|\varrho_{10} \\ -|g|\varrho_{20} & -|g|\varrho_{10} & -\gamma_3 \end{bmatrix}\begin{bmatrix} \delta\varrho_1 \\ \delta\varrho_2 \\ \delta\varrho_3 \end{bmatrix}, \tag{6.84a}$$

$$\frac{d}{dt}\begin{bmatrix} \delta\varphi_1 \\ \delta\varphi_2 \\ \delta\varphi_3 \end{bmatrix} = \begin{bmatrix} -\gamma_1 & -\gamma_1 & \gamma_1 \\ -\gamma_2 & -\gamma_2 & \gamma_2 \\ -|g|\frac{\varrho_{10}\varrho_{20}}{\varrho_{30}} & -|g|\frac{\varrho_{10}\varrho_{20}}{\varrho_{30}} & -\gamma_3 \end{bmatrix}\begin{bmatrix} \delta\varphi_1 \\ \delta\varphi_2 \\ \delta\varphi_3 \end{bmatrix}. \tag{6.84b}$$

The application of the Hurwitz criterion to (6.84a) gives that real parts of all eigenvalues are negative for $|p_3| > p_{\text{th}}$, so that the amplitudes are stable. For (6.84b) and $|p_3| > p_{\text{th}}$ one eigenvalue is zero and the real parts of two remaining eigenvalues are negative. The problem with one zero eigenvalue is not important because it is related to the phase diffusion process. For further discussions we refer the reader to the review by Reynaud, Heidmann, Giacobino and Fabre (1992) and the paper by Martinez-Linares and Garcia-Fernandez (1994).

The linear operator corrections to the stationary solution are obtained similarly solving the equations for the operator corrections $\delta\hat{A}_j$, $\hat{A}_j = A_{jo} + \delta\hat{A}_j$, in the linear approximation,

$$\frac{d\delta\hat{A}}{dt} = \hat{M}\delta\hat{A} + \hat{L}\,, \tag{6.85}$$

where

$$\delta\hat{A} = \begin{bmatrix} \delta\hat{A}_1 \\ \delta\hat{A}_1^\dagger \\ \delta\hat{A}_2 \\ \delta\hat{A}_2^\dagger \\ \delta\hat{A}_3 \\ \delta\hat{A}_3^\dagger \end{bmatrix}, \quad \delta\hat{L} = \begin{bmatrix} \hat{L}_1 \\ \hat{L}_1^\dagger \\ \hat{L}_2 \\ \hat{L}_2^\dagger \\ \hat{L}_3 \\ \hat{L}_3^\dagger \end{bmatrix}, \tag{6.86a}$$

$$\hat{M} = \begin{bmatrix} -\gamma_1 & 0 & 0 & ig^*A_{30} & ig^*A_{20}^* & 0 \\ 0 & -\gamma_1 & -igA_{30}^* & 0 & 0 & -igA_{20} \\ 0 & ig^*A_{30} & -\gamma_2 & 0 & ig^*A_{10}^* & 0 \\ -igA_{30}^* & 0 & 0 & -\gamma_2 & 0 & -igA_{10} \\ igA_{20} & 0 & igA_{10} & 0 & -\gamma_3 & 0 \\ 0 & -ig^*A_{20}^* & 0 & -ig^*A_{10}^* & 0 & -\gamma_3 \end{bmatrix}, \tag{6.86b}$$

which can be solved in the form

$$\delta\hat{A}_j(t) = \sum_{l=1}^{3}\left[U_{jl}(t)\delta\hat{A}_l(0) + V_{jl}(t)\delta\hat{A}_l^\dagger(0)\right] + \hat{F}_j(t)\,, \tag{6.87}$$

where the functions $U_{jl}(t)$ and $V_{jl}(t)$ are determined by the nonlinear dynamics and $\hat{F}_j(t)$ are related to the Langevin forces. It is obvious that the statistical properties of this interaction are described in this approximation by the generalized superposition of coherent fields and quantum noise, as treated in Sec. 4.8.

We have simplified our treatment to three single modes for simplicity. However, a general interaction of three quantum fields with arbitrary spectral composition can be given [Hong and Mandel (1985a)] providing space-time relations.

It is advantageous for generation of nonclassical light to adopt the four-wave mixing described by the third-order susceptibility $\chi^{(3)}$. Two waves of frequencies $\omega_1$ and $\omega_2$ and with wave vectors $\mathbf{k}_1$ and $\mathbf{k}_2$ represent signal waves interacting with the nonlinear medium pumped by two waves of frequencies $\omega_3$ and $\omega_4$ with wave vectors $\mathbf{k}_3$ and $\mathbf{k}_4$. The scattering can be arranged as forward or backward with counterpropagating beams. In the latter case the counterpropagating beams $\mathbf{k}_3(\omega_3)$ and $\mathbf{k}_4(\omega_4)$ are pumping beams and the counterpropagating wave $\mathbf{k}_2(\omega_2)$ is phase conjugated to the incident signal wave $\mathbf{k}_1(\omega_1)$. From the momentum and energy conservation laws it holds that $\mathbf{k}_1+\mathbf{k}_2=\mathbf{k}_3+\mathbf{k}_4$ and $\omega_1+\omega_2=\omega_3+\omega_4$, respectively. Usually the degenerate process is applied fulfilling $\omega_1=\omega_2=\omega_3=\omega_4$. Such a four-wave mixing process was the first to be used for generation of squeezed light, whereas with the optical parametric process the highest squeezing was obtained (Chap. 8).

If the propagation effects are to be considered along the $z$-axis, the Heisenberg equations are derived from the momentum operator

$$\begin{aligned}\hat{G} &= \sum_{j=1}^{4}\hbar k_j\hat{a}_j^\dagger\hat{a}_j+\hbar g\hat{a}_1\hat{a}_2\hat{a}_3^\dagger\hat{a}_4^\dagger+\text{h.c.}\\ &\quad+\left[i\hbar p_3^*\exp(-ik_3z)\hat{a}_3+i\hbar p_4^*\exp(-ik_4z)\hat{a}_4+\text{h.c.}\right], \qquad (6.88)\end{aligned}$$

provided that classical pumping powers of modes 3 and 4 are $p_3$ and $p_4$, respectively. In the interaction picture the Heisenberg-Langevin equations are

$$\begin{aligned}\frac{d\hat{A}_1}{dz} &= -\frac{\gamma_1}{2}\hat{A}_1+ig^*\hat{A}_2^\dagger\hat{A}_3\hat{A}_4+\hat{L}_1,\\ \frac{d\hat{A}_2}{dz} &= -\frac{\gamma_2}{2}\hat{A}_2+ig^*\hat{A}_1^\dagger\hat{A}_3\hat{A}_4+\hat{L}_2,\\ \frac{d\hat{A}_3}{dz} &= -\frac{\gamma_3}{2}\hat{A}_3+ig\hat{A}_1\hat{A}_2\hat{A}_4^\dagger+\hat{L}_3+p_3,\\ \frac{d\hat{A}_4}{dz} &= -\frac{\gamma_4}{2}\hat{A}_4+ig\hat{A}_1\hat{A}_2\hat{A}_3^\dagger+\hat{L}_4+p_4, \qquad (6.89)\end{aligned}$$

with the same meaning of the symbols as above and taking into account the Heisenberg equation for the propagation in the $z$-direction

$$i\hbar\frac{d\hat{a}_j}{dz}=[\hat{G},\hat{a}_j]. \qquad (6.90)$$

In the case that dispersion can be neglected both the formulations in the $t$- and $z$-domains are equivalent and $z=vt$, $v$ being the speed of light in the medium. If moreover the pumping waves $\hat{A}_3$ and $\hat{A}_4$ are strong and classical, the three- and four-wave mixing are described in the same way and both the descriptions are identical using the corresponding denotations. In that case the results for the optical parametric process with classical pumping derived above are equally valid for the four-wave mixing with strong classical pumping. The linear stability analysis can be performed for four-wave mixing [Peřina and Křepelka (1992)] similarly as for the optical parametric process. Under the assumptions that $\arg(g) = \pi/2$ and $\varphi_1 + \varphi_2 - \varphi_3 - \varphi_4 = 2\pi n$, $n = 0, 1, \ldots$, $\varphi_j = \arg(A_j)$, $j = 1, 2, 3, 4$, we arrive at the classical stationary solution

$$\gamma_1 \varrho_{10}^2 = \gamma_2 \varrho_{20}^2 , \qquad \varrho_{30}^2 \varrho_{40}^2 = \frac{\gamma_1 \gamma_2}{4|g|^2} , \tag{6.91}$$

where $\varrho_{jo} = |A_{jo}|$ and $A_{jo}$, $j = 1, 2, 3, 4$ are stationary amplitudes. If additionally $\gamma_1 = \gamma_2 = \gamma_3 = \gamma_4 = \gamma$ and both the pump modes are equally pumped, $p_3 = p_4 = p$, we can solve the system of equations for the stationary solution giving the same solution for forward and backward mixing in the form

$$\varrho_{10} = \varrho_{20} = \left[ \frac{|p|}{|g| \varrho_{30}} - \varrho_{30}^2 \right]^{1/2} , \qquad \varrho_{30} = \varrho_{40} = \sqrt{\frac{\gamma}{2|g|}} . \tag{6.92}$$

Thus it is obvious that above the threshold it holds that

$$|p| \geq \frac{\gamma}{2} \varrho_{30} . \tag{6.93}$$

The eigenvalues of the corresponding matrix $\hat{M}$ in (6.85) are given
i) for forward mixing as

$$\lambda_1^{(\varrho)} = -\gamma , \qquad \lambda_2^{(\varrho)} = -\gamma + \frac{|p|}{\varrho_{30}} ,$$

$$\lambda_{3,4}^{(\varrho)} = -\frac{|p|}{2\varrho_{30}} \left\{ 1 \mp \left[ 1 - \frac{4\gamma^2 \varrho_{10}^2}{|p|^2} \right]^{1/2} \right\} ,$$

$$\lambda_1^{(\varphi)} = 0 , \qquad \lambda_2^{(\varphi)} = -\frac{|p|}{\varrho_{30}} ,$$

$$\lambda_{3,4}^{(\varphi)} = -\left( \gamma - \frac{|p|}{2\varrho_{30}} \right) \left\{ 1 \mp \left[ 1 - \frac{\gamma^2}{8\left(\gamma - |p|/\varrho_{30}\right)^2} \right]^{1/2} \right\} , \tag{6.94a}$$

ii) for backward mixing as

$$\lambda_{1,2,3,4}^{(\varrho)} = \pm i \left\{ \frac{|p|}{2\varrho_{30}} \left( \frac{|p|}{\varrho_{30}} - \gamma \right) \left( 1 \mp \left[ 1 + \frac{4\gamma^3 \varrho_{10}^2}{|p|^2 \left(|p|/\varrho_{30} - \gamma\right)} \right]^{1/2} \right) \right\}^{1/2} ,$$

$$\lambda_{1,2}^{(\varphi)} = 0 , \qquad \lambda_{3,4}^{(\varphi)} = \pm i \left[ \frac{|p|}{\varrho_{30}} \left( \frac{|p|}{\varrho_{30}} - \gamma \right) \right]^{1/2} , \tag{6.94b}$$

where $\lambda^{(\varrho)}$ and $\lambda^{(\varphi)}$ denote the corresponding eigenvalues related to the real amplitude $\varrho$ and to the phase $\varphi$. Additionally to the above threshold condition (6.93) we have for forward scattering from (6.94a)

$$|p| < \gamma\varrho_{30} \tag{6.95}$$

to ensure the stability of the solution. For backward scattering the solution is more complicated. In general, the well-known fact can be observed that the change of the sign in the direction of propagation leads to the change of the quality of the solution, i.e. a monochromatic solution is transformed to oscillating one and vice versa. An important conclusion is [Peřina, Peřinová, Sibilia and Bertolotti (1984)] that the signal-signal quantum correlations (entanglement) are always stronger than signal-pump quantum correlations, as being in the short-length approximation and for the incident coherent state $|\xi_1\rangle|\xi_2\rangle|\xi_3\rangle|\xi_4\rangle$

$$\langle\Delta(\hat{a}_1^\dagger\hat{a}_1)\Delta(\hat{a}_2^\dagger\hat{a}_2)\rangle \approx 2|g|z|\xi_1||\xi_2||\xi_3||\xi_4|\sin\left[\arg(g)+\varphi_1+\varphi_2-\varphi_3-\varphi_4\right], \tag{6.96a}$$

$$\langle\Delta(\hat{a}_1^\dagger\hat{a}_1)\Delta(\hat{a}_3^\dagger\hat{a}_3)\rangle \approx -|g|^2z^2|\xi_1|^2|\xi_3|^2\left(|\xi_2|^2+|\xi_4|^2\right), \tag{6.96b}$$

$$\left\langle\left(\Delta \begin{matrix}\hat{Q}_{12}\\ \hat{P}_{12}\end{matrix}\right)^2\right\rangle \approx 2\left\{1 \pm 2|g|z|\xi_1||\xi_2|\sin\left[\varphi_1+\varphi_2-\arg(g)\right]\right\}, \tag{6.97a}$$

$$\left\langle\left(\Delta \begin{matrix}\hat{Q}_{13}\\ \hat{P}_{13}\end{matrix}\right)^2\right\rangle \approx 2\Big\{1+|g|^2z^2\Big[|\xi_1|^2|\xi_2|^2+|\xi_3|^2|\xi_4|^2 \mp \left(|\xi_2|^2+|\xi_4|^2\right)|\xi_1||\xi_3|\cos\left(\varphi_1+\varphi_3\right)\Big]\Big\}, \tag{6.97b}$$

and similarly for the other mode combinations. Of course, similar conclusions are correct for the three-mode parametric process, as follows from the short-time results [Peřina (1991), Sec. 10.2, Drobný, Jex and Bužek (1993)].

We can conclude that the optical parametric processes and four-wave mixing can create a number of regimes for generation of nonclassical states of radiation. These effects are more pronounced in compound modes as a result of mode coupling. However, if fully quantum description with pump depletion is considered, also nonclassical behaviour in single modes of radiation can be exhibited. Initial weakly nonclassical behaviour can be more pronounced by the nonlinear dynamics whereas strong input nonclassical properties cannot be made stronger. Quantum behaviour of radiation can also be exhibited by oscillations in the photon-number distribution. The quantum properties are rapidly degraded by external noise, lossy mechanism and also they may be degraded during propagation.

We note that similar analysis of optical parametric processes has been presented by Reynaud, Heidmann, Giacobino and Fabre (1992). One can also

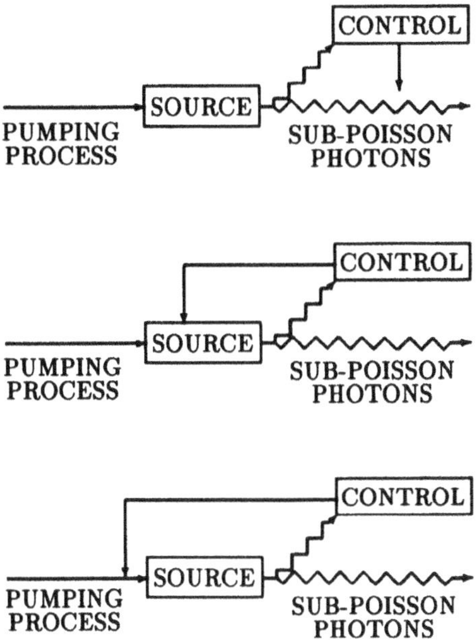


Figure 6.9: Feedback mechanisms for generation of nonclassical light.

derive corrections for quantum pumping compared to classical one [Neumann and Haug (1979), Milburn, Walls and Levenson (1984)]. Especially, the quantum fluctuations in the pumping modes are negligible when the mean photon number of the pump field exceeds $10^4$ photons (corresponding to an intensity of about $10^{-3}$ $\mathrm{Wcm}^{-2}$). As the pump photon number of $10^{12}$ photons is usual, the quantum fluctuations of the pumping field are usually negligible.

The nonlinear process of down conversion is particularly suitable for experimental investigation of optical quantum effects because such a process produces quantum correlated (entangled) states [Hong and Mandel (1985a), Friberg, Hong and Mandel (1985)]. The reason is that one of the sub-frequency beams can be used for feedback or feedforward modification of quantum statistics of the other signal sub-frequency beam. As indicated in Fig. 6.9 the sub-frequency beam can also modify the excitation mechanism inside the source or the pumping process. Such a feedback down conversion was realized by Rarity, Tapster and Jakeman (1987) using an optical fibre, as shown in Fig. 6.10. In this experiment one of the down converted photons provided an electrical feedback signal, which controlled an optical shutter in the signal beam guided by an optical fibre. The optical shutter was closed for a fixed dead-time period following the detection of an event. Poisson

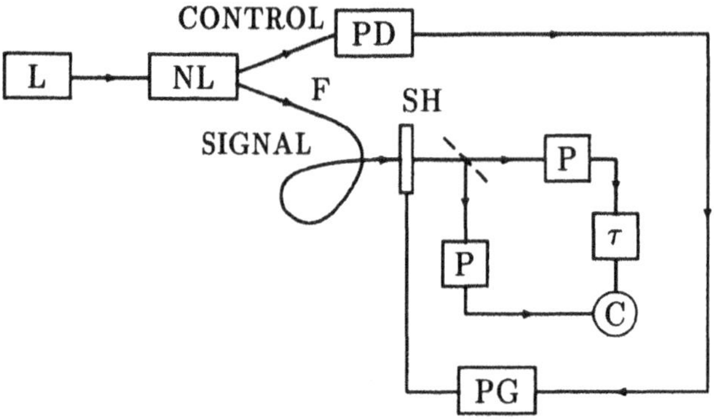


Figure 6.10: Controlled down conversion for generation of sub-Poisson light; L is laser, NL nonlinear crystal, PD photodiode, SH shutter, F optical fibre, P are photodetectors, $\tau$ is time delay, C correlator, PG pulse generator.

photons obtained from a He-Cd laser operating at 325 nm served for pumping of a potassium dihydrogen phosphate (KD*P) crystal generating down converted photons. The correlator C was able to provide the intensity correlation degree $\gamma^{(2)}(\tau)$ and the corresponding Fano factor $F \doteq 0.9998$ indicating slightly sub-Poisson photon statistics.

Instead of using feedback one can use feedforward as suggested by Walker (1986, 1987) based on adding photons to an initially Poissonian beam of photons. One again uses the above scheme with the parametric down conversion in which one of the sub-frequency beams serves for the control. However, photons are now added to the signal beam, instead of being filtered to regularize the signal beam. An estimation of the Fano factor indicates the possibility of significant reduction of photon-number fluctuations in this way. This method could be helpful particularly in optical communication systems, where losses play an important role, because it cannot only serve for the reduction of fluctuations, but also for the compensation of losses in the communication channel.

Heidmann, Horowitz, Reynaud, Giacobino, Fabre and Camy (1987) used a nonlinear KTP-crystal pumped by an $Ar^+$-laser and placed in an optical cavity. The cavity had a high finesse for the signal and idler waves, and a low finesse for the pump. When the pumping power is sufficiently large, the system operates above the threshold of oscillations and emits two stimulated beams with orthogonal polarizations (twin beams). They can be separated at the output of the optical cavity by a polarizing beam splitter. Then their intensities are detected by two photodetectors and the difference noise is analyzed. A half-wave plate is inserted to rotate the polarizations by 45°, which transforms the twin beams into uncorrelated beams giving the shot-noise level. These authors observed up to 86 % noise reduction under shot-noise level [Mertz, Debuisschert, Heidmann, Fabre and Giacobino (1991)].

Using a beam splitter, also one beam can serve as a feedback or a feedforward system, where the correlation is applied after the beam splitter. However, in this case one cannot correct quantum fluctuations because the beam splitter couples vacuum fluctuations to the beam through free ports. In fact Shapiro, Saplakoglu, Ho, Kumar, Saleh and Teich (1987) demonstrated that if a feedback is applied in one channel of a beam splitter, one can obtain sub-Poisson light on an in-loop photodetector, whereas on an out-of-loop photodetector, the photon statistics are super-Poissonian. Therefore it is convenient to use a "closed" arrangement without using beam splitter, as discussed above [see also Machida and Yamamoto (1986)]. It is also possible to obtain squeezed light by using a twin beam generator instead of the beam splitter [Reynaud, Heidmann, Giacobino and Fabre (1992)] to correct fluctuations in the feedback or feedforward configurations. The actual intensity noise reduction depends on the degree of correlation of twin beams and on the intensity of the excess noise in each of the beams. The intensity noise reduction of about 20 % has been observed using twin photons in parametric fluorescence [Tapster, Rarity and Satchel (1988)] or using intense twin beams generated by an optical parametric oscillator [Mertz, Heidmann, Fabre, Giacobino and Reynaud (1990)].

For generation of sub-Poissonian light, we can distinguish two principal ways:

i) The use of the various excitation mechanisms, based on the influence of statistics of primary excitations and individual emissions on coherence and photon statistics of light, with reduced Poisson fluctuations, as investigated by Teich, Saleh and Peřina (1984) [see also Peřina (1991), Chap. 9]. These mechanisms can be modified and made more effective using negative feedbacks and are able, in principle, to prepare sub-Poisson light beams with arbitrary low fluctuations. They can be applied to various dead-time configurations, to the Franck-Hertz experiment [Teich and Saleh (1985)] and to the down-conversion process in order to generate light with sub-Poisson fluctuations.

ii) The use of nonlinear optical processes in which fluctuations are reduced in the beams involved in the nonlinear interaction by nonlinear consumation of photons. In this case, the principle of the generation of nonclassical light is the coincidence decimation of light beams as a consequence of their nonlinear transformations.

An interesting case of excitation feedback is the use of the space-charge-limited Franck-Hertz experiment for generation of sub-Poissonian light on the basis of reduction of Poissonian fluctuations of exciting electrons, as indicated in equation (2.125) (Sec. 2.9.6). Further details can be found in Sec. 8.5 and

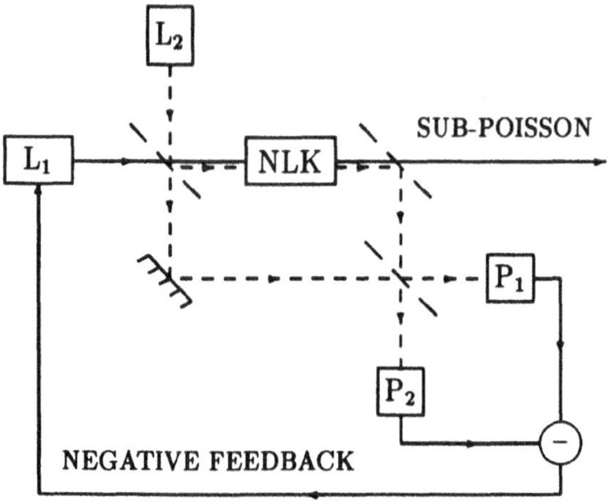


Figure 6.11: Generation of sub-Poisson light using the quantum nondemolition measurement of the laser output photon number to control the laser excitation rate; $L_1$ is a semiconductor laser, $L_2$ a probe laser, NLK nonlinear Kerr medium, P are photodetectors.

in the review by Teich and Saleh (1988).

Recently methods of quantum nondemolition measurements have been developed which make it possible to measure observables without perturbing their free motion [for brief reviews, see Teich and Saleh (1988), Reynaud, Heidmann, Giacobino and Fabre (1992), Meystre (1992)]. As very convenient nonlinear phenomenon for such a purpose the lossless optical Kerr effect can be adopted. It permits to measure the photon number in a signal mode without photon annihilation because the phase of a probe wave passed through the Kerr medium provides information about the change of the nonlinear index of refraction depending on the signal photon number in the medium. Precision in the signal photon number measurement makes an increased uncertainty in the phase variable, as indicated in the photon-number-phase uncertainty relation (5.57). Yamamoto, Imoto and Machida (1986) suggested to use such a quantum nondemolition measurement at the output of a semiconductor diode injection laser for a negative feedback to control the rate of excitation of the laser. This permits to produce sub-Poisson photons, as shown in Fig. 6.11.

A degenerate version of the fourth-wave mixing process represents an anharmonic oscillator. These processes are related to the interaction of radiation with a nonlinear medium possessing the Kerr nonlinearity, i. e. the refractive index of the medium is the intensity dependent. In such a case the hamiltonian of radiation can be written in the form [Peřina, Horák, Hradil, Sibilia and

Bertolotti (1989)]

$$\hat{H} = \sum_j \hbar\omega_j \left(\hat{a}_j^\dagger \hat{a}_j + \frac{1}{2}\right) + \sum_{j,k,m,n} \hbar U_{jkmn} \hat{a}_j^\dagger \hat{a}_k^\dagger \hat{a}_m \hat{a}_n \,, \tag{6.98}$$

where $U_{jkmn}$ are matrix elements of the permittivity coefficient. The Kerr effect in the nonlinear medium is important for realizations of quantum nondemolition measurements (see Fig. 6.11), as discussed in Sec. 3.6. The reason is that the nonlinear dynamics of such an anharmonic oscillator influences only the phase of radiation. Considering two coupled anharmonic oscillators we can write for the radiation hamiltonian

$$\begin{aligned} \hat{H} &= \hbar\omega_1 \left(\hat{a}_1^\dagger \hat{a}_1 + \frac{1}{2}\right) + \hbar\omega_2 \left(\hat{a}_2^\dagger \hat{a}_2 + \frac{1}{2}\right) \\ &\quad + \hbar g_1 \hat{a}_1^{\dagger 2} \hat{a}_1^2 + +\hbar g_2 \hat{a}_2^{\dagger 2} \hat{a}_2^2 + \hbar\kappa \hat{a}_1^\dagger \hat{a}_2^\dagger \hat{a}_2 \hat{a}_1 \,, \end{aligned} \tag{6.99}$$

where $g_1$, $g_2$ and $\kappa$ are the corresponding coupling constants related to the third-order susceptibility $\chi^{(3)}$. Applying the Heisenberg-Langevin or the Fokker-Planck methods (Secs. 4.6 and 4.7), we can obtain the Heisenberg-Langevin equations in the interaction picture

$$\begin{aligned} \frac{d\hat{A}_1}{dt} &= -\frac{\gamma_1}{2}\hat{A}_1 - i2g_1 \hat{A}_1^\dagger \hat{A}_1^2 - i\kappa \hat{A}_1 \hat{A}_2^\dagger \hat{A}_2 - i\sum_l \kappa_{1l} \hat{b}_{1l}(0) \exp\left[-i\left(\psi_{1l} - \omega_1\right)t\right] \,, \\ \frac{d\hat{A}_2}{dt} &= -\frac{\gamma_2}{2}\hat{A}_2 - i2g_2 \hat{A}_2^\dagger \hat{A}_2^2 - i\kappa \hat{A}_1^\dagger \hat{A}_1 \hat{A}_2 - i\sum_l \kappa_{2l} \hat{b}_{2l}(0) \exp\left[-i\left(\psi_{2l} - \omega_2\right)t\right] \,, \end{aligned} \tag{6.100}$$

or the corresponding Fokker-Planck equation for the quasidistribution $\Phi_A(\alpha_1, \alpha_2, t)$ related to the antinormal ordering of field operators

$$\begin{aligned} \frac{\partial \Phi_A}{\partial t} &= \sum_{j=1}^{2} \left[ i2g_j |\alpha_j|^2 \alpha_j \frac{\partial \Phi_A}{\partial \alpha_j} + i g_j \alpha_j^2 \frac{\partial^2 \Phi_A}{\partial \alpha_j^2} + \text{c.c.} \right] \\ &\quad + i\kappa \left[ |\alpha_2|^2 \alpha_1 \frac{\partial \Phi_A}{\partial \alpha_1} + |\alpha_1|^2 \alpha_2 \frac{\partial \Phi_A}{\partial \alpha_2} + \alpha_1 \alpha_2 \frac{\partial^2 \Phi_A}{\partial \alpha_1 \alpha_2} - \text{c.c.} \right] \\ &\quad + \sum_{j=1}^{2} \left[ \frac{\gamma_j}{2} \frac{\partial}{\partial \alpha_j} (\alpha_j \Phi_A) + \text{c.c.} + \gamma_j \left( \langle n_{Rj} \rangle + 1 \right) \frac{\partial^2 \Phi_A}{\partial \alpha_j \partial \alpha_j^*} \right] , \end{aligned} \tag{6.101}$$

where $\gamma_j$ are the damping constants, $\hat{b}_{jl}(0)$ the initial reservoir operators, and $\langle n_{Rj} \rangle = \langle \hat{b}_{jl}^\dagger(0) \hat{b}_{jl}(0) \rangle$ are the mean numbers of reservoir oscillators (Sec. 4.6). The first two terms on the right-hand side of (6.101) represent the effect of nonlinear dynamics whereas the last one is the reservoir (lossy) term.

If we neglect losses (the first and last terms in (6.100) and the last term in (6.101) are omitted since $\kappa_{jl} = 0$, $\gamma_j = 0$ and $\langle n_{Rj} \rangle = 0$ in this case), we can solve the Heisenberg equations (6.100) in the simple form

$$\begin{aligned} \hat{A}_1(t) &= \exp\left(-i2g_1 t\hat{a}_1^\dagger \hat{a}_1 - i\kappa t\hat{a}_2^\dagger \hat{a}_2\right)\hat{a}_1 , \\ \hat{A}_2(t) &= \exp\left(-i2g_2 t\hat{a}_2^\dagger \hat{a}_2 - i\kappa t\hat{a}_1^\dagger \hat{a}_1\right)\hat{a}_2 , \end{aligned} \tag{6.102}$$

clearly expressing only the phase influence of the nonlinear dynamics on the solution. The solution of the Fokker-Planck equation (6.101) can be obtained in the lossless case in the form of non-Gaussian solution

$$\begin{aligned} \Phi_{\mathcal{A}}(\alpha_1, \alpha_2, t) &= \frac{1}{\pi^2}\langle\alpha_1|\langle\alpha_2|\hat{\varrho}(t)|\alpha_2\rangle|\alpha_1\rangle \\ &= \frac{1}{\pi^2}\exp\left(-|\alpha_1|^2 - |\alpha_2|^2 - |\beta_1|^2 - |\beta_2|^2\right) \\ &\quad \times \sum_{k,l,m,n} \frac{(\alpha_1\xi_1^*)^k (\alpha_1^*\xi_1)^l (\alpha_2\xi_2^*)^m (\alpha_2^*\xi_2)^n}{k!\,l!\,m!\,n!} \\ &\quad \times \exp\left[ig_1 tk(k-1) - ig_1 tl(l-1) + ig_2 tm(m-1)\right. \\ &\qquad \left. - ig_2 tn(n-1) + i\kappa tkm - i\kappa tln\right] , \end{aligned} \tag{6.103}$$

provided that the initial field is coherent with the complex amplitudes $\xi_1$ and $\xi_2$.

From here one can obtain all quantum statistical characteristics [Peřina (1991), Sec. 10.8.3], we only mention the short-time effect for squeezing of vacuum fluctuations as being

$$\begin{aligned} \left\langle\left(\Delta \begin{matrix}\hat{Q}_j \\ \hat{P}_j\end{matrix}\right)^2\right\rangle &\approx 1 \pm 4g_j t|\xi_j|^2 \sin(2\varphi_j), \qquad j = 1, 2, \\ \left\langle\left(\Delta \begin{matrix}\hat{Q}_{12} \\ \hat{P}_{12}\end{matrix}\right)^2\right\rangle &\approx 2\Big\{1 \pm 2t\Big[g_1|\xi_1|^2 \sin(2\varphi_1) + g_2|\xi_2|^2 \sin(2\varphi_2) \\ &\qquad + \kappa|\xi_1||\xi_2| \sin(\varphi_1 + \varphi_2)\Big]\Big\} , \end{aligned} \tag{6.104}$$

where $\varphi_j$ are phases of the initial coherent amplitudes $\xi_j$. If, for instance, $\varphi_1 = \varphi_2 = -\pi/4$, the squeezing effect expressed in the quantities $\hat{Q}_j$ and $\hat{Q}_{12}$ is evident.

One can prove that the normal generating function equals $\exp[-\lambda(|\xi_1|^2 + |\xi_2|^2)]$ and, consequently, the photon statistics are Poissonian for all times of the interaction.

Even it is more complicated matter to include losses in this interaction this can be done in an exact way including arbitrary initial photon statistics. For reviews of this subject we refer the reader to works by Peřina, Bajer,

Peřinová and Hradil (1993) (Sec. B1), and by Peřinová and Lukš (1994). Some other aspects of quantum description of nonlinear optical processes have been discussed by Klyshko (1986, 1988).

# Chapter 7

# Quantum optical and Bell's inequalities

The above discussed nonclassical phenomena of photon antibunching, sub-Poissonian photon statistics, squeezing of vacuum fluctuations, etc. represent fully quantum features of light beams having no classical analogues. Consequently all these physical effects are violating various inequalities valid with classical optical fields. Experimental verification of their violation can then support the validity of quantum theory. As a particular case in 1964 Bell [see Bell (1987)] constructed inequalities between probabilities providing a way to test experimentally the predictions of various local hidden variable theories against the predictions of quantum theory. A review of classical and quantum inequalities and suggestions for their optical realizations together with Bell's inequalities and their interpretation has been presented by Reid and Walls (1986). We will follow this treatment.

## 7.1 Inequalities of quantum optics and their interpretation

Here we consider optical fields which are ergodic and stationary, i.e. the ensemble or quantum averages can be replaced by the time average over one realization of the field and the correlation functions are dependent only on time differences. Such assumptions are important for experiments measuring the correlation functions of optical fields and in this way providing tests of classical theory against quantum theory of light.

Adopting the definition of the fourth-order degree of coherence as given in (5.22) written more generally for two intensities $I_1$ and $I_2$,

$$\gamma^{(2)}_{\mathcal{N},12}(\tau) = \frac{\langle I_1(t) I_2(t+\tau)\rangle_{\mathcal{N}}}{\langle I_1\rangle_{\mathcal{N}}\langle I_2\rangle_{\mathcal{N}}}, \tag{7.1}$$

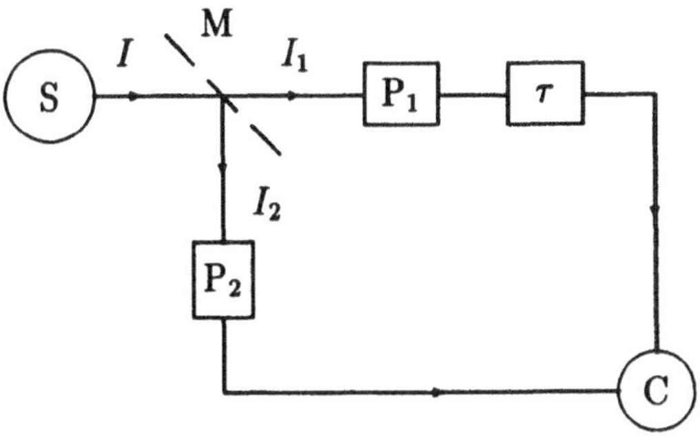


Figure 7.1: Scheme for verification of inequality (7.4b); S is a source, M a splitter, $P_1$ and $P_2$ are photomultipliers, $\tau$ is a delay line and C a correlator.

and assuming that the optical field under consideration has a classical analogy, i.e. that the Glauber-Sudarshan quasidistribution $\Phi_{\mathcal{N}}$ is well behaved as a classical distribution, we easily derive the following classical inequalities:

$$\langle (I - \langle I \rangle_{\mathcal{N}})^2 \rangle_{\mathcal{N}} \geq 0, \tag{7.2a}$$

that is

$$\langle I^2 \rangle_{\mathcal{N}} \geq \langle I \rangle_{\mathcal{N}}^2, \tag{7.2b}$$

and therefore

$$\gamma_{\mathcal{N}}^{(2)}(0) \geq 1. \tag{7.2c}$$

The classical Schwarz inequality directly provides

$$\langle I_1 I_2 \rangle_{\mathcal{N}}^2 \leq \langle I_1^2 \rangle_{\mathcal{N}} \langle I_2^2 \rangle_{\mathcal{N}}. \tag{7.3}$$

In particular for an arbitrary time delay $\tau$ we have

$$[\gamma_{\mathcal{N},12}^{(2)}(\tau)]^2 \leq \gamma_{\mathcal{N},11}^{(2)}(0)\, \gamma_{\mathcal{N},22}^{(2)}(0) \tag{7.4a}$$

and for a single beam

$$\gamma_{\mathcal{N}}^{(2)}(\tau) \leq \gamma_{\mathcal{N}}^{(2)}(0). \tag{7.4b}$$

Obviously, antibunching of photons discussed in Sec. 5.2 violates this inequality. Such an inequality is realized by the Hanbury Brown-Twiss scheme shown in Fig. 7.1. If we use a 50/50 splitter, the correlator will measure, after normalization, the correlation function $\langle I_1(t) I_2(t+\tau) \rangle_{\mathcal{N}} / \langle I_1(0) \rangle_{\mathcal{N}} \langle I_2(0) \rangle_{\mathcal{N}} = \langle I(t) I(t+\tau) \rangle_{\mathcal{N}} / \langle I \rangle_{\mathcal{N}}^2 = \gamma_{\mathcal{N}}^{(2)}(\tau)$ and for $\tau = 0$ we can measure the highest bound value $\gamma_{\mathcal{N}}^{(2)}(0)$. Continuing the arrangement of Fig. 7.1 as shown in Fig. 7.2 we can verify the Schwarz inequality (7.4a). The correlators $C_1$ and $C_2$

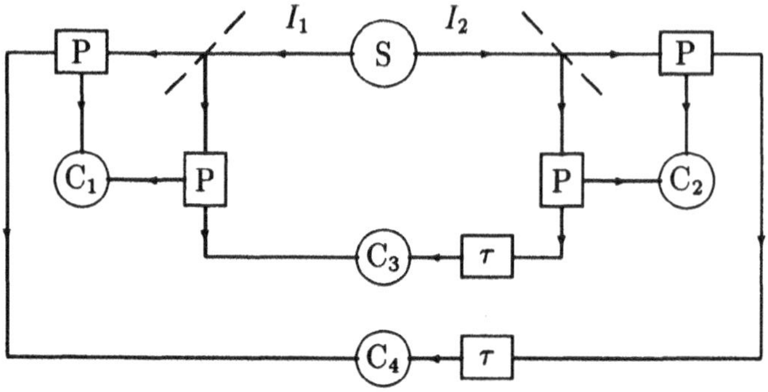


Figure 7.2: Scheme for optical realization of the Schwarz inequality (7.4a); S is a source, P are photodetectors, C correlators and $\tau$ are delay elements.

will provide, similarly as in Fig. 7.1, the degrees of coherence $\gamma_{\mathcal{N},11}^{(2)}(0)$ and $\gamma_{\mathcal{N},22}^{(2)}(0)$, respectively, and the correlators $C_3$ and $C_4$ can determine the cross-correlations $\langle I_1(t)I_2(t+\tau)\rangle_{\mathcal{N}}/\langle I_1\rangle_{\mathcal{N}}\langle I_2\rangle_{\mathcal{N}} = \gamma_{\mathcal{N},12}^{(2)}(\tau)$ and consequently the classical Schwarz inequality (7.4a) can be verified or its quantum violation can be observed.

Rather different interpretation of such inequalities is provided by quantum theory. Considering a single mode for simplicity we clearly have

$$\gamma_{\mathcal{N}}^{(2)}(0) = \frac{\langle \hat{a}^{\dagger 2}\hat{a}^2\rangle}{\langle \hat{a}^\dagger \hat{a}\rangle^2}. \tag{7.5}$$

Using the boson commutation rule $[\hat{a}, \hat{a}^\dagger] = \hat{1}$, we can derive the quantum inequality

$$\langle (\hat{n} - \langle \hat{n}\rangle)^2\rangle = \langle \hat{n}^2\rangle - \langle \hat{n}\rangle^2 = \langle \hat{a}^{\dagger 2}\hat{a}^2\rangle + \langle \hat{a}^\dagger\hat{a}\rangle - \langle \hat{a}^\dagger\hat{a}\rangle^2 \geq 0, \tag{7.6}$$

where $\hat{n} = \hat{a}^\dagger\hat{a}$ is the number operator. Therefore the quantum fourth-order degree of coherence fulfils

$$\gamma_{\mathcal{N}}^{(2)}(0) \geq 1 - \frac{1}{\langle n\rangle}. \tag{7.7}$$

In the double beam arrangement the quantum Schwarz inequality holds for two commuting operators $\hat{a}^\dagger\hat{a}$ and $\hat{b}^\dagger\hat{b}$,

$$[\langle \hat{a}^\dagger\hat{a}\hat{b}^\dagger\hat{b}\rangle]^2 \leq \langle (\hat{a}^\dagger\hat{a})^2\rangle \langle (\hat{b}^\dagger\hat{b})^2\rangle. \tag{7.8}$$

If both the beams are equivalent, then $\langle (\hat{a}^\dagger\hat{a})^2\rangle = \langle (\hat{b}^\dagger\hat{b})^2\rangle$ and $\langle \hat{a}^\dagger\hat{a}\rangle = \langle \hat{b}^\dagger\hat{b}\rangle$, and we arrive at the quantum inequality

$$\gamma_{\mathcal{N},12}^{(2)}(0) \leq \gamma_{\mathcal{N}}^{(2)}(0) + \frac{1}{\langle n\rangle}. \tag{7.9}$$

As we have seen light violating the inequality $\gamma_{\mathcal{N}}^{(2)}(0) \geq 1$ is exhibiting photon antibunching as a fully quantum feature. Similarly fully quantum fields having no classical analogues can violate the Schwarz inequality (7.4a) and in particular the inequality

$$[\gamma_{\mathcal{N},12}^{(2)}(0)]^2 \leq \gamma_{\mathcal{N},11}^{(2)}(0)\, \gamma_{\mathcal{N},22}^{(2)}(0)\,. \tag{7.10}$$

The quantum description of the schemes of Figs. 7.1 and 7.2 can be illustrated writing $\hat{a} = (\hat{a}_1 + \hat{a}_2)/\sqrt{2}$, $\hat{a}_1$ corresponds to the transmitted light and $\hat{a}_2$ is related to the reflected light. Considering a single photon state, we clearly have

$$|\psi\rangle = \hat{a}^\dagger|0\rangle = \frac{1}{\sqrt{2}}\left(\hat{a}_1^\dagger + \hat{a}_2^\dagger\right)|0\rangle|0\rangle = \frac{1}{\sqrt{2}}\left[|0\rangle|1\rangle + |1\rangle|0\rangle\right], \tag{7.11}$$

where $|0\rangle$ denotes the physical vacuum of the field $\hat{a}$, whereas $|0\rangle|0\rangle$ is the vacuum state for the combined modes $\hat{a}_1$ and $\hat{a}_2$. It is clear that

$$\gamma_{\mathcal{N}}^{(2)}(0) = \frac{\langle \hat{a}^{\dagger 2}\hat{a}^2\rangle}{\langle \hat{a}^\dagger \hat{a}\rangle^2} = \frac{\langle \hat{a}_1^\dagger \hat{a}_1 \hat{a}_2^\dagger \hat{a}_2\rangle}{\langle \hat{a}_1^\dagger \hat{a}_1\rangle\langle \hat{a}_2^\dagger \hat{a}_2\rangle}, \tag{7.12}$$

expressing a single particle character of the state implying $I_1 = 1$ or $0$ and $I_2 = 1 - I_1$. In this way new possibilities for photon statistics are arising.

It should be pointed out that violation of classical inequalities is directly related to the non-existence of the Glauber-Sudarshan quasidistribution $\Phi_{\mathcal{N}}$, which is singular and negative in these cases.

In analogy to the inequalities derived in Sec. 2.9.4 for correlation functions, we can obtain from the positive definiteness condition

$$\langle(\alpha_1 + \lambda\alpha_2^*)(\alpha_1^* + \lambda^*\alpha_2)\rangle \geq 0 \tag{7.13}$$

for the complex field amplitudes $\alpha_1$, $\alpha_2$ and a parameter $\lambda$ the following classical inequality

$$\begin{vmatrix} \langle|\alpha_1|^2\rangle & \langle\alpha_1\alpha_2\rangle \\ \langle\alpha_1^*\alpha_2^*\rangle & \langle|\alpha_2|^2\rangle \end{vmatrix} \geq 0\,, \tag{7.14a}$$

i.e.

$$|\langle\alpha_1\alpha_2\rangle|^2 \leq \langle|\alpha_1|^2\rangle\,\langle|\alpha_2|^2\rangle\,. \tag{7.14b}$$

The corresponding quantum analogue reads

$$\langle(\hat{a} + \lambda\hat{b}^\dagger)(\hat{a}^\dagger + \lambda^*\hat{b})\rangle \geq 0\,, \tag{7.15}$$

leading to

$$\begin{vmatrix} \langle\hat{a}\hat{a}^\dagger\rangle & \langle\hat{a}\hat{b}\rangle \\ \langle\hat{a}^\dagger\hat{b}^\dagger\rangle & \langle\hat{b}^\dagger\hat{b}\rangle \end{vmatrix} \geq 0\,, \tag{7.16a}$$

and finally

$$|\langle\hat{a}\hat{b}\rangle|^2 \le \langle\hat{a}\hat{a}^\dagger\rangle\langle\hat{b}^\dagger\hat{b}\rangle = (\langle\hat{a}^\dagger\hat{a}\rangle + 1)\langle\hat{b}^\dagger\hat{b}\rangle = \langle\hat{a}^\dagger\hat{a}\rangle\langle\hat{b}^\dagger\hat{b}\rangle + \langle\hat{b}^\dagger\hat{b}\rangle \tag{7.16b}$$

and

$$|\langle\hat{a}\hat{b}\rangle|^2 \le \langle\hat{a}^\dagger\hat{a}\rangle^2 + \langle\hat{a}^\dagger\hat{a}\rangle\,, \tag{7.16c}$$

provided that $\langle\hat{a}^\dagger\hat{a}\rangle = \langle\hat{b}^\dagger\hat{b}\rangle$ ( $\hat{a}^\dagger\hat{a} - \hat{b}^\dagger\hat{b}$ is a conservation expression for the three-mode interaction $\hbar g\hat{a}\hat{b}\hat{c}^\dagger$+h.c., $g$ being the coupling constant). Although the correlation element $\langle\hat{a}\hat{b}\rangle$ is not directly measured the Schwarz inequality

$$\langle\hat{a}^\dagger\hat{a}\hat{b}^\dagger\hat{b}\rangle \le \langle\hat{a}^{\dagger 2}\hat{a}^2\rangle \tag{7.17}$$

following from the classical arguments is maximally violated since

$$\langle\hat{a}^\dagger\hat{a}\hat{b}^\dagger\hat{b}\rangle = \langle\hat{a}^{\dagger 2}\hat{a}^2\rangle + \langle\hat{a}^\dagger\hat{a}\rangle \tag{7.18}$$

in this case.

It is interesting to note that the classical inequalities are also violated including dissipations as shown by Graham (1984) in deriving the expressions

$$\begin{aligned}\langle\hat{a}^\dagger\hat{a}\hat{b}^\dagger\hat{b}\rangle &= \langle\hat{a}^{\dagger 2}\hat{a}^2\rangle + \frac{1}{2}\langle\hat{a}^\dagger\hat{a}\rangle\,, \\ |\langle\hat{a}\hat{b}\rangle|^2 &= \langle\hat{a}^\dagger\hat{a}\rangle^2 + \frac{1}{2}\langle\hat{a}^\dagger\hat{a}\rangle\end{aligned} \tag{7.19}$$

valid in a steady state.

It should be noted that the parametric interaction $\hbar g\hat{a}\hat{b} +$ h.c. provides a maximally correlated pure quantum states in which a measurement of the photon number in the mode a completely determines the photon number in the mode b, which is related to the violation of Bell's inequalities and to the Einstein-Podolsky-Rosen (EPR) "paradox" (Sec. 7.2).

We can also write that

$$\langle:\hat{n}_1\hat{n}_2:\rangle = \frac{1}{2}\left[\langle\hat{n}_1\rangle + \langle\hat{n}_2\rangle + \langle:\hat{n}_1^2:\rangle + \langle:\hat{n}_2^2:\rangle\right] \tag{7.20}$$

using (7.18), which provides the quantum inequality

$$\langle:\hat{n}_1\hat{n}_2:\rangle > [\langle:\hat{n}_1^2:\rangle + \langle:\hat{n}_2^2:\rangle] \tag{7.21}$$

and it is in contradiction with the corresponding classical inequality

$$\langle I_1 I_2\rangle \le \frac{1}{2}\left[\langle I_1^2\rangle + \langle I_2^2\rangle\right] \tag{7.22}$$

following from the fact that $\langle(I_1 - I_2)^2\rangle \ge 0$. Zou, Wang and Mandel (1991) measured the intensity correlation $\langle I_1 I_2\rangle$ by coincidence counting using two photodetectors. A $LiIO_3$-crystal was used to produce signal and idler beams. Both the beams were separately counted and also their coincidence measurement was performed. The classical inequality was found to be violated by at least 580 standard deviations in the experiment. This was probably the largest violation of classical description in any experiment with light.

Recently an interesting violation of the Schwarz inequality for multiwave mixing was reported by Ansari (1993). Further it has been found by Franson (1991) that two correlated photons incident on two distant interferometers can produce coincidences depending "nonlocally" on the sum of phases in the interferometers. It is also shown that results of experiments violate simple Cauchy inequality $a^2 + b^2 \geq 2ab$, including some classical experiments, which provides the lack of objective realism in the electric field.

Munro and Reid (1993) suggested the use of the quantum macroscopic states generated by the process of parametric amplification to test the violation of a Bell's inequality of the Clauser-Horne-Shimony-Holt type (Sec. 7.2).

## 7.2 Einstein-Podolsky-Rosen "paradox" and Bell's inequalities

We have seen that sub-frequency photons in the optical parametric down conversion are strongly correlated and form the so-called entangled state. Similar highly correlated states, usually spin or polarization states, represent an important tool of investigations of the completeness of quantum mechanics based on the EPR "paradox". Such states allow a violation of another class of classical inequalities called the Bell's inequalities. It is obvious from the argument of Sec. 7.1 that the classical theory of the electromagnetic field is unable to explain completely all properties of light observed in correlation experiments of various orders. The quantum interpretation of the violation of classical inequalities exhibits the evidence for particle properties of optical field. The question of completeness of quantum theory may be reduced to the validity or violation of the Bell's inequalities and to the problem of completing the quantum theory by local hidden variables. Quantum theory describes in general an objective physical reality independent of observation. This raises a question if it is possible to formulate a physical theory in physical quantities representing objective reality independent of observation.

Assume that such objective and real physical quantities have values uniquely determined by certain hidden variables and that statistical character of quantum theory arising from the representation of physical quantities by operators is a consequence of our ignorance of the hidden variables. Now we can pose a question whether it is possible to construct such a hidden variable theory allowing the objective description and being in full agreement with predictions of quantum theory. Bell answered this question showing that a class of hidden-variable local theories cannot be in full agreement with quantum theory. This can be expressed in terms of the Bell's inequalities. This means that locality, reality and completeness are incompatible with some predictions of quantum mechanics. Now it is clear that in the EPR argument there is no doubt about the correctness of quantum predictions, it does not represent a paradox, but rather a program to solve a problem. The Bell's theorem shows, however, that such a program cannot be right and that the EPR propositions with the quantum mechanical predictions lead to contradictions. The contradiction is represented by the Bell's inequalities in terms of statistical predictions expressed by probabilities, because the experimental data cannot agree with both the predictions of quantum theory and with the Bell's inequalities. Recently, Greenberger, Horne and Zeilinger (1989) [see also Greenberger, Horne, Shimony and Zeilinger (1990)] suggested a new way to demonstrate the incompatibility of quantum theory with EPR predictions considering three and more particles, which demonstrates the incompatibility in terms of perfect correlations rather than statistical probability relations. This suggests new types of multiparticle correlation experiments involving triples and quadruples of photons obtained from atomic cascades, or from higher-order down conversion, or a cascade of two-photon down converters.

The essential argument of Einstein, Podolsky and Rosen (1935) is based on precise definitions of reality and completeness of a physical theory. They adopted the following definition of the physical reality: "If, without in any way disturbing a system, we can predict with certainty (i.e. with the probability equal to unity) the value of a physical quantity, then there exists an element of physical reality corresponding to this physical quantity." From this point of view they examined the completeness of quantum theory. They formulated that for a theory to be complete it is necessary that "every element of the physical reality must have a counterpart in the physical theory". Using the properties of quantum correlated states, they conclude that quantum theory is not able to describe completely physical reality.

To illustrate the problem we can consider a singlet state for two half-spin particles

$$|\psi\rangle = \frac{1}{\sqrt{2}}\Big[|a+,\mathbf{e}\rangle|b-,\mathbf{e}\rangle - |a-,\mathbf{e}\rangle|b+,\mathbf{e}\rangle\Big], \tag{7.23}$$

where $|a\pm,\mathbf{e}\rangle$ is a single-particle state with the particle $a$ having spin up (+) or down (−) along the direction of the unit vector $\mathbf{e}$, situated in any direction. The state (7.23) is perfectly correlated (or anticorrelated) and in Schrödinger terminology it represents an entangled state. With respect to the EPR definition, the spin of particle $b$ in any direction can be predicted with certainty from a measurement of the spin of particle $a$ in that direction, i.e. if the particle has spin up (down), then particle $b$ has its spin down (up).

Hence, measuring the spin of one particle we can predict with certainty the spin of the other particle without disturbing it. According to EPR, the spin of particle $b$ is an element of physical reality, independently of any observer, because we can predict with certainty its value without making any measurement on it. If such a measurement is performed in the $x$-direction, the similar measurement can be realized in the $y$-direction. Therefore the spin components of particle $b$ in both the $x$- and $y$-directions must be elements of physical reality and both the spin components have definite values before any measurement on the particle $b$ is performed. However, according to quantum theory $x$- and $y$-spin components cannot have simultaneously precise values because the associated operators do not commute (for the Pauli matrices we have $[\hat{\sigma}_x, \hat{\sigma}_y] = -2i\hat{\sigma}_z$), and consequently, according to EPR, quantum theory cannot be considered as a complete description of physical reality. However, it should be noted that in fact there is no paradox in the framework of quantum theory because the EPR argument does not involve simultaneous measurements. Separate measurements can provide results with arbitrary accuracy in principle and can give elements of reality. EPR "paradox" is not an expression of incompleteness of quantum theory, but it points out the peculiar nature of quantum reality.

A related question is the so-called reduction of the wave function (state vector). If a measurement is performed e.g. on the particle $a$, the state of the system is changed by the measurement and this measurement reduces the state vector (7.23) to $|\phi\rangle = |a+, \mathbf{e}\rangle|b-, \mathbf{e}\rangle$. Of course, this is nothing nonphysical because the measurement provides new information leading to the decrease of uncertainty, the increase of knowledge about the whole system (both the particles) and consequently to the reduction of the wave function; the wave function is a tool containing information about the past of the system relevant for predicting its future behaviour. A similar situation of the reduction of the state vector, containing the full information about the system, by the measurement is not typical only for quantum theory, but it has analogues also in classical statistical theory, because every measurement interrupts sequences of events and introduces new initial and boundary conditions.

Consider now the Bell's inequalities. Suppose a hidden variable theory in which the different components of the spin of each particle represent simultaneous realities in the sense of EPR. Let $\lambda_a$ be a hidden variable determining uniquely the spin component $A(\mathbf{e}_1, \lambda_a)$ of particle $a$ in the direction $\mathbf{e}_1$. Of course, our restriction to one hidden variable is only a technical problem. To be in correspondence with the quantum description we have to restrict values of $A(\mathbf{e}_1, \lambda_a)$ to two possible values labelled as

$$A(\mathbf{e}_1, \lambda_a) = \pm 1 \tag{7.24a}$$

predicted by quantum mechanics. We can similarly write for particle $b$

$$B(\mathbf{e}_2, \lambda_b) = \pm 1 . \tag{7.24b}$$

Now assume locality, i.e. that measurements performed on particle $a$ cannot change in any way the state of particle $b$ and vice versa, therefore $A(\mathbf{e}_1, \lambda_a)$ is independent of $\mathbf{e}_2$ and $B(\mathbf{e}_2, \lambda_b)$ is independent of $\mathbf{e}_1$. Following a close relation to quantum mechanics we assume that all values of $\lambda_a$ and $\lambda_b$ and for all directions $\mathbf{e}_1$ it holds that

$$A(\mathbf{e}_1, \lambda_a) = -B(\mathbf{e}_2, \lambda_b) , \tag{7.25}$$

i.e. if $\mathbf{e}_1 = \mathbf{e}_2$, then the spin components should lie in the opposite directions.

Now let $P(\lambda_a, \lambda_b)$ be the joint probability distribution of hidden variables $\lambda_a$ and $\lambda_b$. Then the expectation value of the product of spin of particle $a$ in the direction $\mathbf{e}_1$ and the spin of particle $b$ in the direction $\mathbf{e}_2$ is equal to

$$E(\mathbf{e}_1, \mathbf{e}_2) = \int\int P(\lambda_a, \lambda_b) A(\mathbf{e}_1, \lambda_a) B(\mathbf{e}_2, \lambda_b)\, d\lambda_a\, d\lambda_b . \tag{7.26}$$

According to Bell we can consider the quantity $E(\mathbf{e}_1, \mathbf{e}_2) - E(\mathbf{e}_1, \mathbf{e}_3)$. Taking into accont (7.25) and assuming $\lambda_a = \lambda_b = \lambda$ for simplicity, we have for three directions

$$\begin{aligned} E(\mathbf{e}_1, \mathbf{e}_2) - E(\mathbf{e}_1, \mathbf{e}_3) &= -\int P(\lambda)[A(\mathbf{e}_1, \lambda)A(\mathbf{e}_2, \lambda) - A(\mathbf{e}_1, \lambda)A(\mathbf{e}_3, \lambda)]\, d\lambda \\ &= \int [-A(\mathbf{e}_1, \lambda)A(\mathbf{e}_2, \lambda)][1 - A(\mathbf{e}_2, \lambda)A(\mathbf{e}_3, \lambda)]P(\lambda)\, d\lambda . \end{aligned} \tag{7.27}$$

Now for all $\lambda$, the quantity $-A(\mathbf{e}_1, \lambda)A(\mathbf{e}_2, \lambda)$ equals $+1$ or $-1$ having the absolute value 1 and $1 - A(\mathbf{e}_1, \lambda)A(\mathbf{e}_3, \lambda)$ is non-negative and therefore equals its absolute value. Therefore taking the absolute value of (7.27) we arrive at the Bell's inequality

$$|E(\mathbf{e}_1, \mathbf{e}_2) - E(\mathbf{e}_1, \mathbf{e}_3)| \leq 1 + E(\mathbf{e}_2, \mathbf{e}_3) , \tag{7.28}$$

using that $\int P(\lambda)\, d\lambda = 1$ and the definition (7.26). This is rather a member of family of such inequalities. Any local theory based on the above assumptions must satisfy this Bell's inequality.

According to quantum theory (7.24a,b) and (7.25) are correct, but instead of (7.26) we have

$$E(\mathbf{e}_1, \mathbf{e}_2) = -\mathbf{e}_1 . \mathbf{e}_2 \tag{7.29}$$

and it can be verified that some values of $\mathbf{e}_1$, $\mathbf{e}_2$ and $\mathbf{e}_3$ in (7.29) are inconsistent with (7.28). For simplicity, if $\mathbf{e}_1$ and $\mathbf{e}_2$ lie in the $x$-$y$ plane and have azimuthal angles $\theta_1$ and $\theta_2$, then

$$E(\mathbf{e}_1, \mathbf{e}_2) = -\cos(\theta_1 - \theta_2). \tag{7.30}$$

Choosing $\mathbf{e}_1$, $\mathbf{e}_2$, $\mathbf{e}_3$ in the $x$-$y$ plane with azimuthal angles 0, $\pi/3$, $2\pi/3$ we have

$$E(\mathbf{e}_1, \mathbf{e}_2) = E(\mathbf{e}_2, \mathbf{e}_3) = -\frac{1}{2}, \qquad E(\mathbf{e}_1, \mathbf{e}_3) = \frac{1}{2} \tag{7.31}$$

and consequently the Bell's inequality (7.28) is clearly violated. Therefore quantum theory is inconsistent with any local realistic theory.

Clauser, Horne, Shimony and Holt (1969) performed very accurate test of Bell's inequalities involving the correlations of polarization of two photons emitted in opposite directions in an atomic cascade decay. The polarization state of such a two-photon system can be written in the form

$$|\psi\rangle = \frac{1}{\sqrt{2}}\Big[|\mathbf{x}, a\rangle|\mathbf{x}, b\rangle + |\mathbf{y}, a\rangle|\mathbf{y}, b\rangle\Big], \tag{7.32}$$

where the single-photon state $|\mathbf{x}, a\rangle$ describes photon $a$ which is linearly polarized in the direction $\mathbf{x}$, and $\mathbf{x}$ and $\mathbf{y}$ are any orthogonal directions transverse to the direction of propagation. Thus a quantum correlated pair of systems is under consideration. Each photon passes through a wavelength filter and a linear polarizer followed by a photomultiplier. On each arm there is a single detector for photon that has passed through the polarizer. The most convincing experiments of this type were performed by Aspect, Grangier and Roger (1981) using the $4p^2\,{}^1S_0 \longrightarrow 4p\,4s\,{}^1P_1 \longrightarrow 4s^2\,{}^1S_0$ cascade of calcium. The $4p^2\,{}^1S_0$ level was excited by two-photon absorption using a krypton ion laser (406 nm wavelength) and a dye laser (580 nm wavelength) focused on a calcium beam, resulting in a cascade rate $\approx 10^7$ s$^{-1}$ and a coincidence rate longer than 100 s$^{-1}$. With such a rate a statistical accuracy of 1 % could be obtained in a data-collection time of only 100 s, compared to times of tens of hours needed for comparable accuracy in earlier experiments. It was found that Bell's inequalities were violated by 9 standard deviations.

In analogy to the scheme of Fig. 7.2 for verification of the Schwarz inequality, one can suggest the similar scheme with double-beam polarizers A and B [Reid and Walls (1986)] as shown in Fig. 7.3. The 50/50 beam splitters in Fig. 7.2 are replaced by double-beam polarizers A and B, which transmit light polarized along a direction forming the angles $\phi$ and $\psi$ with the reference axis $x$ in the $x$-$y$ plane, respectively, and reflect light polarized in the orthogonal direction in the $x$-$y$ plane. Photomultipliers then detect mean intensities $\langle I_{A+}\rangle$, $\langle I_{A-}\rangle$, $\langle I_{B+}\rangle$, $\langle I_{B-}\rangle$ and the correlators provide correlations $\langle I_{A+}I_{B+}\rangle$,

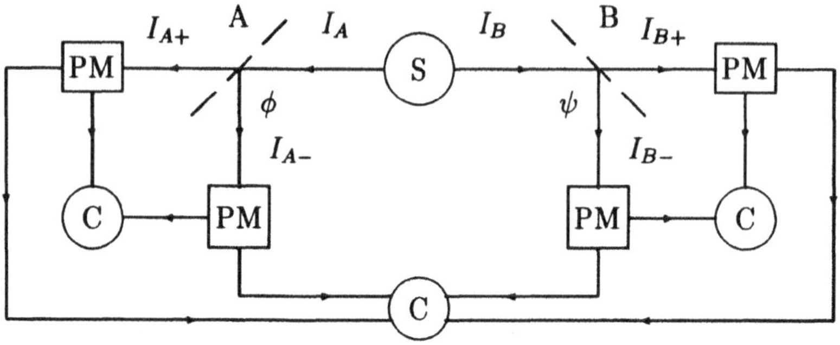


Figure 7.3: Scheme of the test of Bell's inequalities; S is a source of light beams, A, B are polarizer-beam splitters, PM are photomultipliers, C are correlators.

$\langle I_{A-}I_{B-}\rangle$, $\langle I_{A+}I_{B-}\rangle$, $\langle I_{A-}I_{B+}\rangle$, etc. Assume again a hidden variable $\lambda$, which may be discrete or continuous, so that we have similarly as above

$$\langle I_{Aj}I_{Bk}\rangle = \int P(\lambda)I_{Aj}(\phi,\lambda)I_{Bk}(\psi,\lambda)\,d\lambda\,, \tag{7.33}$$

$j$, $k$ being $+$ or $-$ and locality is again assumed, which means that for a given $\lambda$ the quantity $I_A$ ($I_B$) is independent of $\psi$ ($\phi$) because the measurement of a quantity at A (B) cannot causally affect the value of a quantity at B (A) and vice versa.

Now we can introduce the sum quantities

$$\begin{aligned} I_A(\lambda) &= I_{A+}(\phi,\lambda) + I_{A-}(\phi,\lambda)\,, \\ I_B(\lambda) &= I_{B+}(\psi,\lambda) + I_{B-}(\psi,\lambda)\,, \end{aligned} \tag{7.34}$$

which correspond to the full incident intensities on polarizers and are independent of $\phi$ and $\psi$ from the definition. We can therefore introduce the normalization factor

$$N = \int P(\lambda)I_A(\lambda)I_B(\lambda)\,d\lambda\,. \tag{7.35}$$

Now we can further introduce quantities

$$\begin{aligned} C_A(\phi,\lambda) &= \frac{I_{A+}(\phi,\lambda) - I_{A-}(\phi,\lambda)}{I_A(\lambda)}\,, \\ C_B(\psi,\lambda) &= \frac{I_{B+}(\psi,\lambda) - I_{B-}(\psi,\lambda)}{I_B(\lambda)}\,. \end{aligned} \tag{7.36}$$

It is important to note that $|C_{A,B}| \leq 1$. Consider the expectation value

$$E(\phi,\psi) = \frac{\langle I_{A+}I_{B+}\rangle + \langle I_{A-}I_{B-}\rangle - \langle I_{A+}I_{B-}\rangle - \langle I_{A-}I_{B+}\rangle}{\langle I_{A+}I_{B+}\rangle + \langle I_{A-}I_{B-}\rangle + \langle I_{A+}I_{B-}\rangle + \langle I_{A-}I_{B+}\rangle}$$

$$= \frac{\langle (I_{A+} - I_{A-})(I_{B+} - I_{B-}) \rangle}{\langle (I_{A+} + I_{A-})(I_{B+} + I_{B-}) \rangle}, \tag{7.37}$$

which can also be generalized to the quantum formulation if $I_{A\pm}$, $I_{B\pm}$ are replaced by the corresponding operators $\hat{I}_{A\pm}$, $\hat{I}_{B\pm}$ and normal ordering of the operators is adopted in (7.37). Thus including the hidden variable $\lambda$,

$$E(\phi, \psi) = \frac{1}{N} \int P(\lambda)[I_{A+}(\phi, \lambda) - I_{A-}(\phi, \lambda)][I_{B+}(\psi, \lambda) - I_{B-}(\psi, \lambda)]\, d\lambda, \tag{7.38}$$

where the normalization factor $N$ is given by (7.35). Now we can obtain the corresponding Bell's inequality including stochasticity (not only assuming deterministic conditions $|C_{A,B}| = 1$) and for general intensities $I_{A,B}(\lambda) > 0$ (not only assuming a single photon so that $I_{A,B}(\lambda) = 1$). We can write for the expectation values

$$E(\phi, \psi) = \frac{1}{N} \int \varrho(\lambda) C_A(\phi, \lambda) C_B(\psi, \lambda)\, d\lambda, \tag{7.39}$$

where $\varrho(\lambda) = P(\lambda) I_A(\lambda) I_B(\lambda)$, and hence

$$\begin{aligned} E(\phi, \psi) - E(\phi, \psi') &= \frac{1}{N} \int \varrho(\lambda) C_A(\phi, \lambda) C_B(\psi, \lambda)[1 \pm C_A(\phi', \lambda) C_B(\psi', \lambda)] \\ &- \frac{1}{N} \int \varrho(\lambda) C_A(\phi, \lambda) C_B(\psi', \lambda)[1 \pm C_A(\phi', \lambda) C_B(\psi, \lambda)]\, d\lambda . \end{aligned} \tag{7.40}$$

Taking the absolute value we arrive at

$$\begin{aligned} |E(\phi, \psi) - E(\phi, \psi')| \;\le\; & \frac{1}{N} \int \varrho(\lambda)[1 \pm C_A(\phi', \lambda) C_B(\psi', \lambda)]\, d\lambda \\ & + \frac{1}{N} \int \varrho(\lambda)[1 \pm C_A(\phi', \lambda) C_B(\psi, \lambda)]\, d\lambda \\ & = 2 \pm [E(\phi', \psi') + E(\phi', \psi)], \end{aligned} \tag{7.41}$$

or

$$|S| \le 2, \tag{7.42}$$

where

$$S = E(\phi, \psi) - E(\phi, \psi') + E(\phi', \psi') + E(\phi', \psi). \tag{7.43}$$

This is the Bell's inequality, also termed the Clauser-Horne-Shimony-Holt inequality.

From elementary quantum mechanics it follows that

$$E(\mathbf{e}_1, \mathbf{e}_2) = P_{++}(\mathbf{e}_1, \mathbf{e}_2) + P_{--}(\mathbf{e}_1, \mathbf{e}_2) - P_{+-}(\mathbf{e}_1, \mathbf{e}_2) - P_{-+}(\mathbf{e}_1, \mathbf{e}_2) \tag{7.44}$$

in relation to (7.37), where

$$\begin{aligned} P_{++}(\mathbf{e}_1,\mathbf{e}_2) &= P_{--}(\mathbf{e}_1,\mathbf{e}_2) = \frac{1}{2}\sin^2\left[\frac{(\mathbf{e}_1,\mathbf{e}_2)}{2}\right], \\ P_{+-}(\mathbf{e}_1,\mathbf{e}_2) &= P_{-+}(\mathbf{e}_1,\mathbf{e}_2) = \frac{1}{2}\cos^2\left[\frac{(\mathbf{e}_1,\mathbf{e}_2)}{2}\right] \end{aligned} \tag{7.45}$$

and we have (7.29).

In the experiments by Aspect et al. they have obtained [see Aspect (1984)]

$$S_{\text{exper}} = 2.697 \pm 0.015\,, \tag{7.46}$$

in agreement with the quantum prediction

$$S_{\text{quant}} = 2.70 \pm 0.05 \tag{7.47}$$

and in violation of the Bell's inequality $|S| < 2$. The Bell's inequality was violated by about 40 standard deviations. This measurement was performed for the angle $(\mathbf{e}_1,\mathbf{e}_2) = (\mathbf{e}_2,\mathbf{e'}_1) = (\mathbf{e'}_1,\mathbf{e'}_2) = 22.5°$ and $(\mathbf{e}_1,\mathbf{e'}_2) = 67.5°$. Also excellent agreement of the experimental data was obtained with the quantum prediction leading to $\cos[(\mathbf{e}_1,\mathbf{e}_2)]$ for other angles.

Clauser and Horne (1974) derived the modified Bell's inequality, which does not require auxiliary assumption of independence of (7.34) and (7.35) on $\phi$ and $\psi$. Such an inequality is related to an experiment using single channel polarizers allowing detection of the transmitted light in the direction of the polarizer axis only.

In terms of the hidden variables

$$R(\phi,\psi) = \langle I_{A+}I_{B+}\rangle = \int P(\lambda)I_{A+}(\phi,\lambda)I_{B+}(\psi,\lambda)\,d\lambda\,. \tag{7.48}$$

If polarizers are not present, the total intensities measured at photodetectors are independent of $\phi$ and $\psi$, and $I_A(\lambda) \geq I_{A+}(\phi,\lambda)$ and $I_B(\lambda) \geq I_{B+}(\psi,\lambda)$. Therefore

$$R(\phi,\psi) = \int \varrho(\lambda)\frac{I_{A+}(\phi,\lambda)I_{B+}(\psi,\lambda)}{I_A(\lambda)I_B(\lambda)}\,d\lambda\,, \tag{7.49}$$

where $\varrho(\lambda) = P(\lambda)I_A(\lambda)I_B(\lambda)$ again. The expectation values with the removed polarizer are

$$\begin{aligned} r_A(\phi) &= \langle I_{A+}I_B\rangle = \int P(\lambda)I_{A+}(\phi,\lambda)I_B(\lambda)\,d\lambda = \int \varrho(\lambda)\frac{I_{A+}(\phi,\lambda)}{I_A(\lambda)}\,d\lambda\,, \\ r_B(\psi) &= \langle I_A I_{B+}\rangle = \int P(\lambda)I_A(\lambda)I_{B+}(\psi,\lambda)\,d\lambda = \int \varrho(\lambda)\frac{I_{B+}(\psi,\lambda)}{I_B(\lambda)}\,d\lambda\,. \end{aligned} \tag{7.50}$$

Denoting

$$\begin{aligned} x &= \frac{I_{A+}(\phi,\lambda)}{I_A(\lambda)}, & x' &= \frac{I_{A+}(\phi',\lambda)}{I_A(\lambda)}, \\ y &= \frac{I_{B+}(\psi,\lambda)}{I_B(\lambda)}, & y' &= \frac{I_{B+}(\psi',\lambda)}{I_B(\lambda)}, \end{aligned} \tag{7.51}$$

and making use of the number inequality

$$-XY \le xy - xy' + x'y + x'y' + x'Y - yX \le 0, \tag{7.52}$$

if

$$0 \le x, x' \le X, \qquad 0 \le y, y' \le Y, \tag{7.53}$$

we arrive at the Clauser-Horne inequality

$$R = R(\phi,\psi) - R(\phi,\psi') + R(\phi',\psi') + R(\phi',\psi) - r_A(\phi) - r_B(\psi) \le 0. \tag{7.54}$$

This inequality was the basis for the quantum correlation measurements using the frequency down-conversion [Ou and Mandel (1988)] (see Sec. 8.6) where the rate

$$R = 11.5 \pm 2.0 \text{ per minute} \tag{7.55}$$

has been observed for $\phi = 22.5°$, $\psi = 45°$, $\phi' = 67.5°$, $\psi' = 0°$, which violates the Bell's inequality by about 6 standard deviations.

If photons are emitted in opposite directions with respect to the momentum conservation law, the experiment is described by the two-photon state

$$|\psi\rangle = \frac{1}{\sqrt{2}}\left(\hat{a}_+^\dagger \hat{b}_+^\dagger + \hat{a}_-^\dagger \hat{b}_-^\dagger\right)|0\rangle, \tag{7.56}$$

where $\hat{a}_+$ ($\hat{b}_+$) are photon annihilation operators for modes propagating in the $z$ ($-z$) direction with polarization along $x$-axis and similarly $\hat{a}_-$ ($\hat{b}_-$) describe polarization along $y$-axis.

The mode operators $\hat{c}_+$, $\hat{c}_-$, $\hat{d}_+$, $\hat{d}_-$ on the detectors are obtained as orthogonal transformations of $\hat{a}_+$, $\hat{a}_-$, $\hat{b}_+$, $\hat{b}_-$ giving

$$\begin{aligned} \hat{c}_+ &= \hat{a}_+\cos(\phi) + \hat{a}_-\sin(\phi), & \hat{d}_+ &= \hat{b}_+\cos(\psi) + \hat{b}_-\sin(\psi), \\ \hat{c}_- &= -\hat{a}_+\sin(\phi) + \hat{a}_-\cos(\phi), & \hat{d}_- &= -\hat{b}_+\sin(\psi) + \hat{b}_-\cos(\psi), \end{aligned} \tag{7.57}$$

where $\phi$ ($\psi$) is the angle between axis of the polarizer A (B) and $x$-axis. The corresponding correlation functions are

$$\begin{aligned} \langle \hat{c}_+^\dagger \hat{c}_+ \hat{d}_+^\dagger \hat{d}_+ \rangle &= \cos^2(\phi - \psi), \\ \langle \hat{c}_+^\dagger \hat{c}_+ \hat{d}_-^\dagger \hat{d}_- \rangle &= \sin^2(\phi - \psi) \end{aligned} \tag{7.58}$$

and

$$E(\phi,\psi) = \cos[2(\phi-\psi)]\,. \tag{7.59}$$

If we choose $\phi-\psi=\phi'-\psi=\phi'-\psi'=(\phi-\psi')/3$, we obtain

$$R = 3\cos[2(\phi-\psi)] - \cos[6(\phi-\psi)] \tag{7.60}$$

and for $\phi-\psi=22.5°$, $R=2\sqrt{2}$, which is the value violating the Bell's inequality (7.42). Experiments by Aspect (1976) (and references therein) have directly verified the predictions of quantum theory. Further violation of Bell's inequality was experimentally proved and full agreement with quantum theory was found by Fry and Thompson (1976) and Lamehi-Rachti and Mitting (1976).

It is important to point out that the violations of the Bell's inequalities are again related to the non-existence of the Glauber-Sudarshan quasidistribution $\Phi_{\mathcal{N}}$. Consider, for instance, transformations (7.57) where the operators are replaced as follows: $\hat{a}_\pm \to \alpha_\pm$, $\hat{b}_\pm \to \beta_\pm$, $\hat{c}_\pm \to \gamma_\pm$, $\hat{d}_\pm \to \delta_\pm$. For the quantum analogue of (7.37) we therefore have [Reid and Walls (1986)]

$$\begin{aligned} E(\phi,\psi) &= \frac{1}{N}\int \Phi_{\mathcal{N}}(\alpha_+,\alpha_-,\beta_+,\beta_-)(|\gamma_+|^2-|\gamma_-|^2)(|\delta_+|^2-|\delta_-|^2) \\ &\quad \times d^2\alpha_+ d^2\alpha_- d^2\beta_+ d^2\beta_- \\ &= \frac{1}{N}\int \Phi_{\mathcal{N}}(\alpha_+,\alpha_-,\beta_+,\beta_-)(|\alpha_+|^2+|\alpha_-|^2)(|\beta_+|^2+|\beta_-|^2) \\ &\quad \times C(\gamma_+,\gamma_-)C(\delta_+,\delta_-)\, d^2\alpha_+ d^2\alpha_- d^2\beta_+ d^2\beta_-\,, \end{aligned} \tag{7.61}$$

where

$$N = \int \Phi_{\mathcal{N}}(\alpha_+,\alpha_-,\beta_+,\beta_-)(|\alpha_+|^2+|\alpha_-|^2)(|\beta_+|^2+|\beta_-|^2)\, d^2\alpha_+ d^2\alpha_- d^2\beta_+ d^2\beta_-\,, \tag{7.62}$$

and

$$C(\gamma_+,\gamma_-) = \frac{|\gamma_+|^2-|\gamma_-|^2}{|\alpha_+|^2+|\alpha_-|^2}\,, \qquad C(\delta_+,\delta_-) = \frac{|\delta_+|^2-|\delta_-|^2}{|\beta_+|^2+|\beta_-|^2}\,. \tag{7.63}$$

As $\gamma_\pm$ ($\delta_\pm$) are the functions of only $\phi$ ($\psi$), the Glauber-Sudarshan representation is a local one. Comparing (7.61) and (7.37) we see that for the prove of the Bell's inequality the quasidistribution $\Phi_{\mathcal{N}}$ has to be regular and non-negative. For the quantum optical fields having no classical analogues the $\Phi_{\mathcal{N}}$-quasidistribution may take on negative values and may be singular, which means that the violation of the Bell's inequalities is possible and it is related to quantum properties of optical fields. Recently Agarwal, Home and Schleich (1992) and Agarwal (1993) illustrated that when using the Wigner function related to the symmetric ordering of field operators one can develop a description of the EPR-correlated states similar to the model based on local hidden variables in such a way that quantum results are reproducible allowing the corresponding probability distribution functions become negative. Such an unphysical feature illustrates deep incompatibility between quantum theory and local classical realistic theory.

Further we note that Rauch (1993) suggested to interpret EPR correlations and "nonlocal" behaviour of quantum systems in terms of interferences in momentum-frequency space where modulations occur when interferences in the real space-time are missing (Sec. 6.1). This indicates that coherent wave packets remain coupled even in the case of wide spatial separations.

Note also that a review of original experiments for verifications of Bell's inequalities and their critical analysis were presented in the book by Selleri (1990) (Chap. 5).

Reid and Walls (1986) suggested the non-degenerate optical parametric processes and four-wave mixing described by the effective hamiltonians, as treated in Sec. 6.6, to obtain various regimes violating the Bell's inequalities. Using the non-degenerate parametric process and the counterpropagating four-wave mixing with the strong classical pumping they have shown that the Bell's inequalities are violated if

$$C = \frac{\langle \hat{a}_+^\dagger \hat{a}_+ \hat{b}_+^\dagger \hat{b}_+ \rangle - \langle \hat{a}_+^\dagger \hat{a}_+ \hat{b}_-^\dagger \hat{b}_- \rangle}{\langle \hat{a}_+^\dagger \hat{a}_+ \hat{b}_+^\dagger \hat{b}_+ \rangle + \langle \hat{a}_+^\dagger \hat{a}_+ \hat{b}_-^\dagger \hat{b}_- \rangle} > \frac{1}{\sqrt{2}} \tag{7.64}$$

and the maximum violation occurs for $C = 1$. More generally they have calculated that

$$C = \frac{1}{1 + 2\tanh^2(|\kappa|t)}, \tag{7.65}$$

where $\kappa$ is the coupling constant of the nonlinear process involving the pumping amplitudes. Thus for small interaction times when $|\kappa|t \to 0$, the Bell's inequality is violated, whereas in the asymptotic case $|\kappa|t \to \infty$, $C \to 1/3$ and the inequality is satisfied.

Rigorous description of the state reduction for correlated photon pairs in the signal mode when a measurement is performed on the idler mode was presented by Lukš, Peřinová and Křepelka (1994).

It is also interested to reconsider the two-photon quantum state (7.56) as a superposition of states with phases $\varepsilon_1$ and $\varepsilon_2$,

$$\begin{aligned}
\hat{c}_+ &= \hat{a}_+ \cos(\phi) + \hat{a}_- \sin(\phi) \exp(i\varepsilon_1), \\
\hat{d}_+ &= \hat{b}_+ \cos(\psi) + \hat{b}_- \sin(\psi) \exp(i\varepsilon_2), \\
\hat{c}_- &= -\hat{a}_+ \sin(\phi) + \hat{a}_- \cos(\phi) \exp(i\varepsilon_1), \\
\hat{d}_- &= -\hat{b}_+ \sin(\psi) + \hat{b}_- \cos(\psi) \exp(i\varepsilon_2);
\end{aligned} \tag{7.66}$$

then we obtain

$$E(\phi, \psi) = C[\cos(2\phi)\cos(2\psi) + \sin(2\phi)\sin(2\psi)\cos(\varepsilon_1 + \varepsilon_2)]\,. \qquad (7.67)$$

Therefore the optimal situation is for $\varepsilon_1 + \varepsilon_2 = 0$, whereas for $\varepsilon_1 + \varepsilon_2 = \pi/2$ we obtain the factorization of the expression and the Bell's inequality cannot be violated.

Also in two-beams experiments Reid and Walls (1986) have demonstrated for

$$\frac{\langle \hat{a}^\dagger \hat{a} \hat{b}^\dagger \hat{b} \rangle}{\langle \hat{a}^\dagger \hat{a} \hat{b}^\dagger \hat{b} \rangle + \langle \hat{a}^{\dagger 2} \hat{a}^2 \rangle} \geq \frac{1}{\sqrt{2}} \qquad (7.68)$$

that there is strong violation of the Schwarz and Bell's inequalities. They have also investigated the question of the violation of these inequalities in the optical parametric system and in the process of four-wave mixing including dissipations.

All inequalities discussed till now involved stochastic quantities, i.e. probabilities and correlation functions. However, it was shown by Greenberger, Horne and Zeilinger (1989) [see Greenberger, Horne, Shimony and Zeilinger (1990) for a review] that it is possible to consider correlations between three or four particles and to establish Bell's theorem without inequalities leading to deterministic contradictions.

If four 1/2-spin particles are produced by a source and particles 1 and 2 are propagating freely in the positive $z$-direction and particles 3 and 4 are propagating in the negative $z$-direction, they enter different Stern-Gerlach analyzers with orientations $\mathbf{e}_1$, $\mathbf{e}_2$, $\mathbf{e}_3$ and $\mathbf{e}_4$. If the four particles result from a decay of a single particle into a pair of particles with spin 1, each of which then decays into a pair of particles with spin 1/2, with the $z$-component of spin initially zero and remaining so during the decay process, then the quantum spin state of these four particles is

$$|\psi\rangle = \frac{1}{\sqrt{2}}\Big[|+\rangle_1|+\rangle_2|-\rangle_3|-\rangle_4 - |-\rangle_1|-\rangle_2|+\rangle_3|+\rangle_4\Big]\,, \qquad (7.69)$$

leading to the expectation value of the product of outcomes

$$E(\mathbf{e}_1, \mathbf{e}_2, \mathbf{e}_3, \mathbf{e}_4) = \cos(\phi_1)\cos(\phi_2)\cos(\phi_3)\cos(\phi_4) - \sin(\phi_1)\sin(\phi_2)\sin(\phi_3)\sin(\phi_4)\cos(\vartheta_1 + \vartheta_2 - \vartheta_3 - \vartheta_4)\,, \qquad (7.70)$$

where $(\phi_j, \vartheta_j)$ are the polar and azimuthal angles of $\mathbf{e}_j$, $j = 1, 2, 3, 4$. If the $\mathbf{e}_j$ lie in the $x$-$y$ plane, we have

$$E(\mathbf{e}_1, \mathbf{e}_2, \mathbf{e}_3, \mathbf{e}_4) = -\cos(\vartheta_1 + \vartheta_2 - \vartheta_3 - \vartheta_4)\,. \qquad (7.71)$$

The particularly interesting cases are if

$$\vartheta_1 + \vartheta_2 - \vartheta_3 - \vartheta_4 = 0 \,,$$
$$E(\mathbf{e}_1, \mathbf{e}_2, \mathbf{e}_3, \mathbf{e}_4) = -1 \,, \tag{7.72a}$$

and

$$\vartheta_1 + \vartheta_2 - \vartheta_3 - \vartheta_4 = \pi \,,$$
$$E(\mathbf{e}_1, \mathbf{e}_2, \mathbf{e}_3, \mathbf{e}_4) = 1 \,. \tag{7.72b}$$

From here we see that the knowledge of the outcomes for any three particles permits the prediction of the outcome for the fourth one. Since at the time of measurement the four particles are arbitrarily far apart, they do not interact and no real change of their state can take place of any one of them as a consequence of the change of states of the other three, i.e. the locality assumption is fulfilled. From this it can be inferred the existence of four functions $A(\vartheta_1, \lambda)$, $B(\vartheta_2, \lambda)$, $C(\vartheta_3, \lambda)$, $D(\vartheta_4, \lambda)$ having values $\pm 1$. These are the outcomes of spin measurements along the respective directions when the complete state of the four particles is $\lambda$. Therefore instead of (7.72a,b) we have

$$A(\vartheta_1, \lambda) B(\vartheta_2, \lambda) C(\vartheta_3, \lambda) D(\vartheta_4, \lambda) = -1 \,,$$
$$\vartheta_1 + \vartheta_2 - \vartheta_3 - \vartheta_4 = 0 \,, \tag{7.73a}$$
$$A(\vartheta_1, \lambda) B(\vartheta_2, \lambda) C(\vartheta_3, \lambda) D(\vartheta_4, \lambda) = 1 \,,$$
$$\vartheta_1 + \vartheta_2 - \vartheta_3 - \vartheta_4 = \pi \,. \tag{7.73b}$$

Considering these equations for angles $\pi$ and 0, it is possible to deduce a contradiction that $A(\pi, \lambda) = A(0, \lambda) = -A(0, \lambda)$. Thus the EPR program is not realizable.

Similar considerations can be performed with three particles (photons). Assume a decaying particle with zero momentum and that emitted particles have the same energy, and with respect to the momentum conservation, they must be emitted 120° apart. The central source can be surrounded by an array of six apertures a, b, c at 120° separation and shifted apertures a', b', c' also at 120° separation. Three emitted particles 1, 2, 3 must emerge either through a, b, c or a', b', c'. Therefore the resulting quantum state beyond the apertures is given by

$$|\psi\rangle = \frac{1}{\sqrt{2}} \left[ |a\rangle_1 |b\rangle_2 |c\rangle_3 + |a'\rangle_1 |b'\rangle_2 |c'\rangle_3 \right] , \tag{7.74}$$

where $|a\rangle_1$ denotes particle 1 in beam a, etc. Moreover predictions can be modified substituting phase factors into (7.74). Then beams a, a', b, b' and c, c' can be superimposed with some phases $\varphi_1$, $\varphi_2$, $\varphi_3$ and the quantum expectation value can be obtained as [Greenberger, Horne, Shimony and Zeilinger (1990)]

$$E(\varphi_1,\varphi_2,\varphi_3) = \sin(\varphi_1 + \varphi_2 + \varphi_3)\,. \tag{7.75}$$

Perfect correlations are obtained for

$$E(\varphi_1,\varphi_2,\varphi_3) = 1\,,$$

$$\varphi_1 + \varphi_2 + \varphi_3 = \pi/2\,, \tag{7.76a}$$

and

$$E(\varphi_1,\varphi_2,\varphi_3) = -1\,,$$

$$\varphi_1 + \varphi_2 + \varphi_3 = 3\pi/2\,. \tag{7.76b}$$

Similarly as above for four particles, a contradiction can be deduced.

Thus the most significant feature of such deterministic arguments is that the EPR program cannot realize even perfect quantum correlations with three or four particles (photons).

Such experiments could be realized using correlations among three or more photons emitted in the process of parametric down conversion as a generalization of two-photon down conversion [Burnham and Weinberg (1970), Sec. 6.6]. Also generalizations may be adopted of the atomic cascade two-photon experiments using an atom cascading through two intermediate levels to produce three photons. Or it may be easier to utilize a two-step cascade of two photon down conversions to produce a four-photon entangled state [see Zou, Wang and Mandel (1991), Zou, Grayson, Barbosa and Mandel (1993), Sec. 8.6].

A two photon interferometer with a momentum correlated source for observation of "nonlocal" quantum effects, as suggested by Rarity and Tapster (1990), is shown in Fig. 7.4. If a detector is placed in one arm to provide path information for a photon, the interference fringes cannot be observed; they can be observed only if both photons are detected after recombination at their respective beam splitters. The similar three-photon experiment using a momentum correlated source suggested by Greenberger, Horne, Shimony and Zeilinger (1990) is shown in Fig. 7.5. Three-photon correlations can be measured there. Detectors inserted in any path to provide information about the photon position will smooth out interference fringes in all detectors. All three photons have to be detected after recombination at their respective beam splitters to observe interference fringes as a result of quantum correlated three-photon state. Multiparticle interferences were discussed by Rice, Osborn and Lloyd (1994).

Note that experiments with multiport devices are under way at the Innsbruck University. Such experiments should demonstrate new features of quantum mechanics for more complex entangled systems.

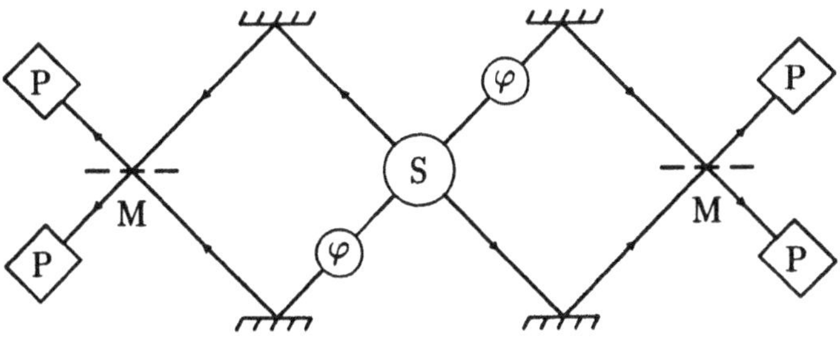


Figure 7.4: A scheme of a two-photon interferometer with correlated photons for observation of "nonlocal" quantum effects, S is a two-photon source, M are splitters, P photodetectors, $\varphi$ phase elements.

Recently great attention has been devoted to the relation between quantum and classical theories [see e.g. Zurek (1991), Cini and Levy-Leblond (1990)]. It may be assumed that quantum theory is universal and also valid in macroworld, i.e. also classical world can be described by quantum superpositions of states; however such classical systems are not isolated and as results of dissipation effects arising during the interaction with surroundings, the quantum interference terms are missing as a result of decoherence effects. Such reduced states then can behave like classical states. When we adopt the Wigner function for the description for quantum states it is characteristic that it also takes on negative values obtained with respect to quantum coherence effects. The environment-induced decoherence reduces such off-diagonal effects, which may lead to classical probability distribution. Or some quantum interference effects can occur in the Wigner function with changing signs excepting some classical non-negative parts of the quasidistribution. As a result of decoherence the positive and negative oscillating parts can be reduced and cancelled out, which leads to the classical description of macroscopic systems, so that in appropriate limits both the classical structure of the phase space and classical dynamics naturally coincide.

In quantum optics a number of interesting investigations have been performed following the effect of dissipations and quantum decoherence [Daniel and Milburn (1989), Peřinová and Lukš (1990), Milburn (1991), Kim and Bužek (1992a,b), Brune, Haroche, Raimond, Davidovich and Zagury (1992), Bužek, Moya-Cessa, Knight and Phoenix (1992), Zurek, Habib and Paz (1993)].

There exist various suggestions to use the EPR correlations for superluminal communications [Bennett and Wiesner (1992), Squires (1992)]. However, such interpretations are overlooking the information content of the wave func-

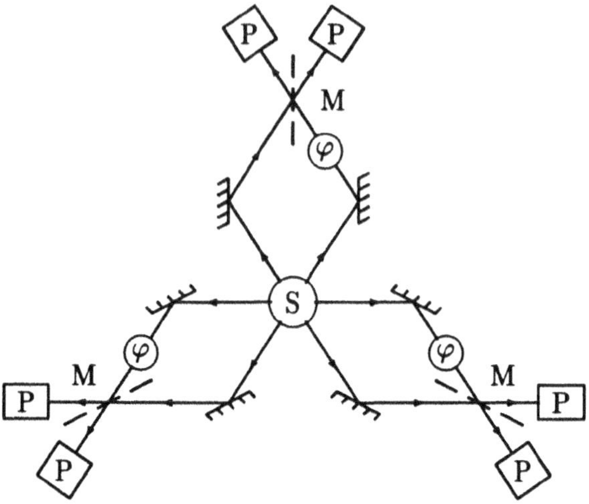


Figure 7.5: A scheme of a three-photon interferometer with correlated photons for observation of "nonlocal" quantum effects, denotation as in Fig. 7.4.

tion and of its collapse and seem to be impossible [Glauber (1986), Milonni and Singh (1991)]. On the other hand quantum interference effects and their missing if a which-path information is obtained can be used to save information during communication and a new direction is arising here called quantum cryptography [Ekert, Rarity, Tapster and Palma (1992)]. Recently Muller, Buguet and Gisin (1993) experimentally demonstrated quantum cryptography using polarized photons in an optical fibre for transmission more than 1 km. Townsend, Rarity and Tapster (1993) observed single-photon interferences in 10 km long optical fibre interferometer with the visibility higher than 90 % for future quantum cryptography systems. Models of two-photon EPR interference experiments were discussed by Rubin and Shih (1992).

Yurke and Stoler (1992) discussed the EPR effects from independent widely separated particle sources. Reid and Munro (1992) discussed the Greenberger-Horne-Zeilinger correlations. Busch, Kienzler, Lahti and Mittelstaedt (1993) considered a test of quantum theory against a full set of Bell's inequalities. Kärtner and Haus (1993) examined the "collapse" of the wave function in relation to reduction of coherence. Klyshko (1993) and Belinsky and Klyshko (1993) investigated the relation of "nonlocality" and its content to the Bell and Greenberger-Horne-Zeilinger theorems for cases of three-photon and many-particle interference. A two-particle interferometry using down-conversion process for quantum entanglement of states was discussed by Horne, Shimony and Zeilinger (1989). Lyubomirsky, Shirasaki,

Kärtner and Haus (1993) suggested an experiment using Sagnac fibre ring in a Mach-Zehnder interferometer producing squeezed light to test Bell's inequalities, which can be possible for small photon numbers employing quantum correlations. Some general questions of quantum theory of measurements can be found in proceedings by Wheeler and Zurek (1983). A discussion of wave-particle duality can be found in the book edited by Selleri (1992) and questions of multiparticle interferometry and quantum superposition principle were considered by Greenberger, Horne and Zeilinger (1993), Reck and Zeilinger (1993) and Zeilinger, Zukowski, Horne, Bernstein and Greenberger (1993). A thorough discussion of the interference phenomena in relation to the Bell's theorem has been presented in the review by Belinsky and Klyshko (1993a) where other references to the relevant literature can be found. Further discussion of hidden variables and Bell's theorems was provided by Mermin (1993). Shih and Rubin (1993) discussed a four-photon interference experiment for testing the Greenberger-Horne-Zeilinger correlations using an arrangement based on two Franson interferometers (Sec. 8.6) involving two down-frequency converters. Kiess, Shih, Sergienko and Alley (1993) realized the EPR state using photon pairs from parametric down conversion demonstrating Bell's inequality violation of 22 standard deviations. Franson (1993) showed that violation of Bell's inequalities and "nonlocal" quantum effects can be demonstrated not only using measurements on individual particles, but also using high-intensity fields containing large numbers of photons producing macroscopic bursts of energy in two distant absorbers with "nonlocal" quantum correlations between them. This provides new point of view for interpretation of the correspondence principle. Żukowski, Zeilinger, Horne and Ekert (1993) suggested to use two down converters as independent sources for realization of Bell-Einstein-Podolsky-Rosen experiment as another test of quantum "nonlocality". Greater details and further experiments related to quantum correlated states and "nonlocal" quantum effects are presented in Chapter 8.

# Chapter 8

# Quantum optical experiments supporting quantum theory

In this chapter we review the most important experiments performed in quantum optics supporting quantum theory. We deal with one-photon as well as two-photon experiments, resonance fluorescence experiments, experiments in cavity quantum electrodynamics, experiments with various feedbacks, and experiments with squeezed light and sub-Poisson light generated in various nonlinear optical processes, such as optical parametric processes, particularly frequency down conversion, four-wave mixing, phase conjugation, optical Kerr effect as well as Raman scattering. In particular the process of the interaction of single atoms with light can provide effective tests of quantum behaviour. Some experiments related to violation of classical inequalities, Bell's inequalities and quantum "nonlocality" were discussed in Chap. 7.

## 8.1 Single-photon interference experiments

It is verified in single particle experiments (as discussed in Sec. 6.5) that many cumulative realizations of single photon interference lead to the classical picture, in agreement with the probabilistic interpretation of the wave function and the principle of correspondence of quantum and classical descriptions. The most interesting experiments of this kind with single photons were performed by Grangier, Roger and Aspect (1986), Aspect, Grangier and Roger (1989) and Aspect and Grangier (1991). In the last two papers one can also find a review of experiments with very weak light starting with the Taylor experiment [Taylor (1909)], in which the diffraction pattern of a needle illuminated by an attenuated discharge lamp was recorded on a photographic plate with exposure of six months. As the photon flux was about $10^6$ photons/s so that an average distance between two photons was about 300 m, one could say that in average only one photon was present during the interaction with the photographic plate.

In the above papers Aspect et al. first tested the operation of a beam splitter. A source was emitting light pulses incident on the beam splitter and the reflected and transmitted light was detected by two photodetectors providing the number of pulses $N_r$ and $N_t$. Moreover they were able to measure the number $N_c$ of coincidences in both the channels and the total number of events $N$. It is clear that in the case of single photons anticorrelation between responses of both the photodetectors has to be observed, whereas in the case of attenuated classical wave packets there is some probability of coincidences. If we define the probabilities $p_r = N_r/N$, $p_t = N_t/N$ and $p_c = N_c/N$ it follows from quantum optics that for a single-photon wave packet a perfect anticorrelation is obtained, i.e. there is no coincidence between transmitted and reflected photons. For fields having classical analogues it holds that

$$p_c \geq p_r p_t \,. \tag{8.1a}$$

Denoting the corresponding transmitted and reflected intensities $I_t$ and $I_r$, respectively, then the coincidence probability is proportional to $\langle I_r I_t \rangle$. In the case of attenuated classical wave packets there is a non-zero probability $p_c$ of coincidences fulfilling (8.1a) and in such a case it also holds for the following parameter $\beta$ that

$$\beta = \frac{N_c N}{N_r N_t} \geq 1 \,. \tag{8.1b}$$

This is clear quantitative criterion experimentally accessible allowing to distinguish between an attenuated classical wave packet and a single-photon wave packet. The violation of the inequality (8.1b) can also express a failure of semiclassical theories.

The above authors first realized an experiment with light from a pulsed photodiode with a rise time 1.5 ns and a fall time about 6 ns. The gates triggered by the electric pulses driving the photodiode were 9 ns wide and they had an almost complete overlap with the light pulses. The source was attenuated to a level corresponding to one detection per 1000 pulses emitted. The detector quantum efficiency was about 10 % and the average energy per pulse was estimated to correspond to about 0.01 photon. It has been found that $\beta = 1$ in this case and the experiment verified that attenuated light of a classical source does not exhibit single-photon behaviour on a beam splitter under the condition of very low energy in the pulse of the order of $10^{-2}$ photon.

The above authors further investigated fluorescent light emitted from a single atom. To have only one atom is quite necessary because in natural sources, such as discharge lamps, many randomly excited atoms are producing a macroscopic light field seen by a photodetector so that resulting light is a superposition of single-photon wave packets with random initial times [Teich, Saleh and Peřina (1984)]. Taking into account fluctuations of the number of excited atoms quantum behaviour of light is rapidly smoothed out [see discussions in the book by Peřina (1991), Secs. 9.1 and 11.1]. The authors used radiative cascade in calcium to produce single-photon pulses. The atom is excited to its upper level by a two-photon excitation using two lasers. Then two photons of frequencies $\upsilon_1$ and $\upsilon_2$ are emitted. The time intervals between pulses of frequencies $\upsilon_1$ and $\upsilon_2$ are distributed according to the exponential law. Choosing suitably the decaying times it is possible to obtain cascades well separated in time. Using photons of the frequency $\upsilon_1$ as a trigger for a gate, the probability of detecting a photon of the frequency $\upsilon_2$ coming from the same atom emitting a photon of the frequency $\upsilon_1$ is much higher than the probability of detecting a photon of the frequency $\upsilon_2$ coming from any other atom inside the source. Thus they have approached an ideal single-photon pulse and the corresponding anticorrelation of the transmitted and reflected pulses can be expected. Denoting the gate width as $w$, the authors measured the coefficient $\beta$ in dependence on $Nw$, i.e. as a function of the excitation rate of the cascade. Of course, for an ideal single atom $\beta = 0$. Taking into account the possibility of occurring several atoms excited inside the source during the same gate $w$, $\beta > 0$ is violating (8.1b). The maximum violation of more than 13 standard deviations has been observed for an integration counting time of 5 hours. The value $\beta = 0.18 \pm 0.06$ is corresponding to a total number of 9 coincidences, instead of the minimum value 50 expected for a classical wave packet. Of course, with increasing $Nw$ the coefficient $\beta$ is increasing to the value 1 corresponding to the classical field. Therefore such experiments confirm the anticorrelation on a beam splitter and represent good criterion for distinguishing between single-photon pulses and classical attenuated pulses. The experiments with single photons provide photon anticorrelation supporting quantum treatments and representing evidence against semiclassical theories of light, which never predict any violation of inequalities (8.1a,b).

The same authors also used the Mach-Zehnder interferometer (Fig. 6.6) with one photon. The gate system was again adopted and the path difference was discretized to 256 points of the distance $\lambda/50$. The integration time was 1 s and 15 s in every channel. The source with $\beta = 0.18$ was again employed and the visibility $C = 98.7 \pm 0.5$ % of the interference fringes was achieved with increasing counting time, clearly representing the classical interference pattern in strong intensity limit, in the sense of the discussion presented in Chap. 6. As an exhibition of particle character of single photons the anticorrelation of responses from the photodetectors occurs which is reflected by the fact that at places of the maxima at the detector $P_1$ there are minima at the

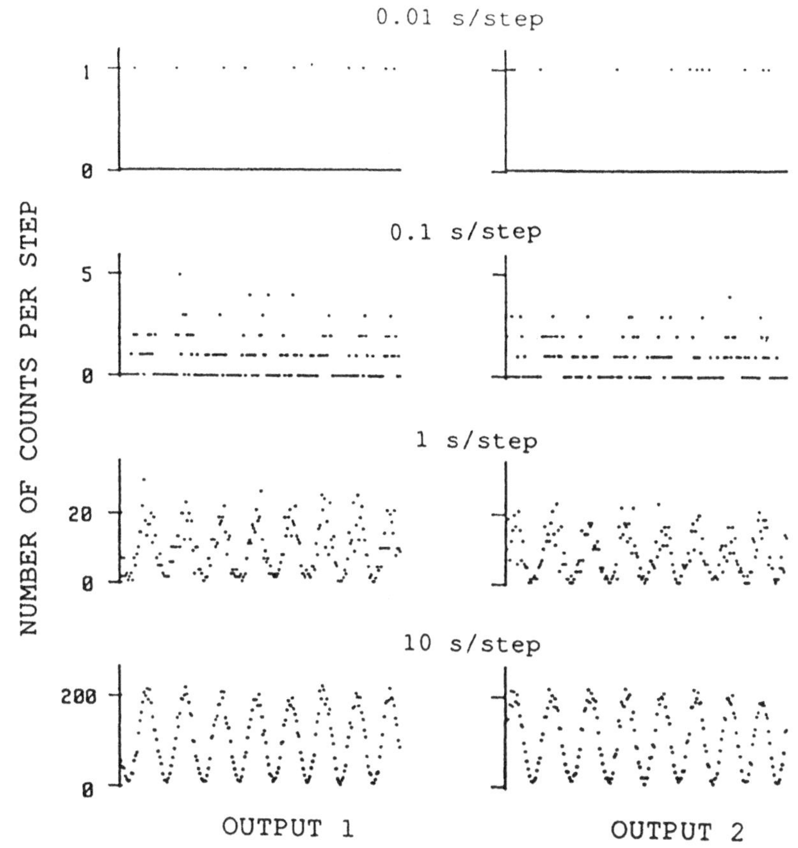


Figure 8.1: Successive creation of the interference pattern with single photons [after A. Aspect, P. Grangier and G. Roger, *J. Optics (Paris)* **20** (1989) 119; ©MASSON, Paris].

detector $P_2$ and vice versa. As a demonstration of single-photon interferences we present Fig. 8.1 showing the cumulative creation of the interference fringes in the experiment of Aspect et al. at the detectors $P_1$ and $P_2$, respectively with increasing integration counting times from 0.01 s to 10 s (in this case $\beta = 0.2$ and the channel distance was $\lambda/25$).

The above experimental results demonstrate the wave-particle duality for a single photon. They verify predictions of quantum theory, in relation to the other nonclassical effects such as sub-Poisson photon statistics, photon antibunching, squeezing of vacuum fluctuations, violation of classical and Bell's inequalities, oscillations in photocount distributions, collapses and revivals of atomic coherent oscillations and quantum nondemolition measurements.

## 8.2 Polarization correlation experiments

In relation to the EPR "paradox" we mention older optical experiments reported by Kocher and Cummins (1967) and Freedman and Clauser (1972) who adopted 6S–4P–4S cascade decay of Ca atoms. An atomic beam of calcium was excited by light of a hydrogen or deuterium arc producing a population of the 6S level of calcium. This represents the initial level for the 6S–4P–4S cascade decay producing two visible light pulses at wavelengths 551.3 nm and 422.7 nm. The light pulses were detected by photomultipliers placed on opposite sides where coincidences in photoelectric emissions were observed (Fig. 8.2). Interference filters and linear polarizers, whose orientations could be changed, were used to filter light before detection. It was found that the rate of coincidence counting was greater when the linear polarizers were aligned and close to zero when the polarizers were crossed. This represents a general disproof of hidden variable theories as discussed by Clauser, Horne, Shimony and Holt (1969), Clauser and Horne (1974) and Bell (1987). This also supports quantum theory and contradicts any neoclassical theory. As in Chap. 7 for the joint detection probability we obtain

$$p_{12} = \frac{1}{2} \cos^2 (\phi_1 - \phi_2) , \tag{8.2}$$

which vanishes for orthogonal position of polarizers $P_1$ and $P_2$. If we consider

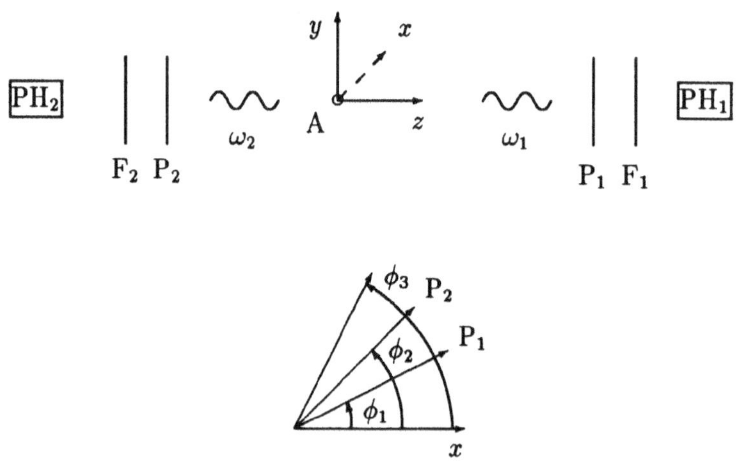


Figure 8.2: Polarization correlation experiment; two photon wave packets of frequencies $\omega_1$ and $\omega_2$ are emitted by the atom A; $P_1$ and $P_2$ are polarizers, $F_1$ and $F_2$ filters transmitting frequencies $\omega_1$ and $\omega_2$, respectively, and $PH_1$ and $PH_2$ are photodetectors; the corresponding angles are also indicated.

a linear polarization forming the angle $\phi_3$ with the axis $x$ (Fig. 8.2), wave packets passing through the two linear polarizers emerge with amplitudes $\cos(\phi_3 - \phi_1)$ and $\cos(\phi_3 - \phi_2)$, respectively. The probability amplitude for the joint detection of the two pulses is then

$$p_{12} \sim \cos(\phi_3 - \phi_1)\cos(\phi_3 - \phi_2) \,. \tag{8.3}$$

From the point of view of the classical theory the result is the same, only the probability amplitude is replaced by the actual field amplitude. If the light is unpolarized we have to perform the average over the direction $\phi_3$. According to quantum theory we must average (8.3) over $\phi_3$ as a quantum superposition, which leads to (8.2). On the other hand in the framework of the classical theory we detect the intensity obtained by squaring (8.3) and then we average over $\phi_3$, which provides the classical result

$$p_{12} \approx 1 + \frac{1}{2}\cos^2(\phi_1 - \phi_2) \,. \tag{8.4}$$

Hence, quantum theory predicts zero coincidences for crossed polarizers, whereas for classical theory $p_{12} = 1$ in this case. The measurements by Freedman and Clauser (1972) were in perfect agreement with quantum theory and in disagreement with the classical theory.

This process was employed by Aspect, Grangier and Roger (1981, 1982) and Aspect, Dalibard and Roger (1982) to demonstrate strong violation of Bell's inequalities. Saleh and Teich (1985) [for a review, see Teich and Saleh (1988)] suggested to use the cascaded photon emission from $^{40}Ca$ atoms for generation of sub-Poisson light. Green photons (551.3 nm wavelength) are collected by a lens and again passed through a polarizer and green-transmitting filter to a photomultiplier and discriminator, which produces the control signal (Fig. 8.3). The violet photons are collected from the other side of the source, passing through a polarizer and violet-transmitting interference filter, a delay path and optical gate that is opened for a period determined by green photon pulses. In this way the violet beam is regularized to obtain sub-Poisson nonclassical light.

## 8.3 Experiments with feedbacks

In general feedbacks repesent a tool for quieting light fluctuations. For this purpose Walker and Jakeman (1985) suggested to use photon pairs from parametric down conversion where one sum-frequency photon decays to two subfrequency photons. It was proved by Hong and Mandel (1985, 1985a) and Mandel (1986) that such pairs of sub-frequency photons arise simultaneously and they are strongly quantum correlated (entangled states). Therefore, one

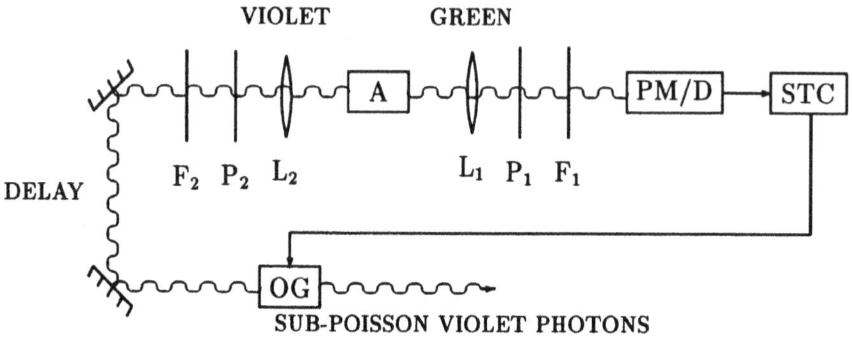


Figure 8.3: Scheme for generation of sub-Poisson light based on cascaded photon polarization correlated emissions of $^{40}Ca$ atoms (A); $L_1$, $L_2$ are lenses, $P_1$, $P_2$ polarizers, $F_1$, $F_2$ filters, PM/D is photomultiplier/discriminator (green), STC selected trigger circuit, OG optical gate.

of the sub-frequency photons from such a photon pair can be used to control the pumping process, emission mechanism, or photon detection process to produce sub-Poisson light.

Walker and Jakeman (1985) employed one photon from the photon pair arising in the down-conversion experiment using an ADP crystal to realize an electrical feedback signal, which controlled an optical shutter in the pumping beam. Poisson photons obtained from a He-Cd laser operated at 325 nm wavelength served as the excitation process. This UV radiation was passed through an acousto-optical shutter which served as a control gate and it was introduced to the ADP nonlinear crystal, which produced red photon pairs by means of the parametric down conversion. The authors observed a small but statistically significant increase in the fourth-order degree of coherence $\gamma_{\mathcal{N}}^{(2)}(\tau)$ in dependence on $\tau$ in the vicinity of $\tau = 0$. This indicated antibunching effect in the sense of the positive-derivative definition, although the values of $\gamma_{\mathcal{N}}^{(2)}(\tau)$ were higher than unity so that the observed radiation was super-Poissonian. In the later experiments, Rarity, Tapster and Jakeman (1987) and Tapster, Rarity and Satchell (1988) succeded in producing sub-Poisson light using this feedback method. The scheme of experimental arrangement is shown in Fig. 6.10 and finally they were able to produce sub-Poisson light with the Fano factor $F_n = 0.78$.

As we have mentioned in Sec. 6.6 (Fig. 6.11) Yamamoto, Imoto and Machida (1986) suggested the scheme for generating sub-Poisson light based on the quantum nondemolition measurement performed with the help of a lossless optical Kerr effect operating without photon destruction. Precision in the photon-number measurements is achieved at the expense of increased uncertainty in phase variable, as discussed in Sec. 5.5. These authors also

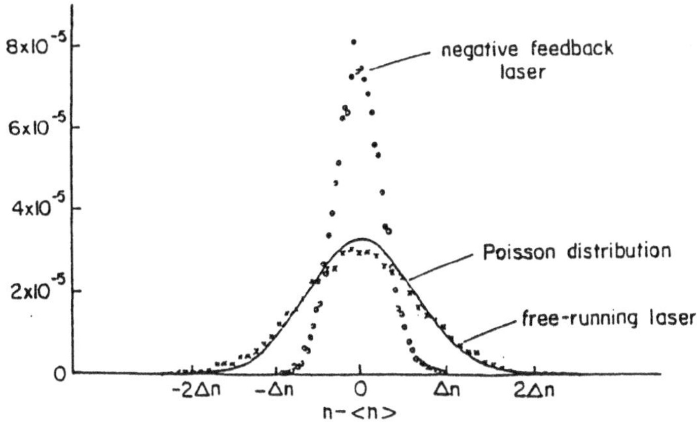


**Figure 8.4:** Photocount distribution for semiconductor laser with and without negative feedback [after Y. Yamamoto, N. Imoto and S. Machida, *Phys. Rev.* **A 33** (1986) 3243].

proposed that the results of such a quantum nondemolition measurement at the output of a semiconductor diode injection laser could be negatively fed back to control the excitation rate of the laser and to produce sub-Poisson photon beams.

Yamamoto, Imoto and Machida (1986) investigated, both theoretically and experimentally, the photon statistics of a semiconductor laser with and without negative feedback. They experimentally verified that the photon statistics in the feedback loop could be substantially reduced up to the Fano factor $F_n \approx 0.26$, whereas in the free-running regime of the semiconductor laser, the photon statistics were slightly super-Poissonian (see Fig. 8.4). Further Machida and Yamamoto (1988, 1989) have used high impedance suppression for pump-amplitude fluctuations to obtain the constant-current-driven semiconductor laser that produced field with its amplitude noise reduced by 32 % to below the standard quantum limit over a broad frequency range from near dc to 1.1 GHz. Quantum nondemolition measurements were investigated and applied by Yamamoto, Imoto and Machida (1986) and Kitagawa, Imoto and Yamamoto (1987).

Mertz, Heidmann, Fabre, Giacobino and Reynaud (1990) used the optical parametric oscillator to generate high-intensity sub-Poisson light with 24 % reduction below the shot-noise limit using one of the twin beams for feed-forward corrections. The corresponding analysis of feedback and feedforward control mechanisms for generation of sub-Poisson light, including the use of twin beams from an optical parametric oscillator, was presented by Mertz, Heidmann and Fabre (1991).

More details about applications of various feedbacks can be found in reviews by Shapiro, Saplakoglu, Ho, Kumar, Saleh and Teich (1987) and Teich and Saleh (1988).

## 8.4 Experiments with squeezed light

It should be noted that relatively strong effort to obtain substantially nonclassical sub-Poisson or antibunched light suitable for applications was less successful than the effort to generate quadrature squeezed light. In the beginning experiments [Slusher, Hollberg, Yurke, Mertz and Valley (1985)] nonclassical effects of tens per cent were achieved in reducing light fluctuations below fluctuations in the coherent field or physical vacuum. The first experiments with optical parametric oscillator performed by Wu, Kimble, Hall and Wu (1986) and Wu, Xiao and Kimble (1987) provided 63 % squeezing of vacuum fluctuations in the sub-threshold regime of the degenerate parametric down conversion including a number of degradation effects being a consequence of linear loss mechanism present in the measuring apparatus. They were also able to show that the generated optical field was in the minimum uncertainty state. Now experiments with optical parametric processes are sources of squeezed light with 90 % reduction of vacuum fluctuations in a quadrature component.

Now we can describe experiments based on the theory presented in Sec. 6.6. A schematic arrangement of the experiment realized by Wu et al. (1986, 1987) is outlined in Fig. 8.5. They have used the degenerate parametric oscillator in a cavity with a strong pumping. The parametric oscillator contained a nonlinear crystal of $MgO{:}LiNbO_3$ of dimensions $25 \times 9 \times 9$ mm$^3$, heated to phase matching temperature of 98°C. The pump field enters the optical parametric oscillator from a ring laser at frequency $2\omega$ (dashed line). The green pump light at the 0.53 $\mu$m wavelength is generated with a crystal of $Ba_2NaNb_5O_{15}$ inside the cavity of a Nd:YAG ring laser by the frequency doubling in the nonlinear crystal. The infrared emission at the 1.06 $\mu$m wavelength derived from the ring-laser beam, serves as a strong local oscillator beam in a balanced homodyne detector. The pump light is converted in the optical parametric oscillator to sub-frequency correlated photons with the wavelength of 1.06 $\mu$m, which exhibits squeezing of vacuum fluctuations, as discussed in Secs. 5.4 and 6.6. The balanced homodyne detector (Sec. 6.4) was formed by two InGaAs photodiodes of quantum efficiency $\eta = 0.9$. The photocurrents were amplified and subtracted to obtain phase-dependent noise variance, as indicated in (5.6) and (5.7). In the analysis it was also incorporated that real lasers do not produce exactly coherent states, but mostly produce coherent light with a small noise component, i.e. the superposition of the coherent field and chaotic field. Also the excess local-oscillator fluctuations were incorporated to the analysis involving the spectral correlation function of the local-oscillator beam. The square root of the variance $V$ (cf. (5.7)), depending on the local oscillator phase $\theta$, is demonstrated in Fig. 8.6. When changing the phase $\theta$ of the local oscillator, the square root $\sqrt{V}$ of

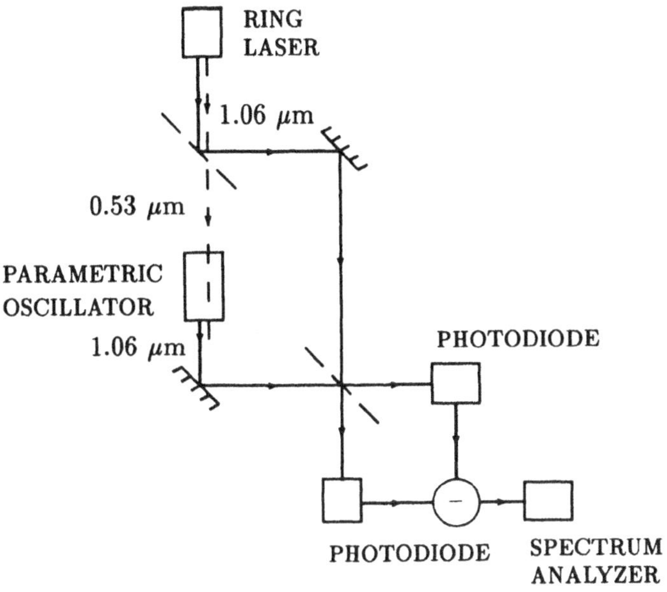


Figure 8.5: A scheme of the experiment for generation of squeezed light by means of optical parametric oscillator.

the variance $V$ follows the noise ellipse (Sec. 5.1), as indicated in Fig. 8.6 by the solid curve. The dashed circle and dashed line of $\sqrt{V}$ shown in Fig. 8.6 correspond to isotropic vacuum fluctuations when the signal from the optical parametric oscillator is blocked. The square-root variance was displayed as a function of the local oscillator phase $\theta$ at a fixed frequency of 1.8 MHz and a bandwidth of 100 kHz in the spectral distribution of photocurrent fluctuations. If the optical parametric oscillator is in operation, the dips below the vacuum level of the fluctuations represent about a 60 % (−4 dB) reduction of noise power relative to the vacuum noise level. The authors calculated the spectrum of squeezing showing that the intracavity field in their experiment was about 10 times squeezed, but the effect was masked by losses, which were substantially removed in later experiments.

Hence, Wu et al. (1986, 1987) were successful in applying the nonlinear process of sub-threshold optical parametric down conversion to reduce the noise level in homodyne detection by more than 60 % (−4 dB) relative to the vacuum noise level. The spectrum of squeezing extracted from their measurements indicated that the observed noise reduction resulted from light in which fluctuations were squeezed more than tenfold in the case of lossless light propagation and detection. The uncertainty product $\langle(\Delta\hat{Q})^2\rangle\langle(\Delta\hat{P})^2\rangle$

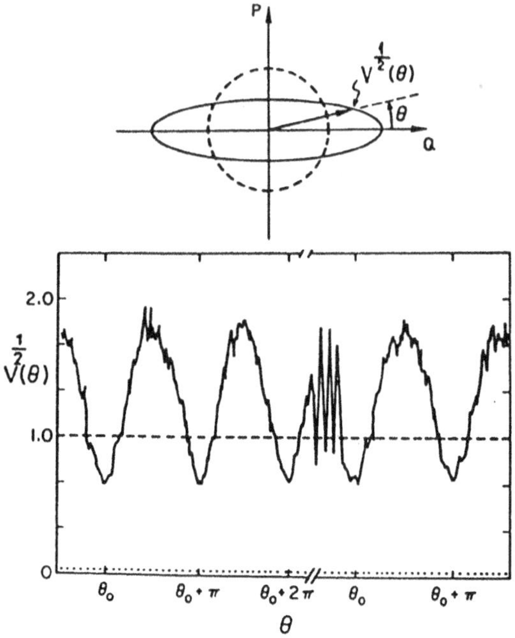


Figure 8.6: The noise ellipse for squeezed vacuum fluctuations in optical parametric oscillator and the phase dependence of fluctuations [after L.-A. Wu, H. J. Kimble, J. L. Hall and H. Wu, *Phys. Rev. Lett.* **57** (1986) 2520].

(given by (5.3)) derived from the measurements suggested that the optical field state produced in the lossless conditions is a minimum uncertainty state (squeezed vacuum state). Degradation of squeezing of vacuum fluctuations as a quantum effect was caused by various linear attenuation mechanisms, which were eliminated in later experiments. This permits to obtain sufficient conditions for generation of squeezed light for practical uses in high precision measurements, optical communications and high resolution spectroscopy. For instance such squeezed light can be used to achieve signal detection in an interferometer with substantial improvement in its sensitivity beyond the shot-noise limit (Sec. 3.7). Such an experiment with nonclassical light represents the manipulation with the basic commutation relations of quantized light fields.

The process of optical parametric down conversion in $KTiOPO_4$ (KTP) crystals was used by Slusher, Grangier, LaPorta, Yurke and Potasek (1987) to generate squeezed light in the form of trains of pulses, approximately of a duration of 100 ps. The arrangement of the experiment was similar to that desribed above. These authors used the frequency doubled output (532 nm wavelength) of a mode-locked Nd:YAG laser as a pump to the KTP squeezer. The peak of green pump power was 50 W. A portion of the initial 1.064 $\mu$m

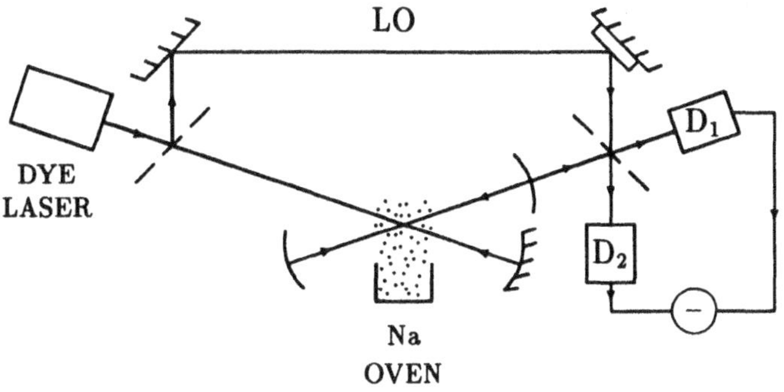


Figure 8.7: A scheme of backward four-wave mixing experiment for generation of squeezed light; $D_1$ and $D_2$ are detectors and LO is a local oscillator beam.

wavelength pulse train from the mode-locked laser was split off and delayed to form the local oscillator pulse train with pulses of 140 ps at width of 5 ns intervals. Parametric down conversion from green to infrared was obtained in a pair of KTP crystals. Linearly polarized squeezed light was formed by a superposition of photon signal and idler pairs. The squeezed pulse was again combined with the local oscillator beam at a beam splitter where their spatial and temporal forms, as well as their phase fronts, were matched. The usual balanced detection was then applied, ensuring the coincidence of the signal and local oscillator pulses. In this experiment the authors achieved a noise reduction of $-0.6$ dB in the pulse trains below the shot-noise limit. This source of squeezed light combined with an interferometer can increase the sensitivity of picosecond time-scale measurements. Such results are important particularly for quantum noise control of soliton light pulses in optical fibres [Rosenbluh and Shelby (1991), Singer, Potasek, Fang and Teich (1992)]. Also, optical communication, processing and computing involve pulse sequences where the noise is determined by quantum effects and it is requested to be suppressed as possible. It may be expected that in optical quantum computers pulsed squeezed light will play an important role in the information processing in the computer. We refer the reader to a discussion of this subject in papers by Yurke, Grangier, Slusher and Potasek (1987), LaPorta, Slusher and Yurke (1989), Debuisschert, Reynaud, Heidmann, Giacobino and Fabre (1989), Mertz, Heidmann, Fabre, Giacobino and Reynaud (1990) and Hirano and Matsuoka (1990).

The first successful experiment to observe squeezing of vacuum fluctuations was performed by Slusher, Hollberg, Yurke, Mertz and Valley (1985) using the backward four-wave mixing on Na atoms in a cavity. They have used the arrangement schematically shown in Fig. 8.7. A cw ring dye laser was frequency and phase stabilized and used as a pumping beam of frequency $v_p$ for Na atoms placed in a cavity. A mirror serves to obtain the counter-propagating pumping beams. The pumping beams nonlinearly interact with cavity modes $v_p \pm Nv_{SC}$, $v_{SC} = 140.5$ MHz being the cavity mode-spacing frequency and $N$ is an integer ($N = 3$ in this case). The signal beams are generated from the vacuum fluctuations, one of the signal beams is amplified and the other is deamplified in relation to the phase of the pump field relative to the phase of the cavity modes, in agreement with the discussion in Sec. 6.6. All four waves are phase matched over the sodium interaction region. A part of the pumping laser beams is used as a strong local oscillator beam for the balanced homodyne detection operating as described before. The registered variance behaved similarly as that in Fig. 8.6 in dependence on the phase of the local oscillator which was piezoelectrically controlled. In the original measurements 7 % squeezing of vacuum fluctuations was observed and after correcting for the phase jitter, detector efficiency and spontaneous emission, the squeezing effect was 28 %. In the later experiments Slusher, Yurke and Mertz (1987) and Slusher, Yurke, Grangier, LaPorta, Walls and Reid (1987) obtained direct squeezing of vacuum fluctations of about 20 % (−1 dB) and of about 44 % after the phase jitter and detection efficiency corrections. More explicit comparison of the experimental data for the squeezing of vacuum fluctuations with the theory was performed by Slusher, Yurke, Grangier, LaPorta, Walls and Reid (1987).

Maeda, Kumar and Shapiro (1987, 1987a) observed squeezed vacuum fluctuations produced by forward four-wave mixing in sodium vapour. The two pumping beams and the two signal beams are forward propagating and intersecting each other in a sodium cell (Fig. 8.8). The reflected-conjugate and transmitted signal beams were isolated and combined on a 50 % beam splitter to produce the composite output beam. They were further combined with the local-oscillator beam on a second beam splitter, and the resulting output beams were focused onto detectors for balanced homodyne detection. They observed about 4 % (−0.2 dB) squeezing of vacuum fluctuations corresponding to actual squeezing of −0.6 dB after elimination of losses. These authors also demonstrated that nonclassical behaviour cannot be observed in the signal beam homodyne detection experiment or in the signal-conjugate beam direct detection.

The third group of pioneering experiments to observe squeezing of vacuum fluctuations was realized by Shelby, Levenson, Perlmutter, DeVoe and Walls (1986), Shelby, Levenson, Walls, Aspect and Milburn (1986), Levenson, Shelby, Reid and Walls (1986), Levenson and Shelby (1987) and Milburn, Levenson, Shelby, Perlmutter, DeVoe and Walls (1987). In this case two-mode squeezed state generation in optical fibres was realized by means of the forward four-wave mixing based on the optical Kerr effect. Strong pump waves

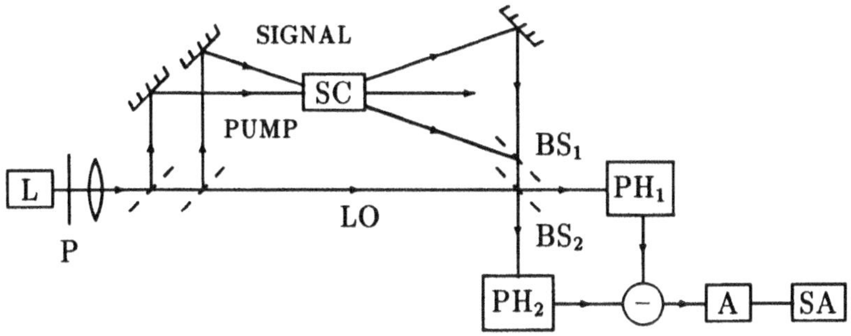


Figure 8.8: Forward four-wave mixing for generation of squeezed light; L is dye laser, P polarizer, SC sodium cell, LO local oscillator beam, $BS_1$ and $BS_2$ are beam splitters, $PH_1$ and $PH_2$ photodetectors, A is amplifier, SA spectrum analyzer.

at two different wavelengths of 647 and 676 nm from a krypton-ion laser were combined in a prism. Then they propagated through 114 m of a single-mode optical fibre and were separated by a second prism. One wave was incident on a photodiode, which determined its amplitude fluctuation noise, while the other wave was shifted in an interferometer and detected by another photodiode. The dc current from both the photodiodes were measured by digital voltmeters, and ac noise currents were delayed and combined. The spectrum analyzer displayed the electric noise power in the combined signal. The best experimental results were obtained with the optical fibre cooled to below 2 K to minimize the phase noise imposed by the fibre. At such low temperatures stimulated Brillouin scattering in the fibre was very low (about 5 mW for each frequency component when the power of pumping was 150 mW). The four-wave mixing in the single-mode fibre generated squeezed light with a 20 % reduction of vacuum fluctuations. Application to a quantum nondemolition measurement by Levenson, Shelby, Reid and Walls (1986) led to a 37 % reduction of quadrature fluctuations on one wave at the expense of the increase of quantum fluctuations of the other wave. It was shown by Milburn, Levenson, Shelby, Perlmutter, DeVoe and Walls (1987) that absorbing media act as a source of vacuum fluctuations adding noise to squeezed states of light, i.e. they absorb the correlated signal and idler photons and add vacuum fluctuations. The squeezing effect is reduced in the way like the quantum efficiency of the detector was reduced.

Removing scattered light effects and effects of polarization decorrelation in the optical fibre, further improvement can be gained and it is possible to go beyond the standard quantum limit using the contemporary optical fibres and detection techniques, and at power levels accessible with current technology one may obtain noise suppression of about 6 dB. Using nondemolition (back-action-evading) measurements [for a review see Meystre (1992)] (Sec. 3.6) and two-frequency homodyne detection systems, it was estimated that noise levels up to 200 times below the vacuum noise levels may be realized.

There are other interesting physical phenomena suitable for generation of squeezed light. For instance we can mention the experiment by Yurke, Kaminsky, Miller, Whittaker, Smith, Silver and Simon (1988) who reported the observation of the equilibrium noise squeezing at the temperature 4.2 K using the Josephson parametric amplifier. They achieved 42 % squeezing of vacuum fluctuations at 19.4 GHz. Orozco, Raizen, Xiao, Brecha and Kimble (1987) reported the squeezed state generation in optical bistability [Gibbs (1985)]. A collection of two-level Na atoms was placed in a high-finesse cavity. They have used a frequency-stabilized ring dye laser as a source of atomic excitations. Part of the beam was split to serve as a local oscillator beam, while the other part of the beam has driven the cavity for squeezed-state generation as a signal beam. Further laser beams in addition to the signal and local oscillator beams were employed to prepump optically the sodium atoms into $3^2S_{1/2}$, F=2, $m_F$=2 state of the $D_2$ line (589 nm). One of the beams is frequency shifted with an acousto-optic modulator to permit a fixed detuning between the optical pumping and the signal beam, the other pumping beam is collinear with the frequency shifted one and is used for initial alignment and for measurements of the input-output characteristics in absorptive bistability. Another dye laser transferred the atomic population from the F=1 to the F=2 ground state by the F=2 excited state to increase the optical density of the atomic beam entering the squeezing cavity. The weak-field coupling of atoms to the cavity mode produced a splitting in the normal mode structure of the atom-field system, which is large compared with the atomic linewidth. Reductions of 30 % (−1.55 dB) below the noise level of vacuum fluctuations were observed in a balanced homodyne detector. A degree of about 50 % reduction was inferred for generated field state in the absence of propagation and detection losses. The observed spectrum of squeezing extended over a very broad range of frequencies ($\sim \pm 75$ MHz).

Kürz, Paschotta, Fiedler and Mlynek (1993) reported squeezed light generation of the considerable power of 3.2 mW with 52 % squeezing of vacuum fluctuations under the shot-noise level with about 10 s stabilization using second-harmonic generation in a monolithic resonator.

Interesting effects of squeezing of vacuum fluctuations can occur in quantum soliton propagation [Drummond and Carter (1987), Kodama and Hasegawa (1992)]. Classical solitons represent noiseless nonlinear waves propagating along single-mode optical fibres thousands of kilometers without any change, which is an important feature needed for optical communications. The generation and propagation of the soliton is determined by the nonlinear

Schrödinger equation involving the cubic nonlinearity in the field amplitude [see e.g. Saleh and Teich (1991), Sec. 19.8] derived from the intensity dependence of the refractive index (cf. Sec. 6.6) of the optical fibre (optical Kerr effect). If such nonlinear Schrödinger equation is quantized, we have the equation for propagation of quantum solitons involving fluctuations of the field physical vacuum. If a linear approximation for the quantum propagation equations is adopted, it is possible to demonstrate that fluctuations of one quadrature of the soliton field can be reduced below the value of the vacuum fluctuations at the expense of an increase of fluctuations of the other quadrature of the soliton field, thus obtaining squeezed solitons [Drummond and Carter (1987)], and also indicating reduced quantum fluctuations at different spatial directions. Squeezing of quantum fluctuations below the vacuum level in a propagating soliton represents a general feature of nonlinear Schrödinger equation for soliton propagation in the quantum description. Further details including possibilities of experimental realizations can be found in the review article by Kodama and Hasegawa (1992).

## 8.5 Experiments with sub-Poisson light

We can now pay attention to experiments permitting to generate nonclassical light exhibiting photon antibunching and sub-Poisson photon statistics, which also represent fundamental quantum features of light supporting quantum theory and also having important applications in optical communications, high-precision measurements and high sensitivity spectroscopy involving new states of optical fields.

The first experiment to observe nonclassical behaviour was performed by Kimble, Dagenais and Mandel (1977), more precise measurements were performed later on by Kimble, Dagenais and Mandel (1978), Dagenais and Mandel (1978) and Cresser, Häger, Leuchs, Rateike and Walther (1982). As shown by Carmichael and Walls (1976), Kimble and Mandel (1976) and Cohen-Tannoudji (1977), the fourth-order (intensity) correlation function of light emitted by a single atom undergoing resonance fluorescence exhibits photon antibunching because it has the following factorized form

$$\langle \mathcal{T} : \hat{I}(t)\hat{I}(t+\tau) :\rangle = \langle \hat{I}\rangle\langle \hat{I}_G(\tau)\rangle , \tag{8.5}$$

where $\mathcal{T}$ denotes the time-ordering operator, : : means again the normal ordering of field operators, $\hat{I} = \hat{A}^{(-)}\hat{A}^{(+)}$ represents the intensity operator, $\langle \hat{I}\rangle$ is the steady-state mean intensity, and $\langle \hat{I}_G(\tau)\rangle$ is the mean intensity of light that is radiated by the atom driven by an external field at time $\tau$ if it starts in the ground state at $t = 0$. As the atom cannot radiate in its ground state, it follows that $\langle \hat{I}_G(\tau)\rangle$ always starts from zero at $\tau = 0$ and then grows

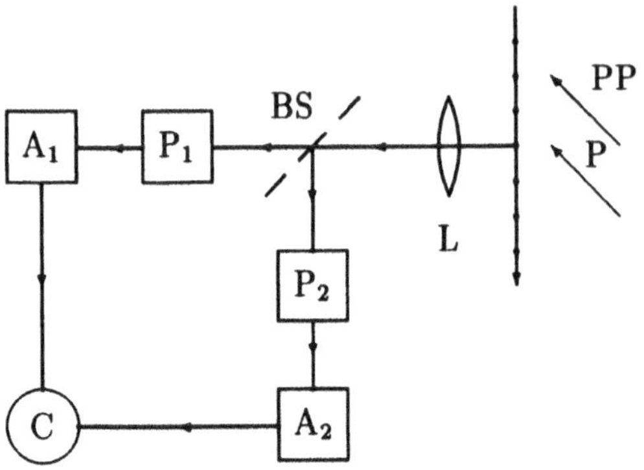


Figure 8.9: A scheme of the experiment for observation of photon antibunching in resonance fluorescence light; a sodium atomic beam is prepumped by a dye laser beam PP, P is the pumping beam, L lens, BS beam splitter, $P_1$, $P_2$ are photodetectors, $A_1$, $A_2$ amplifiers and C is coincidence device.

with $\tau$ and reaches its steady-state value $\langle \hat{I} \rangle$ after a time longer than the natural life time. Denoting the Rabi frequency as $\Omega = 2P\mathcal{E}/\hbar$, $P$ being the atomic dipole matrix element and $\mathcal{E}$ the driving field amplitude and writing $\beta$ for the half of the Einstein A-coefficient for the transition, then

$$\langle \hat{I}_G(\tau) \rangle = \langle \hat{I} \rangle [1 + \lambda(\tau)] , \tag{8.6}$$

where $\lambda(\tau)$ represents the normalized correlation function of fluctuations from both the detectors of the Hanbury Brown-Twiss correlation arrangement,

$$\lambda(\tau) = \frac{\langle \mathcal{T} : \hat{I}_1(t) \hat{I}_2(t+\tau) : \rangle}{\langle \hat{I}_1 \rangle \langle \hat{I}_2(t+\tau) \rangle} - 1 \tag{8.7}$$

and

$$1 + \lambda(\tau) = 1 - \exp\left(-\frac{3}{2}\beta\tau\right) \left[\cos(\Omega'\beta\tau) + \frac{3}{2\Omega'} \sin(\Omega'\beta\tau)\right] ,$$

$$\Omega' = \left(\frac{\Omega^2}{\beta^2} - \frac{1}{4}\right)^{1/2} . \tag{8.8}$$

A scheme of the experimental arrangement of Kimble, Dagenais and Mandel (1977) and Dagenais and Mandel (1978) is outlined in Fig. 8.9. The authors used beam of sodium atoms optically pumped by a dye laser in order to prepare a pure two-level system. The atomic beam was irradiated at right angles with a highly stabilized dye laser tuned on resonance with the $3^2S_{1/2}$,

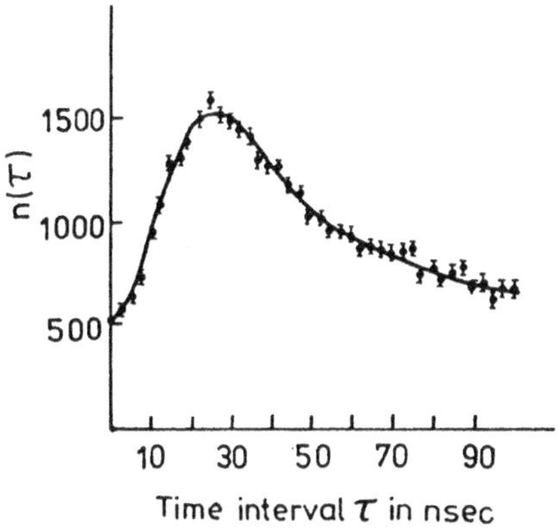


Figure 8.10: Comparison of unnormalized measured values of $n(\tau)$ with the theory for $\Omega/\beta = 3.3$ [after M. Dagenais and L. Mandel, *Phys. Rev.* **A 18** (1978) 2217].

F=2, $m_F$=2 to $3^2P_{3/2}$, F=3, $m_F$=3 transition in sodium. The intensity of the atomic beam was reduced so that, on average, no more than one atom was present in the observation region at a time. The fluorescent light from a small observation volume was obtained in a direction orthogonal to both the atomic and pump laser beams. Further, the standard Hanbury Brown-Twiss coincidence arrangement was used, i.e. fluorescent light was divided by a beam splitter and the arrival of photons in each beam was detected by two photomultipliers. The pulses from the two detectors were fed to the start and stop inputs of a time-to-digital converter where the time intervals $\tau$ between the start and stop pulses were digitized and stored. The number of events $n(\tau)$ stored at the address $\tau$ is a measure of the joint probability density $\eta^2\gamma_N^{(2,2)}(\tau)$ of the separation of photons by $\tau$ seconds provided that the efficiency of the photodetector is $\eta$.

The results of the measurement of Dagenais and Mandel (1978) are shown in Fig. 8.10. The growing part of the curve really violates the classical inequality $\gamma_N^{(2,2)}(\tau) \leq \gamma_N^{(2,2)}(0)$. This demonstrates antibunching of photons in resonance fluorescence light in the sense of positive derivative of the curve in the corresponding region. If the observed results shown in Fig. 8.10 are normalized, scattered light is eliminated and also transit time effects and atom number fluctuations are taken into account, we can obtain a normalized curve giving evidence for photon antibunching in the phenomenon of resonance fluorescence, since $\lambda(0) = -1$ and $1 + \lambda(0) = 0$, whereas for classical system this correlation function takes on its maximum value. However, as a result of atom number fluctuations the light was not sub-Poissonian.

Similar demonstrations have been presented by Cresser, Häger, Leuchs, Rateike and Walther (1982) who measured the normalized intensity correlation function $\gamma_N^{(2,2)}(\tau)$ for atomic beams in dependence on the time delay $\tau$.

They clearly proved on the basis of the experimental data that $\gamma_{\mathcal{N}}^{(2,2)}(\tau) < 1$ in the vicinity of $\tau = 0$ and that $\gamma_{\mathcal{N}}^{(2,2)}(0) = 0$ and they nicely demonstrated photon antibunching from a single atom. They were also able to follow the smoothing effect of photon antibunching caused by increase of the number of atoms leading to the diminishing of antibunching. Later on Diedrich and Walther (1987) investigated the resonance fluorescence of a single atomic ion in a radio-frequency trap. The effect of antibunching around $\tau = 0$ exhibited a strong dependence on the ion number. For a single ion it was exactly $\gamma_{\mathcal{N}}^{(2,2)}(0) = 0$, whereas photons emitted by independent ions were not correlated, thus leading successively to classical correlation function because the number of photons in the field increases. If there are more than one ion stored in the trap, the Coulomb repulsion keeps them at a distance of several micrometers and fluctuations in their positions are of the same order of magnitude. Therefore coherent beating of the light emitted by different ions was negligible, and also background radiation provided no coherent contribution since several coherence zones were observed. The authors also demonstrated the sub-Poisson behaviour of emitted photons obtaining 28 standard deviations from the Poisson statistics, although the reduction of the Fano factor below the unity was very small of the magnitude $10^{-4}$. Photon antibunching in multiphoton resonance fluorescence was observed by Grangier, Roger, Aspect, Heidmann and Reynaud (1986).

Thus, it has been experimentally demonstrated that fluorescent photons from a single atom exhibit antibunching in time, which may be regarded as a reflection of the fact that the excited atom makes a quantum jump to the ground state in the process of emitting a photon and is unable to radiate again afterwards because a relaxation process of the atom has to occur first. Hence, the photon coincidence technique makes it possible to test directly the behaviour of the atom and to observe the antibunching effect, which is a direct manifestation of the quantum nature of light, in full agreement with predictions od quantum electrodynamics.

Later on Short and Mandel (1983) observed also sub-Poisson photon behaviour in resonance fluorescence light with the Fano factor $F_n = 0.998$ slightly reduced below the coherent-state value 1, nevertheless the statistical confidence was very high. They supplemented the experimental arrangement shown in Fig. 8.9 by an additional control system opening the measurement when only one atom was present in the scattering volume. Thorough discussion of the photocount statistics of resonance fluorescence light was presented by Mandel (1979), Cook (1980, 1981), Smirnov and Troshin (1981, 1987) and Lenstra (1982). Furthermore Mandel (1979) derived that the bound value of the quantity $F_n - 1$ in the resonance fluorescence light of a single atom is $-3/4$.

Both the quantum effects, i.e. photon antibunching and sub-Poisson photon statistics in resonance fluorescence light are fastly smoothed out if more atoms are present in the scattering volume as a result of fluctuations in the number of atoms [Jakeman, Pike, Pusey and Vaugham (1977), Carmichael, Drummond, Walls and Meystre (1980), Schubert, Süsse, Vogel, Welsch and Wilhelmi (1982)].

An interesting quantum effect in the system of interacting atom and light is the occurrence of collapses and revivals of atomic oscillations (Sec. 5.8). This was observed by Rempe, Walther and Klein (1987), Diedrich, Krause, Rempe, Scully and Walther (1988) and Rempe, Schmidt-Kaler and Walther (1990), which is the phenomenon very frequently discussed in the literature now.

These experiments with resonance fluorescence light of a single atom have fundamental physical meaning for verification of quantum electrodynamics. Hovewer, such light is of small value for applications. Therefore Teich, Saleh and Stoler (1983) suggested using the classical Franck-Hertz experiment, used originally for the first demonstration of quantum nature of matter, as a source of antibunched and sub-Poisson light at the 253.7 nm wavelength. The basic equation for such an experiment is the Burgess formula (2.125) and the corresponding theory of the effect of statistics of primary excitations and individual emissions was elaborated by Teich, Saleh and Peřina (1984). In the Franck-Hertz tube a quiet electron beam with sub-Poisson fluctuations with the Fano factor $F_m < 1$ occurring in the Burgess formula (2.125) interacts with Hg atoms and produces by spontaneous emissions a sub-Poisson photon beam with the Fano factor $F_n < 1$.

Teich and Saleh (1985) used a specially constructed 25 mm-diameter UV-transmitting Franck-Hertz tube containing 0.75 g of Hg in their experiment. The tube was heated in an oven to permit the vapor pressure of the Hg to be temperature controlled. By means of external voltage, the cathode was heated to an appropriate temperature to produce thermoionically emitted electrons with Coulomb repulsions and to form a space-charged cloud. The cathode was specially formed and the grid voltages were adjusted to provide a desired space-charge-limited electron current with controlled electron distance. In Hg atoms, the transition $6^3P_1 \longrightarrow 6^1S_0$ was used to obtain 253.7 nm Franck-Hertz light by inelastic collisions of Hg atoms with the space-charge-limited quiet electron beam. The most pronounced sub-Poissonian behaviour was obtained at the temperature 26.6°C. The Poisson fluctuations of the electron beam were reduced to the value of the Fano factor $F_m \approx 0.1$ (but values about 0.01 are achievable). The counting time interval was $T = 1$ $\mu$s and from 10 to 15 $\times$ $10^6$ samples were realized to construct the photocount distribution, while 50 photocount distributions provided a value of photon Fano factor $F_n$, giving a global measure of the sub-Poissonian behaviour of generated Franck-Hertz light. As the efficiency $\eta$ of the conversion of the electron beam to the

Franck-Hertz photon beam was very small (about 0.0025), the sub-Poisson effect was also very small, with $F_n \approx 0.998$, but the statistical confidence was high. The observed quantum effect was by 2 to 3 standard deviations. The small effect is a result of losses in the experimental apparatus. If the efficiency of the conversion $\eta$ and the electron Fano factor $F_n$ were 0.5 and 0.01 respectively, one could obtain for the photon Fano factor $F_n = 0.525$, i.e. the photon beam can be quietened by about a factor 2, compared to the Fano factor for a coherent laser source. The output was fed into photocounting equipment which provided the photocount distribution $p(n,T)$. The mean number of counts $\langle n \rangle$ and the Fano factor $F_n$ were then calculated from the photocount distribution $p(n,T)$. The measurements demonstrated that quiet sub-Poisson light is less affected by the photodetection dead-time effect than is coherent light. In principle, this method may be adopted to use atomic, ionic, or molecular beams to produce sub-Poisson X-rays.

Small quantum effect occurring in the Franck-Hertz light is related to very low efficiency of conversion, as mentioned above. Therefore Capasso and Teich (1986) and Teich, Capasso and Saleh (1987) [see Teich and Saleh (1988) for a review] suggested the solid-state version of the experiment using a semiconductor device structure in which sub-Poissonian electron excitations are attained through space-charge-limited current flow. Single photon emissions are then obtained through recombination radiation. Thus sub-Poisson electrons are directly converted into recombined sub-Poisson photons. Of course, the efficiency $\eta$ of the conversion in the semiconductor is substantially higher and significantly sub-Poisson light may be achieved in this way. For instance, for the surface-emitting GaAs/GaAlAs configuration and Si p-i-n photodetectors, the quantum efficiency $\eta \approx 0.1125$, yielding to the post-detection Fano factor $F_n \approx 0.899$. Therefore the space-charge-limited light emitting devices are suitable for generation of significantly nonclassical light. These properties can further be improved by conditions for stimulated emissions as suggested by Yamamoto, Machida and Nilsson (1986). The stimulated emissions can improve the directionality of nonclassical light, high switching speed, spectral properties, etc. Another interesting way to produce sub-Poisson light is to reduce fluctuations of primary excitations in lasers, i.e. to reduce fluctuations in laser pumping below the shot noise as suggested by Yamamoto, Imoto and Machida (1986), Smirnov and Troshin (1985, 1987), Caroll (1986), Golubev (1987) and Marte, Ritsch and Walls (1988) among others. Sub-Poisson light with photon Fano factor $F_n \approx 0.96$ was generated in the broad frequency region by high-efficiency light-emitting diodes with reduced fluctuations of pumping current in the experiment by Tapster, Rarity and Satchell (1987). Machida, Yamamoto and Itaya (1987) generated sub-Poisson light using high-impedance suppression for pump amplitude fluctuations in a highly saturated semiconductor laser. They observed that the noise level of 7.3 % (31 % after

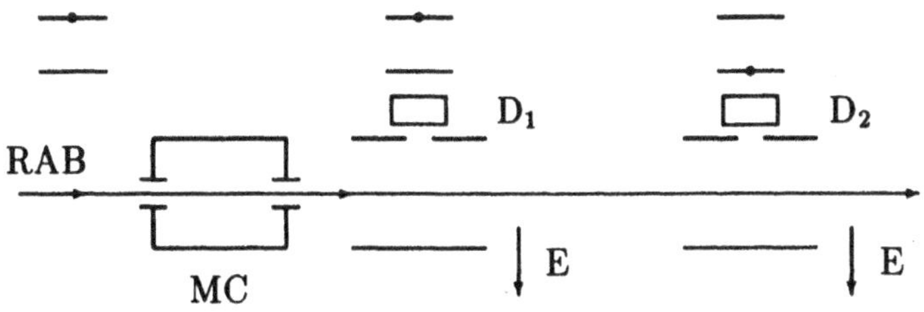


Figure 8.11: A scheme of the experimental arrangement to generate a number state; RAB denotes Rydberg atomic beam, MC is a micromaser cavity, $D_1$ and $D_2$ are detectors of upper-level atoms and lower-level atoms, respectively and E is a ionizing electric field.

corrections) was reduced in power below the standard quantum limit in the frequency range of 350–450 Hz.

Interesting experiments on observation of sub-Poisson photon statistics were performed with the help of micromaser cavities and Rydberg atoms, which have high principal quantum number $n$ and can easily be ionized [Filipovicz, Meystre, Rempe and Walther (1985)], as discussed by Krause, Scully and Walther (1987), Diedrich, Krause, Rempe, Scully and Walther (1988), Krause, Scully, T. Walther and H. Walther (1989), Rempe, Schleich, Scully and Walther (1989) and Rempe, Schmidt-Kaler and Walther (1990).

These authors used Rydberg two-level atoms which were excited to their upper states and were injected into a micromaser cavity. Such a micromaser (one-atom maser) was realized by Meschede, Walther and Müller (1985) using a single Rydberg atom (e. g. rubidium atoms pumped in the transition $5s_{1/2} \longrightarrow 63p_{3/2}$ and transitions $63p_{3/2}$ to $61d_{3/2}$ and $61d_{5/2}$ were investigated) and a single mode of a superconducting cavity with the quality factor $Q = 8 \times 10^8$ at 2 K. Signals were still detected with an average number of 0.06 atoms being present in the cavity. After they leave the cavity, they are probed by a static electric field which ionizes all atoms in their upper level. All the atoms which are not ionized emit a photon in the cavity. Then the atoms in the lower level which have emitted a photon in the cavity can be counted and the total number of photons in the maser cavity can be determined, i.e. in lossless cavity the number of atoms leaving the cavity in the lower state is equal to the number of photons in the cavity (Fig. 8.11). This means that the number state of the outgoing atoms is determined. As no radiation is present in the cavity at the beginning, the generated field will be in a number state when atoms enter the cavity. By the interaction of the atom with the field being in the Fock state $|n\rangle$, the field will change to a superposition of pure states $|n\rangle$ and $|n+1\rangle$, which is reduced by the measurement of the atomic state to the state $|n\rangle$ or $|n+1\rangle$ in dependence on the result of the measurement. For these measurements providing the number-state field, it is important to have long life-time $\tau$ of photons in the cavity. Since the quality factor of the cavity $Q$ can have value $5 \times 10^{10}$, for the frequency $v = 2 \times 10^{10}$ s$^{-1}$ we can have $\tau \approx Q/v = 2.5$ s. The fluxes of atoms used in these experiments were of the order of $10^{-3}$ s$^{-1}$. To obtain a pure number state it is necessary to suppress losses as far as possible, otherwise the pure state $|n\rangle$ is transformed to a mixed state according to the Bernoulli distribution (2.121).

Using a liguid $^3$He cryostat as a cooling environment the mean number of thermal photons was $\langle n_R \rangle = 0.016$ at T = 0.25 K and $\langle n_R \rangle = 3.3 \times 10^{-5}$ at T = 0.1 K for the frequency $v = 2 \times 10^{10}$ s$^{-1}$. The experiment has to be performed repeatedly with a constant number of atoms providing a large number of different number states, from which the probability distribution for photon numbers can be deduced. It is possible to determine the probability $p_M(n)$ to have $n$ photons in the field after $M$ atoms have passed the cavity [Krause, Scully and Walther (1987)]. If the initial field is in the state with $m$ photons, then the time-evolution operator $\hat{U}(t)$ for the interaction of one two-level atom with the field is given by [see e.g. Meystre and Sargent (1991)]

$$\hat{U}(t) = \begin{bmatrix} \cos\left(\sqrt{m+1}\,gt\right) & i\sin\left(\sqrt{m+1}\,gt\right) \\ -i\sin\left(\sqrt{m+1}\,gt\right) & \cos\left(\sqrt{m+1}\,gt\right) \end{bmatrix}, \tag{8.9}$$

where $g$ is the coupling constant of the interaction and $t$ is its duration. Thus the initial atom-field statistical operator is $\hat{\varrho}(0) = |a,m\rangle\langle a,m|$, $|a,m\rangle \equiv |a\rangle|m\rangle$ denoting the upper level of the atom $|a\rangle$, and after the interaction time $t$ we arrive at

$$\begin{aligned} \hat{\varrho}(t) &= \hat{U}(t)|a,m\rangle\langle a,m|\hat{U}^\dagger(t) = \cos^2\left(\sqrt{m+1}\,gt\right)|a,m\rangle\langle a,m| \\ &+ \sin^2\left(\sqrt{m+1}\,gt\right)|b,m+1\rangle\langle b,m+1| + i\sin\left(\sqrt{m+1}\,gt\right) \\ &\times \cos\left(\sqrt{m+1}\,gt\right)\Big[|b,m+1\rangle\langle a,m| - |a,m\rangle\langle b,m+1|\Big], \end{aligned} \tag{8.10}$$

where $|b\rangle$ denotes the lover atomic level. If the atom is found in the upper state $|a\rangle$, then the statistical operator (8.10) reduces to

$$\hat{\varrho}(t) = \cos^2\left(\sqrt{m+1}\,gt\right)|a,m\rangle\langle a,m|, \tag{8.11a}$$

and if the atom is found to be in the lover state $|b\rangle$, the statistical operator is reduced to

$$\hat{\varrho}(t) = \sin^2\left(\sqrt{m+1}\,gt\right)|b,m+1\rangle\langle b,m+1|. \tag{8.11b}$$

Thus the probability for the field to remain in the state $|m\rangle$ is equal to $C(m) \equiv \cos^2\left(\sqrt{m+1}\,gt\right)$, whereas the probability that it makes a transition to the state $|m+1\rangle$ equals $S(m) \equiv \sin^2\left(\sqrt{m+1}\,gt\right)$. When $M-1$ atoms have passed, the field is in the state $|n\rangle$ with the probability $p_{M-1}(n)$ and in the state $|n-1\rangle$ with the probability $p_{M-1}(n-1)$. The probability for the field to be in the state $|n\rangle$ after $M$ atoms have been present in the cavity is then determined by

$$p_M(n) = C(n)p_{M-1}(n) + S(n-1)p_{M-1}(n-1)\,, \tag{8.12}$$

where for the field in the initial vacuum state $|0\rangle$ it holds that $p_0(0) = 1$. The recursive relation (8.12) determines uniquely the probability $p_M(n)$, which can be derived in the following form [Krause, Scully and Walther (1987)]

$$p_M(n) = \prod_{j=0}^{n-1} S(j) \sum_{\substack{\{j_k\}=0,\\ j_{M-1}\le\cdots\le j_n}}^{n} \prod_{k=n}^{M-1} C(j_k)\,, \tag{8.13}$$

where

$$\sum_{\substack{\{j_k\}=0,\\ j_{M-1}\le\cdots\le j_n}}^{n} \prod_{k=n}^{M-1} C(j_k) = \begin{cases} \sum_{j_n=0}^{n}\sum_{j_{n+1}=0}^{j_n}\cdots \\ \times \sum_{j_{M-1}=0}^{j_{M-2}}\prod_{k=n}^{M-1} C(j_k)\,, & M>n\,, \\ 1\,, & M=n\,, \\ 0\,, & M<n \text{ or } n<0\,. \end{cases} \tag{8.14}$$

The above authors analyzed numerically the probability (8.13) for up to $10^3$ atoms. This probability is strongly dependent on the value $gt$, which can be changed by changing the velocity of the atomic beam. It was found that for $gt < 1$, the peak of the photon distribution moves towards higher photon numbers and it is getting narrower with increasing number of atoms going to the $\delta$-function in the steady-state limit. In this way the photon number state $|n\rangle$ can be prepared.

Another possibility how to prepare number states is to control the velocity of the atom beam [Krause, Scully, T. Walther and H. Walther (1989)] using the expression for the probability that the atom emits a photon to the cavity provided that the field is in a number state $|n\rangle$, $\sin^2\left(\sqrt{n+1}\,gt\right)$. To have the probability of unity, giving $g$ and $n$, we have to control the interaction time, i.e. the time of flight of the atom through the cavity. It is difficult to control neutral atoms. Therefore it is more suitable to use a beam of ions (e.g. alkaline earth atoms) which have lost one electron and the outer electron is excited to the upper level of a Rydberg state pair. These two-level Rydberg ions can be accelerated with the help of adjustable dc voltage. For the first atom, which enters the cavity where the radiation in the vacuum state $|0\rangle$ is present, one adjusts the time of flight to $t = \pi/2g$, so that the probability of

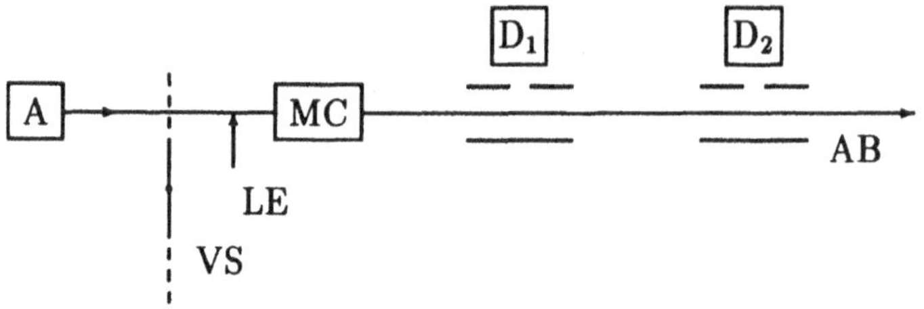


Figure 8.12: A scheme for the experiment to determine photon statistics in micromasers; A is an atomic oven, VS velocity selector, LE cw ring dye laser excitation, MC micromaser cavity, $D_1$, $D_2$ are detectors, AB is atomic beam.

photon emission is unity. When the next atom enters the cavity, the field is in the state $|1\rangle$, and the atom velocity (i.e. the dc voltage) can be changed to $t = \pi/2\sqrt{2}\,g$ adding another photon to the radiation field. Continuing, we arrive at the time value $t = \pi/2\sqrt{n}g$ for the $n$-th ion, adding the $n$-th photon to the field. In this way the Fock photon number state $|n\rangle$ with higher $n$, i.e. nonclassical radiation field with stronger intensities can be generated. The experiments demonstrated that photon statistics of micromasers can be reduced up to 70 % of Poisson statistics. These experiments proved that
i) micromaser represents a unique device exhibiting oscillations with less than one atom on average,
ii) it is suitable for generation of nonclassical light, particularly of sub-Poisson light,
iii) it provides understanding of the generation process of pure quantum light.
A scheme of the experimental arrangement of Rempe, Schmidt-Kaler and Walther (1990) is shown in Fig. 8.12. A highly collimated rubidium atomic beam is used, which is passed through a Fizeau velocity selector. Before entering the superconducting micromaser cavity, the atoms are excited into the upper maser level $63^2p_{3/2}$ of $^{85}$Rb atoms by means of the frequency-doubled light of a cw ring dye laser. The laser frequency is stabilized onto the atomic transition $5^2s_{1/2}$, F=3 $\longrightarrow$ $63^2p_{3/2}$, which has a width of a few MHz determined by the laser linewidth and transit-time broadening. In this way very stable beam of excited atoms is prepared. The superconducting niobium maser cavity is cooled down to the temperature 0.5 K by means of a $^3$He cryostat. At this temperature the mean number of thermal photons is reduced to about 0.15 at the maser frequency 21.5 GHz. The cavity quality factor $Q = 3 \times 10^{10}$ corresponds to a photon storage time $T_{cav} = 0.2$ s. The cavity is shielded against magnetic fields by several layers. Also the Earth magnetic field is compensated. Two maser transitions from $63^2p_{3/2}$ to $61^2d_{3/2}$ and $61^2d_{5/2}$ are investigated. The Rydberg atoms in the upper

and lower maser levels are detected in two separate field-ionization detectors. The field strength is adjusted to ensure that in the first detector the atoms in the upper level are ionized, while those in the lower level are not ionized. As atoms in the lower state already emitted a photon to the cavity, they can be counted in the second detector and consequently the number of maser photons can be deduced. Moreover, the variance of the photon-number distribution can be derived from the number fluctuations of the lower-level atoms, which represents the main interest in the experiment. In steady-state conditions, the number $N$ of atoms in the lower maser level is obtained for a fixed time interval $T$ which is approximately equal to the storage time of photons in the cavity. Many realizations of this measurement permit to derive the probability $p(N,T)$ of finding $N$ atoms in the lower level in the standard way. Then the Fano factor $F_N = \langle(\Delta N)^2\rangle/\langle N\rangle$ or the parameter $Q_a = F_N - 1 = \langle:(\Delta\hat{N})^2:\rangle/\langle N\rangle$ can be calculated from the distribution $p(N,T)$, whose values are negative for sub-Poisson statistics and $Q_a = 0$ for the Poisson statistics. In that experiment the detection efficiency was 10 %. About $2 \times 10^4$ measurements were needed to have the accuracy of $Q_a$ better than 1 %. At the atom-field coupling frequency $\Omega = 10$ kHz for the transition $63^2\mathrm{p}_{3/2} \longrightarrow 61^2\mathrm{d}_{3/2}$ the number of atoms $N > 10$ was necessary to drive the micromaser above threshold. The nonclassical photon statistics were observed for higher flux of atoms during the atom-field interaction time 50 $\mu$s. In the transition $63^2\mathrm{p}_{3/2} \rightarrow 61^2\mathrm{d}_{5/2}$ it was $\Omega = 44$ kHz and the atom-field interaction time was 35 $\mu$s. In this case very low flux of atoms $N < 10$ was sufficient to generate nonclassical maser fields, although such fields were again generated for higher atom fluxes. For $N = 30$ the authors deduced $Q_a = -0.04$ and $F_n = 0.3$, which means that they were able to reduce photon fluctuations to 70 % below the shot-noise level. The possibility to reduce the atomic Poisson fluctuations in this way may have interesting applications in atomic interferometry.

Various effects in the micromaser system and their relation to generation of entangled states and the EPR "paradox" were discussed by Wagner, Brecha, Schenzle and Walther (1993). Phoenix and Barnett (1993) demonstrated "nonlocal" interatomic correlations in the micromaser.

Recently Koashi, Kono, Hirano and Matsuoka (1993) realized antibunched and sub-Poisson light along the original Stoler's suggestion [Stoler (1974)]. They used the nonlinear process of second-harmonic generation as a pump to generate the second sub-harmonic light in another nonlinear crystal. The sub-harmonic light exhibits photon antibunching and sub-Poisson behaviour (Sec. 5.4). For a review of properties of one-atom maser and its use to test quantum theory we refer the reader to the paper by Walther (1993).

# 8.6 Experiments with frequency down conversion

First we mention that the original experiments using for instance atomic cascade decay to produce polarization-correlated photon pairs for tests of Bell's inequalities have a number of shortcomings and may be criticized from various points of view. These difficulties can be removed using generation of quantum correlated photon pairs by the process of optical parametric down conversion, as will be discussed in this section. The first experiment to observe sub-Poisson light in the optical process of parametric down conversion was suggested by Stoler (1974).

The nonlinear optical process of the frequency down conversion, as discussed in Secs. 5.4 and 6.6, proved to be one of the most efficient one for a direct investigation of quantum properties of light. There are several experimental groups which devoted the systematic many-year attention to the research in this field. Many theoretical and experimental results for the second- and fourth-order interference (in the complex amplitudes), including properties of physical vacuum, nonclassical and "nonlocal" properties of quantum system, using two-photon down conversion process have been gained by the Rochester group directed by L. Mandel [see e.g. brief reviews by Ou and Mandel (1990) and Mandel (1992), and the book by Mandel and Wolf (1994/5)]. Many problems of generation of twin photons using down conversion in cavities with particular attention to reduction of noise, nondemolition measurements and quantum limits of measurements were solved by the Paris group directed by Elisabeth Giacobino [for a review, see Reynaud, Heidmann, Giacobino and Fabre (1992)]. Questions of generating sub-Poisson light in optical parametric down conversion, fourth-order interferences, quantum mechanical "nonlocality", communication with correlated photon pairs (twin photons) and noise reduction below the shot noise were discussed by J. G. Rarity, P. R. Tapster, S. F. Seward, J. G. Walker, J. S. Satchell and E. Jakeman in the Malvern group. Their close cooperation with M. C. Teich, T. S. Larchuk and R. A. Campos from Columbia University and B. E. A. Saleh from the University of Wisconsin made it possible to realize very interesting two-photon interference experiments using a Mach-Zehnder interferometer [see Larchuk, Campos, Rarity, Tapster, Jakeman, Saleh and Teich (1993)]. Interesting experiments on "nonlocality" using higher-order splitters are realized by A. Zeilinger and his group at the University of Innsbruck. There is a number of stimulating papers in this field, e.g. by Franson (1991a), Smithey, Beck, Belsley and Raymer (1992), Shih, Sergienko and Rubin (1993), Kwiat, Steinberg and Chiao (1993), etc. [for a review, see Han, Kim and Zachary (1992), Sec. I], which will be discussed in the following.

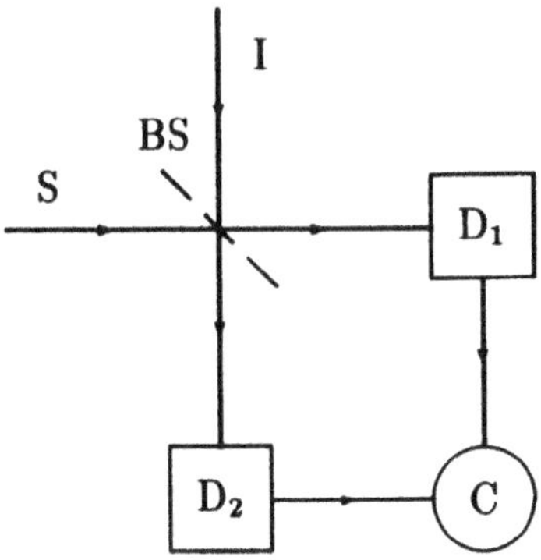


Figure 8.13: Interference of signal (S) and idler (I) beams on the beam splitter BS and their detection; $D_1$, $D_2$ are photodetectors and C is coincidence counter.

First it is interesting to discuss the effect of a beam splitter on photon pairs generated by the parametric down conversion [Ou and Mandel (1988a), Rarity and Tapster (1989)]. When photon wave functions overlap the coincidence count rate in the intensity correlations can be reduced when pairs of sub-frequency photons are combined in the beam splitter, which permits to perform measurements of the time interval between the two photons with sub-optical period accuracy in the sub-picosecond range. The width of the coincidence reduction is related to the bandwith of down-converted light and to the phase matching conditions in the nonlinear crystal. Since the photon position in space-time can be specified with the accuracy of a few wavelengths or a few light periods in time, this method makes it possible to perform measurements in space and time with better accuracy and optical period does not represent a definite limit of the measurement accuracy. The experiment is based on the use of the beam splitter as shown in Fig. 8.13. The signal and idler beams interfere on the beam splitter, the resulting beams are detected beyond the splitter by two photodetectors and their responses are correlated in the coincidence counter. In the experiment by Rarity and Tapster (1989) the arrangement shown in Fig. 8.14 was employed. A crypton-ion laser operating at 413.4 nm wavelength is focused on a 15 mm-long crystal of deuterated potassium dihydrogen phosphate (KD*P), using a 1 m-focal length lens. A pair of small apertures $A_1$ and $A_2$ ensured that the phase matching condition is satisfied and the matched pairs of photons are selected. The two created beams of the wavelength 826.8 nm are reflected by mirrors and right-angle prisms onto a beam splitter. One of the prisms can be translated to introduce a path difference between the beams. The outputs of the beam splitter are focused onto two photocounting avalanche photodiode de-

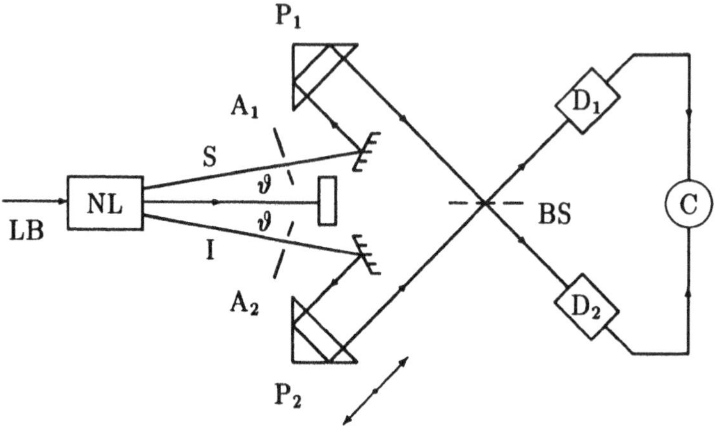


Figure 8.14: Scheme of the experiment for interference of down-converted beams; LB is pump laser beam, NL nonlinear crystal, S and I are signal and idler beams, $P_1$ and $P_2$ prisms, BS is beam splitter, $D_1$ and $D_2$ are detectors, C is coincidence counter.

tectors and the resulting photodetection pulse trains are fed to a single-bit correlator operating as a multichannel coincidence counter with 10 ns resolution. A computer acquires the correlator data and calculates the two-photon coincidence rate. The coincidence rate can be expressed in this case in the form

$$R_{12} \sim 1 - C \exp\left[-\frac{(\delta x)^2}{2\sigma^2 \sin^2(\vartheta)}\right] \tag{8.15}$$

where $C$ is the visibility, $\delta x = c\delta t$ is the displacement and $\sigma$ is the halfwidth. The authors obtained the coincidence dip of the halfwidth 19 $\mu$m (63 fs) for a filter-free case with the visibility $C = 0.8$. They also performed modified experiment in which the measurement is performed in the signal beam only, when another beam splitter is situated and responses from two photodetectors are again correlated. Clearly enhanced coincidence rate was observed. With small apertures and negligible crystal dispersion effects the width of the effect is related to the path-length uncertainty introduced by the finite beam width of the pump beam. Similar results were obtained by Shih and Sergienko (1994) with the visibility $C = 0.97$.

Similar experiments were performed by Ou and Mandel (1988b) and Rarity and Tapster (1990a) for nondegenerate optical parametric process. The arrangement was the same as outlined in Fig. 8.14, only the signal and idler modes have different wavelengths (beams of two colors are used). In this way additional beating oscillations occur in the coincidence rate demonstrating nonclassical and "nonlocal" effects. In this case the coincidence rate (8.15) can be written, provided that the resolving time of the coincidence counter is larger than the coherence time $1/\Delta\omega$ of the pump radiation, as

$$R_{12} \sim 1 - C \exp\left[-\frac{(\delta x)^2}{2\sigma^2 \sin^2(\vartheta)}\right] \cos\left[\frac{\omega \Delta \delta x}{2c \tan(\vartheta)}\right], \tag{8.16}$$

where $\omega$ is the angular pump freguency and $\Delta$ is the angular width of the slit. Fitting modulated experimental data the above authors were able to improve time resolution to about 40 fs with the visibility $C = 0.84$. The sharp central minimum defines the centre of the photon wave packet. Its position can be determined with the accuracy of about 1 fs, regardless of the positional spread of the photon, so that the average position of the photon can be determined with the better time accuracy than the optical period. If one of the apertures is blocked, the effect is destroyed due to lack of overlap between the spectra of the photon pair. Thus if the twin photons are incident on a beam splitter, the destructive interference occurs between quantum amplitudes corresponding to the final state consisting of one photon transmitted and the other one reflected, and consequently neither two photons are transmitted or reflected. This means, that in the case of Fig. 8.13 both pair photons are propagating through one or the other arm to the first or the second detector (cf. eq. (8.22)). Steinberg, Kwiat and Chiao (1992) demonstrated a "nonlocal" dispersion-cancellation effect in propagation of single photons travelled in glass using a two-photon interferometer and pairs of photons from an optical parametric process. The single-photon propagation time was measured to within approximately 4 fs. "Nonlocal" effect of the spectral filtering of one sub-frequency beam in the other sub-frequency beam was demonstrated by Chiao, Kwiat and Steinberg (1992). Zeilinger, Zukowski, Horne, Bernstein and Greenberger (1993) and Zeilinger, Bernstein, Greenberger, Horne and Zukowski (1993) discussed various kinds of higher-order multiport beam splitters (tritters, etc.) providing new measurement possibilities. Mattle (1993) and Aschenwald, Dopfer, Mattle, Weinfurter and Zeilinger (1993) realized the interferometric measurement shown in Fig. 8.14 when the beam splitter was substituted by a tritter. They have demonstrated, as discussed in Sec. 7.2, that multiple correlations are more deterministic by observing deeper dips in coincidences than in the case of the use of a simple beam splitter.

Modifying the above experiment as shown in Fig. 8.15, Rarity and Tapster (1990) performed a test of the Bell's inequality (Sec. 7.2), which can be written in the form

$$E(\vartheta_a, \vartheta_b) - E(\vartheta_a, \vartheta_b') + E(\vartheta_a', \vartheta_b) + E(\vartheta_a', \vartheta_b') \leq 2\sqrt{2}\, C, \tag{8.17}$$

and which is violated for $C > 0.71$; the observed value was $C = 0.74$. As discussed in Sec. 7.2 Ou and Mandel (1988) and Alley and Shih (1988) demonstrated much stronger violation of the Bell's inequality using the arrangement

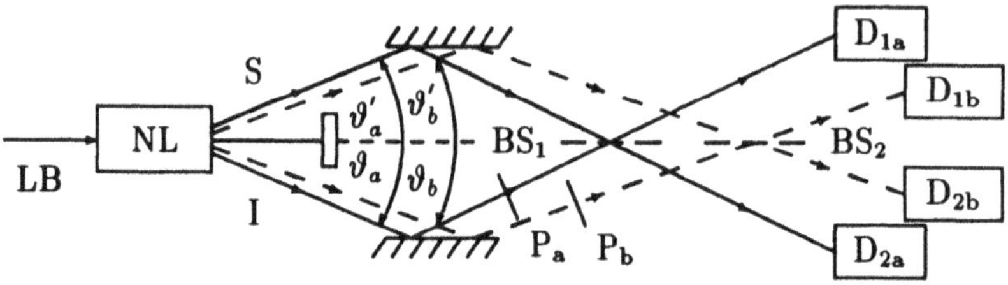


Figure 8.15: Scheme of arrangement for a test of Bell's inequality; LB is pump laser beam, NL nonlinear crystal, S and I are the signal and idler beams, respectively, $BS_1$ and $BS_2$ beam splitters, $D_{1a,b}$–$D_{2a,b}$ photodetectors, $P_a$ and $P_b$ phase plates.

shown in Fig. 8.16. In this experiment two sub-frequency photons are emitted with parallel polarizations from a nonlinear crystal of KDP. A 90° rotator placed in the idler beam, with a compensating glass plate C placed in the signal beam to produce an equal time delay, gives rise to a pair of orthogonally polarized photons. These are mixed by the beam splitter and the resulting beams are fed to two photodetectors after passing polarizers and interference filters. The counters $C_1$ and $C_2$ provide single numbers of counts and the counter C provides the number of incidences. If the compensator, rotator and polarizers are removed and the position of the beam splitter is changed, the above discussed modulated coincidence rate can be observed [Ou and Mandel (1988b)].

Tapster, Seward and Rarity (1991) used a lithium iodate nonlinear crystal as a source of two quantum correlated (twin) beams at 826.8 nm wavelength. The experiment belongs to the group of experiments for generation of nonclassical light using a feedback derived from the pump beam of the wavelength 413.4 nm beyond the crystal to regularize the pump beam again with the help of an electro-optic modulator. Low-noise high efficiency photodiodes again detect the light in each beam and the spectrum of the photocurrents, and the spectrum of the difference between the photocurrents were measured. The authors developed a method of fractional subtraction of the noise in one channel from the noise of the other channel, which is a very effective method because the noise in both the channels is quantum correlated. In this way they were able to perform the noise reduction of 2 dB below the shot-noise level in the single beam and the noise reduction of 4 dB below the shot-noise level in both the beams. The method can be improved to provide near 10 dB noise reduction.

Such light was used for investigation of small fluctuations in the absorption coefficient of highly transmissive samples and for low-light-level com-

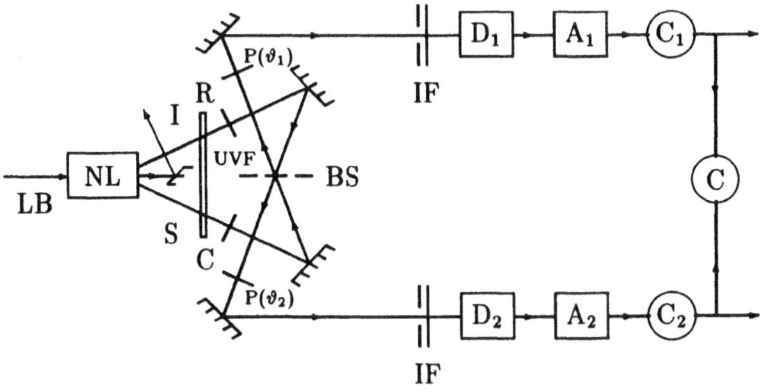


Figure 8.16: Scheme of arrangement for exhibiting violation of Bell's inequality; LB is laser beam, NL nonlinear crystal, S and I are signal and idler beams, C is compensator, R rotator, BS beam splitter, $P(\vartheta_1)$ and $P(\vartheta_2)$ are polarizers, UVF is filter of UV radiation, IF are interference filters, $D_1$, $D_2$ detectors, $A_1$, $A_2$ amplifiers, $C_1$, $C_2$ counters, C is coincidence counter.

munications using correlated photon pairs [Seward, Tapster, Walker and Rarity (1991)]. These authors demonstrated day-light communication at the distance 200 m using quantum correlated photon pairs produced by non-degenerate parametric down conversion. Using 7 % quantum efficient detectors and a correlator sample time of 10 ns, the received signal intensities were typically 0.14 % of the detected background light level. They used He-Cd laser as a pump beam of 441.6 nm wavelength to produce down-converted twin beams in lithium iodate crystal having 883.2 nm wavelength. One of the sub-frequency beams is collimated and sent 200 m to a detector in the receiver unit. The other beam is used for acousto-optic modulator acting as a shutter in a He-Ne beam falling on another detector in the receiver unit. For each sub-frequency photon registered by the second detector, a delayed pulse from the He-Ne laser is emitted. Two different delays may then be used to transmit a binary signal. The second detector in the receiver unit collects this light and a correlator acting as a multichannel coincidence counter together with a computer are used to decode the detector outputs. Thus the photon traversing the high-background channel can be detected on using the time stamps to open the short-coincidence gate. The information is coded in the time delay between the time stamps and coincident photodetections in the receiver.

Franson (1989) suggested to use quantum correlated pairs of photons (from an atomic cascade or a down frequency converter) as inputs into two

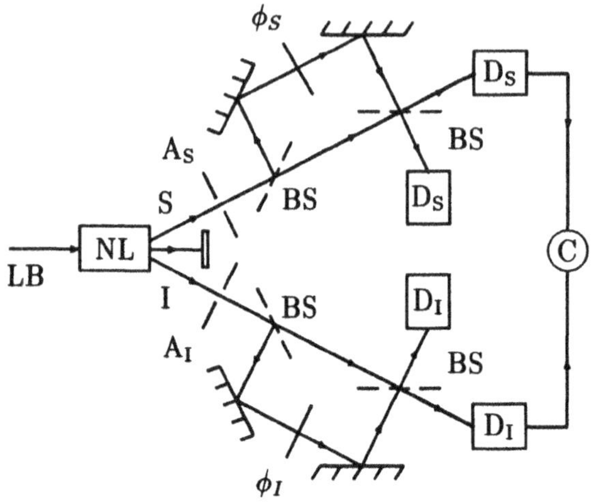


Figure 8.17: A two-photon interferometer; LB is laser pumping beam, NL nonlinear crystal, S and I are signal and idler beams, $A_S$ and $A_I$ signal and idler apertures, BS beam splitters, $D_S$ and $D_I$ signal and idler detectors, $\phi_S$ and $\phi_I$ signal and idler phase plates, C is coincidence counter.

separate Mach-Zehnder imterferometers as demonstrated in Fig. 8.17, representing a two-photon interferometer. Frequency and phase-matched photon pairs selected by signal and idler apertures pass through out-of-balance Mach-Zehnder interferometers to photon counting detectors. Variable phase shifts $\phi_S$ and $\phi_I$ are introduced in the long arms of the interferometers. Such two-photon interferometer makes it possible to observe "nonlocal" quantum effects because the energy and momentum conservation induce a phase coherence of the photon pairs, which extends beyond the coherence length of the individual beams. This quantum entanglement is manifested in the above interferometer. The second-order interference effects visible in single detectors rates are not seen when individual interferometer path-length differences are greater than the coherence length of individual pair-photon beams (removing the splitter in one of the interferometers, one can obtain the path information), while the fourth-order coincidence interferences can be observed. As such a device can exhibit interference pattern with the 100 % visibility, it allows a test of "local realistic" theories (Sec. 6.2). Such large nonclassical oscillations with the visibility greater than 0.5 have been really observed [Ou, Zou, Wang and Mandel (1990), Kwiat, Vareka, Hong, Nathel and Chiao (1990), Rarity, Tapster, Jakeman, Larchuk, Campos, Teich and Saleh (1990), Brendel, Mohler and Martienssen (1991, 1992), Shih, Sergienko and Rubin (1993, 1993a), Kwiat, Steinberg and Chiao (1993, 1993a)]. Rarity and Tapster

(1992) modified this two-photon interferometer introducing into the signal arm, before the Mach-Zehnder interferometer, a multimode fibre-optic cable of 170 m length. Such a modification may be interesting as a more secure form of coding in terms of "nonlocal" correlations for optical communications provided that the identical Mach-Zehnder interferometers are placed in the transmitter and receiver. They have used a lithium iodate crystal pumped by a helium-cadmium laser operating at 441.6 nm wavelength. The crystal axis was tilted to produce 883.2 nm wavelength pair beams separated by an angle of about 30°. The authors observed the coincidence rate modulations when changing the path-length difference in the Mach-Zehnder interferometer with the visibility about 0.11 less than 0.5, but they assumed that the reduced visibility was caused by spatial averaging in the interferometers and mode scrambling in the optical fibre. This experiment does not prove absolutely that the entangled states exist over the 170 m detector separation, even if Franson (1991) demonstrated that such results are in agreement with the quantum description and that a classical description is inadequate. Whereas the coherence length of single beams was a few micrometers, actual path-length differences to observe interference pattern were changed up to a few centimeters. However using the arrangement in Fig. 8.17 Rarity and Tapster (1990) clearly demonstrated the visibility a bit less than 1 but sufficient to show a violation of Bell's inequality. Rarity, Burnett, Tapster and Paschotta (1993) realized high-visibility two-photon interference in a single-mode-fibre interferometer ($C = 90$ %) using entangled photon pairs from a down converter, with a possibility of demonstration of violation of "local realism" at km distance.

Franson (1991a) used the two-photon interferometer in which the optical path-length between the two Mach-Zehnder interferometers was 102 m. The results of this measurement were in good agreement with quantum theory and are inconsistent with any semiclassical theory. Even if the visibility of modulation of coincidence rate $R_c$ was about 0.31, for close positions of the two Mach-Zehnder interferometers, the visibility 0.27 was observed and no modulation was observed in single detector rates. This indicates that quantum correlations are not reduced due to a collapse of the wave function over longer distances, which is in agreement with the information content of the wave function. Although the visibility was not greater than 0.5 to violate Bell's inequality, Franson (1991) derived a weaker nonclassical inequality

$$C \leq \frac{R_{c0}(\tau)}{R_{c0}(0) + R_{c0}(\tau)}, \tag{8.18}$$

where $\tau$ is the difference in travel times in the interferometer and $R_{c0}$ is the coincidence rate provided that the beam splitters are removed, or more generalized form was derived when large coincidence windows are used in the interferometric measurements. These classical inequalities were violated by the experimental data in agreement with quantum theory. Such inequalities are consequencies of the uncertainty relations arising from the commutation relations in space and time.

It should be noted that the coincidence counting rate in this interferometer is equal to [Franson (1989)]

$$R_c = \frac{1}{4} R_{c0} \cos^2 \left( \frac{\phi_1 + \phi_2 + \omega\tau}{2} \right), \tag{8.19}$$

where $\phi_1$ and $\phi_2$ are phase shifts introduced in two long paths of the Mach-Zehnder interferometers and $\omega$ is the frequency of pumping. Expression (8.19) violates Bell's inequality, but it needs the resolution of the coincidence measurements better than $\tau$. For time resolutions worse than $\tau$ this leads to visibility only 0.5. The above results by Franson (1991) and Ou and Mandel (1990a) demonstrated that also such measurements with the coincidence visibility less than 0.5 can be consistent with quantum theory and in disagreement with a classical description. Thus two-photon interferometer experiments with sufficiently large visibilities of coincidence rates (fourth-order correlations) will violate Bell's inequality and are inconsistent with any local hidden-variable theory, proving "nonlocal" effects in quantum description. However, also experiments with smaller coicidence visibilities may violate classical inequalities if the degree of coincidence of the photon counts is taken into account, as in the case of the Franson experiment violating the classical limit by 4 standard deviations, indicating that the observed effects were quantal in nature.

Clear quantum correlations of photon pairs from the nonlinear process of down conversion were experimentally demonstrated by Rarity, Tapster, Jakeman, Larchuk, Campos, Teich and Saleh (1990), Shih, Sergienko and Rubin (1993), and Kwiat, Steinberg and Chiao (1993). Rarity et al. used the arrangement as outlined in Fig. 8.18. The thorough theory of the fourth-order interference of joint single-photon wave packets in lossless optical systems and of the above experiment was presented by Campos, Saleh and Teich (1990). The coincidence probabilities were determined at the output ports of a beam splitter of a Mach-Zehnder interferometer, where two photons were incident at each port and spectral composition of wave packets was arbitrary. For uncorrelated and distinguishable input photons, the fourth-order interferences are standardly understood in terms of particle or wave behaviour of each photon acting independently. For highly correlated photon pairs, e.g. arising in the spontaneous process of down conversion, the fourth-order interference effects at the output of the interferometer are determined by the joint quantum nature of pairs and cannot be described by the particle or wave properties of single photons. In the experiment by Rarity et al. (1990) a krypton-ion laser

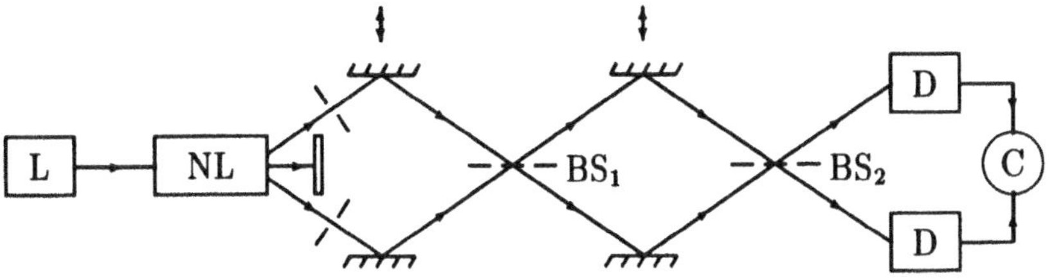


Figure 8.18: A scheme of the two-photon experiment; L is krypton laser, NL nonlinear crystal, BS are beam splitters, D detectors, C is coincidence counter.

producing light of 413.4 nm wavelength was used for the pumping of a KD*P nonlinear crystal which produced sub-frequency photons of the wavelength 826.8 nm wavelength incident on the input ports of the Mach-Zehnder interferometer. The photons were directed to a beam splitter cube $BS_1$, and then redirected onto a lower portion of the same beam splitter cube $BS_2$ by two right-angle prisms, thus forming a folded Mach-Zehnder interferometer. The beam splitter cube and prisms were mounted together as a unit on a translation stage. The distance between one of the prisms and the beam splitter cube can be adjusted within the unit by a combined micrometer and piezoelectric positioner to introduce an adjustable path-length difference time delay between the arms of the interferometer. The photons at the output ports of the interferometer were then focused by two lenses onto two Si avalanche photodiodes. After discrimination, amplification, and shaping, the outputs of the photodiodes were fed to digital correlator that provided the coincidence rate and the individual count rates. A computer controlled the experiment and collected the data. The authors observed nonclassical coincidence interferences with a spatial period equal to the wavelength of pump photons of the visibility 0.62 for path length differences both smaller and greater than the single photon coherence length. This demonstrates the quantum correlated nature of two-photon states produced in parametric down-conversion nonlinear process. The interferences are expected to disappear when the path-length difference in the interferometer exceeds the coherence length of the pump beam. Of course, for independent photons the interference pattern is lost for path-length differences exceeding the coherence length of either photon.

Also in the experiment by Shih, Sergienko and Rubin (1993) the coincidence rate was modulated with the visibility greater than 0.5. The scheme of their arrangement is shown in Fig. 8.19. Signal and idler photons from the nonlinear crystal pass throught independent Michelson interferometers and are detected by the coincidence counter. If the interferometers are set so that

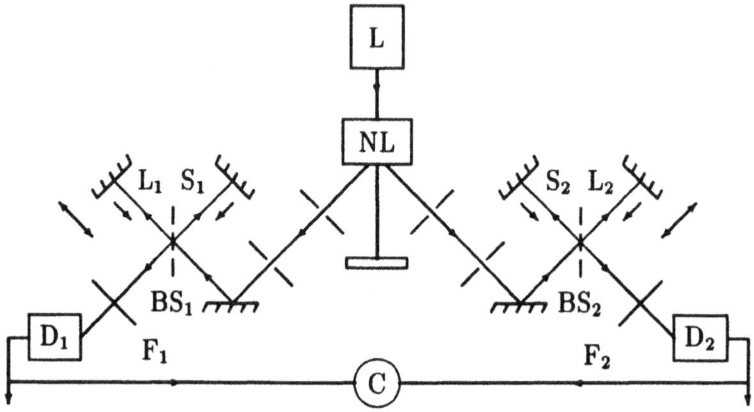


Figure 8.19: A scheme of arrangement to test violation of Bell's inequality, L is cw argon laser (351 nm), NL nonlinear crystal, $BS_1$ and $BS_2$ are beam splitters of two Michelson interferometers, $F_1$ and $F_2$ spectral filters, $D_1$ and $D_2$ detectors, C is coincidence counter, $L_1$, $L_2$ and $S_1$, $S_2$ denote the long and short paths, respectively.

the optical path differences are longer than the coherence length of the beams, no second-order interference is possible. However, if the optical paths of both the interferometers are approximately equal, the fourth-order interference can be observed. This interference arises from the frequency and wave-number correlations in a given photon pair generated under the frequency and phase matching conditions, $\omega = \omega_1 + \omega_2$, $\mathbf{k} = \mathbf{k}_1 + \mathbf{k}_2$. The visibility of the interference pattern can be described in a classical or a quantum way. The classical calculation predicts the maximum visibility 0.5, whereas in idealized conditions quantum theory predicts the visibility 1 (Sec. 6.2) when the coincidence time window is shorter than the optical path difference. In this case, the registration time of one photon traversing the long path and of the other one following the short path of the interferometers is outside of coicidence window and it will not be registered by the coincidence counter. The quantum correlated state in space and time has the form

$$|\psi\rangle = |\psi(L_1, L_2)\rangle + |\psi(S_1, S_2)\rangle\,, \tag{8.20}$$

which leads to the 100 % visibility of coincidence interference pattern. But in the above interferometer rather the state

$$|\psi\rangle = |\psi(L_1, L_2)\rangle + |\psi(S_1, S_2)\rangle + |\psi(L_1, S_2)\rangle + |\psi(L_2, S_1)\rangle \tag{8.21}$$

is observed, which gives only 50 % visibility and cannot be distinguished from the classical description. However, using the coincidence time window shorter than the optical path difference of the interferometers, the last two terms can be cancelled. A 351.1 nm single-mode cw argon laser beam was used as a pumping of KDP nonlinear crystal for optical parametric down conversion. Both the degenerate and non-degenerate photon pairs were used with wavelengths 702.2 nm, and 632.8 and 788.7 nm, respectively. Single rates and their correlation were measured. When both path differences $\Delta L_1$ and $\Delta L_2$ were shorter than the coherence length of the beams, the visibilities 0.97 and 0.82 for the fourth- and second-order interferences were observed, respectively. If $\Delta L_1$ and $\Delta L_2$ are greater than the coherence length, no second-order interference can be observed. The fourth-order coincidence measurements provided the interference patterns with the visibilities 0.59 with the period 632.8 nm and 0.59 with the period of 788.7 nm. When both $\Delta L_1$ and $\Delta L_2$ were changed, the visibility 0.58 with the period 351.1 nm was observed.

Chiao, Kwiat and Steinberg (1992) tested also Bell's inequality in space and time using quantum correlated photon pairs from the optical down-conversion process. They have used a single-mode ultraviolet argon ion laser operating at 351.1 nm wavelength as a pumping source for KDP nonlinear crystal. The signal and idler beams were centered at 702.2 nm wavelength. The idler photon was detected directly, while the signal photon was sent to a Michelson interferometer, inside one arm of which two quarter wave plates were placed. After leaving the Michelson interferometer, the signal beam was split at another beam splitter, where it was transmitted to two detectors through the corresponding filters. Then coincidences between idler and signal beams were observed, providing the coincidence interference effect with the quite high visibility $0.60 \pm 0.05$ for a sufficiently narrow filter in the idler beam, whereas the interference fringes disappear for the broad-band filter. Further Kwiat, Steinberg and Chiao (1993) realized the Franson experiment with the visibility 0.8 violating the Bell's inequality by 16 standard deviations, examing energy and time variables instead of spin components.

Interesting interference effects of quantum correlated photons of different colors with center-wavelength differences about 40 nm (800 and 840 nm wavelengths) were demonstrated by Larchuk, Campos, Rarity, Tapster, Jakeman, Saleh and Teich (1993) using the same arrangement, as shown in Fig. 8.18. The difference frequency of the photons appears directly in the spectrum of the coincidence rate. Such a phase-sensitive behaviour can reveal center-frequency differences from zero to hundreds of THz. They have observed a continuation of pump frequency oscillations for path-length differences exceeding substantially the second-order coherence lengths, which is again fully quantum "nonlocal" effect observed for strong non-degenerate process. By adjusting the angle of one of the input mirrors, the photon paths within the interferometer can be made to have cases of fully spatial overlapping or no spatial overlapping equivalent to the case that each photon is sent into a se-

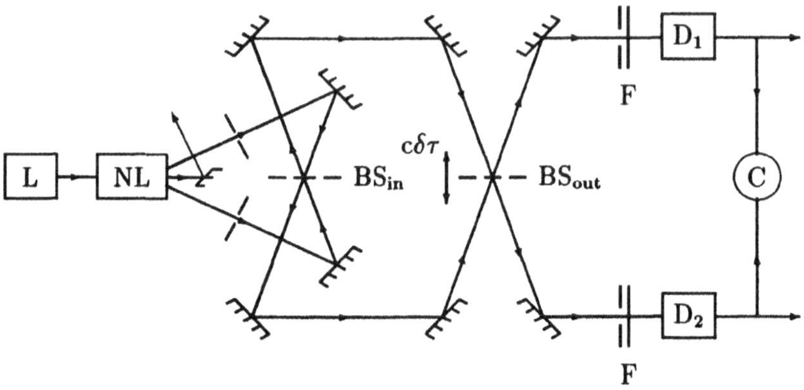


Figure 8.20: A scheme of the interferometric arrangement with down converted photons; L is laser, NL nonlinear crystal, $BS_{in}$ input beam splitter, $BS_{out}$ output adjustable beam splitter, F are interference filters, D detectors, C is coincidence counter.

parate interferometer. The output pulses from the detectors were counted for 1 s to provide a measure of the rate of photon detections at each of the ports, and the sequence of standardized pulses from the two detectors were passed through a 10 ns gate and counted for 1 s to provide the coincidence rate. High visibility of coincidence interference was again observed.

Another theoretical and experimental analysis of the second- and fourth-order interference experiment employing down-converted photons in a Mach-Zehnder interferometer and using coincidences of two detectors at the output was presented by Ou, Zou, Wang and Mandel (1990). The results were again in the agreement with quantum descriptions and in disagreement with any classical analysis. For comparison two coherent light beams were used as inputs to the interferometer. The authors observed the fourth-order interference pattern in the case when classical optics predicts no interference at all so that this experiment violates the laws of classical optics. A scheme of this experiment is shown in Fig. 8.20. The two photons entering at the input to the Mach-Zehnder interferometer are produced by down conversion in a nonlinear crystal $LiIO_3$, which is pumped by an incident light beam at the wavelength of 351.1 nm from an argon-ion laser. Down-converted signal and idler photons of about 700 nm wavelength emerge at relative angles $\pm 7°$ from the crystal and provide the inputs to the beam splitter $BS_{in}$. This beam splitter is mounted on a micrometer allowing changes in the position. If it is in the symmetric position, the two photons emerging from the output sides of $BS_{in}$ appear almost always together on one or the other side, and almost never on both sides simultaneously, because of destructive interference [Hong,

Ou and Mandel (1987), Ou, Hong and Mandel (1987)]. This is given by the fact that the outgoing state $|\psi\rangle$ from the beam splitter is determined by

$$|\psi\rangle = (|t|^2 - |r|^2)|1,1\rangle + i\sqrt{2}\,rt\,(|2,0\rangle + |0,2\rangle)\,, \tag{8.22}$$

which leads to the above conclusion concerning propagation of single photons beyond the splitter if the transmittivity and reflectivity coefficints are equal, $t = r = 1/\sqrt{2}$. The light emerging from $\mathrm{BS_{in}}$ passes through the interferometer and is mixed by the output beam splitter $\mathrm{BS_{out}}$, whose outputs are detected by detectors $D_1$ and $D_2$. The path difference through the interferometer is varied over a range of a few wavelengths by means of the beam splitter $\mathrm{BS_{out}}$ mounted on a piezoelectric transducer varying the applied voltage. After amplification and pulse shaping a coincidence counter yielded the coincidence rate. The experimental results provide no interference effect in dependence on the optical path difference $c\delta\tau$ in single counting rates of detectors $D_1$ and $D_2$ or if one of the inputs to the beam splitter $\mathrm{BS_{in}}$ is blocked. This reflects the fact that the interferometer was a bit unbalanced so that the path difference $c\delta\tau$ exceeded the coherence length $c\tau_c$ and therefore the second-order coherence could not appear. The coincidence rate $R_{12}$ exhibits quantum interference with the visibility slightly less than 100 %. This is in full contradiction with the classical result predicting no fourth-order interference in this case. For comparison the authors modified this experiment sending on the beam splitter $\mathrm{BS_{in}}$ two coherent beams derived by splitting a beam from He-Ne laser so that both the beams were mutually coherent. They observed second-order interferences in dependence on $c\delta\tau$ in single photodetectors, however there was no fourth-order interference in the coincidence rate, as has to be for a classical coherent field, nicely contrasting with the above quantum effect. This complementary behaviour of classical and quantum fields is a result of one-photon interferences in the classical case and two-photon interferences in the quantum case of down-converted beams.

Ou, Zou, Wang and Mandel (1990a) also realized the above described Franson experiment, where they used unbalanced Michelson interferometers instead of the Mach-Zehnder interferometers. If two sub-frequency beams with centre frequencies $\omega_1$ and $\omega_2$ are emitted by the nonlinear crystal, having the spectral bandwidth $\Delta\omega$, the photons propagate through channels 1 and 2 with time differences $T + \tau_1$ in channel 1 and $T + \tau_2$ in channel 2, with $T \gg 1/\Delta\omega$ and $\tau_1$, $\tau_2 \ll 1/\Delta\omega$. Thus no second-order coherence can be expected. However, for the joint probability $p_{12}$ that a photon is detected by $D_1$ and by $D_2$ in coincidence we have

$$p_{12} \sim [1 + C\cos(\omega_1\tau_1 + \omega_2\tau_2 + \mathrm{const})]\,, \tag{8.23}$$

where $C = 1$ for the resolving time $T_R$ sufficienntly short, $T \gg T_R \gg 1/\Delta\omega$, whereas $C = 0.5$ for $T_R \gg T \gg 1/\Delta\omega$. This is understood in terms of the interference of photon pairs. The following cases can occur: i) both photons follow the short paths in the interferometers and arrive simultaneously at the two detectors; ii) both photons follow the long paths and arrive simultaneously at the detectors; iii) one photon follows the long path and one photon follows the short part, the time differences lie within the coincidence resolving time $T_R$ and photons arrive "simultaneously". As these probabilities cannot be distinguished, this leads to (8.23). As the outcome measurement at $D_1$ depends on $\tau_2$ even if the two detectors and interferometers may be far apart, some "nonlocal" features are exhibited. The authors used an argon-ion laser to pump $LiIO_3$ nonlinear crystal to observe the fourth-order interference effect. Even if the visibility was less than 50 %, Ou and Mandel (1990a) proved that although a classical field can give rise to such interference effects, no classical field can account for the observed results in terms of physical ergodic processes.

Note that Wang, Zou and Mandel (1991) demonstrated experimental results for testing the de Broglie guided-wave theory for photons. Such a theory assumes that both photons and electromagnetic waves exist as physical reality and the wave serves as a pilot for photons. It represents some combination of quantum and classical theories. Croca (1987) and Croca, Garuccio and Selleri (1988) suggested an experiment that should exhibit coincidence interferences according to the de Broglie guided-wave theory and no interferences according to quantum theory. A scheme realized by Wang et al. is shown in Fig. 8.21. Pairs of signal and idler photons are produced simultaneously in the down-conversion process within a $LiIO_3$ crystal optically pumped by argon-ion laser oscillating at the 351.1 nm wavelength. Signal S and idler I photons at the 702.2 nm wavelength enter the interferometer at the beam splitter $BS_1$ and $BS_3$. Two similar interference filters IF restrict the bandwidth of detected light to have the coherence length about 0.3 mm. Light beams from $BS_1$ and $BS_2$ are detected by the photocounting detectors $D_1$ and $D_2$ and their responses are correlated in coincidence counter C with the resolving time $T_R = 7$ ns. It is evident that only signal photons can be detected by the detector $D_2$, whereas both signal and idler photons can be registered by the detector $D_1$. When the path difference is much shorter than 0.3 mm (coherence length), the interference of probability amplitudes, or light waves, occurs. Therefore the detector $D_1$ will exhibit interferences when changing the displacement $\Delta x$ of $BS_2$, whereas the counting rate of $D_2$ is independent of $\Delta x$. As the de Broglie waves are supported to have a reality independent of the existence of photons, the corresponding fourth-order correlation function factorizes and the coincidence rate must be modulated as the rate of $D_1$. On the other hand, from quantum theory modulation of coincidence rate is zero because the idler photons cannot reach $D_2$. The experimental results show clearly no modulation of the coincidence rate in agreement with quantum

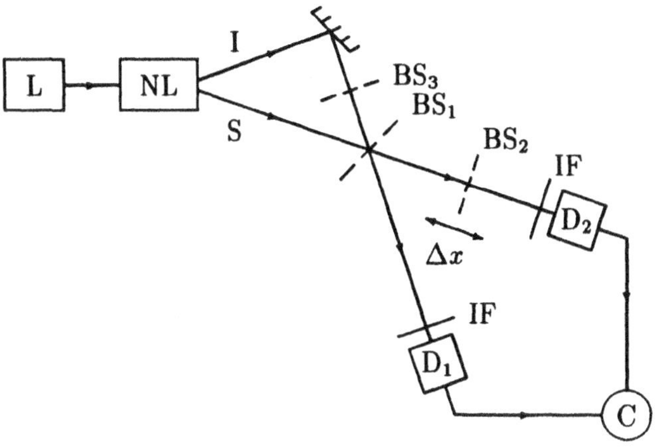


Figure 8.21: A scheme of the experiment to test the de Broglie guided-wave theory; L is laser source, NL nonlinear crystal, I and S are idler and signal beams, BS beam splitters, IF interference filters, $D_1$ and $D_2$ detectors, C is coincidence counter.

theory and disagreement with de Broglie guided-wave theory.

A suggestion was made by Tan, Walls and Collett (1991) to realize an experiment for demonstration of "nonlocal" properties od a single-photon field and a violation of Bell's inequality.

A number of interesting down conversion experiments was suggested and performed by L. Mandel and collaborators at the Rochester University using two identical nonlinear crystals. In one of the experiments (cf. Fig. 8.22) two coherent pump waves fall on two identical nolinear crystals, down-converted signal and idler beams from the two crystals are mixed by two beam splitters and the coincidence counting rate is measured. The counting rate depends on the phase difference between the two coherent pump waves and interference of the physical vacuum with the down-converted photons can be examined, which can be used for testing Bell's inequality and "locality" violation.

Grangier, Potasek and Yurke (1988) pointed out that the down-converted photon pair carries phase information about the pump field. They suggested that such an information may be extracted when the signal and idler photons are mixed with coherent oscillator fields performing a photoelectric correlation measurements. This may provide new tests of Bell's inequality in phase measurements. Practical difficulties arise from the fact that the coherent reference beam has to oscillate at the down-converted frequency, rather than at the pump frequency. Therefore Ou, Wang and Mandel (1989) suggested another experiment for testing the phase coherence and the Bell's inequality without involving any coherent reference beam. For this, down-converted

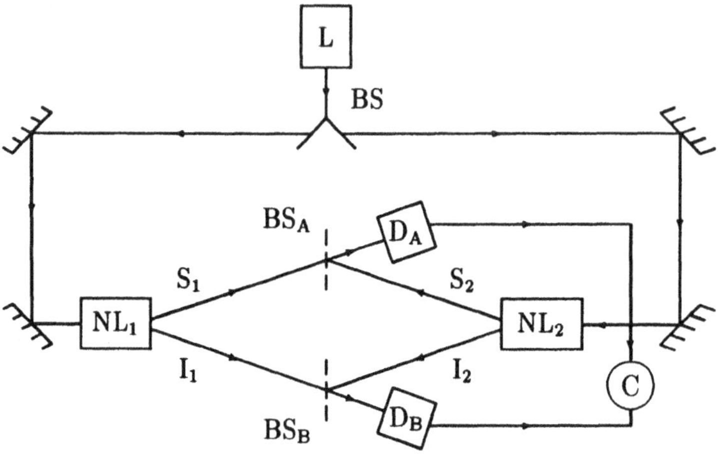


Figure 8.22: Scheme for observing interference of down-converted photons from two nonlinear crystals; L is laser source, $NL_1$ and $NL_2$ are nonlinear crystals, $S_1$, $S_2$ and $I_1$, $I_2$ the corresponding signal and idler beams, $D_A$ and $D_B$ detectors, BS, $BS_A$, $BS_B$ beam splitters, C is coincidence counter.

photons from two similar nonlinear crystals are mixed and the coincidence correlations are measured. This explains the role of the vacuum field in the interference phenomena. A scheme of the arrangement of the experiment suggested by Ou, Wang and Mandel (1989) is shown in Fig. 8.22. A coherent beam of frequency $\omega$ is split into two beams by the beam splitter BS, and the two beams are used to pump two identical nonlinear crystals $NL_1$ and $NL_2$. The generated signal beams $S_1$ and $S_2$ and idler beams $I_1$ and $I_2$ are mixed by the two beam splitters $BS_A$ and $BS_B$, where they are mixed and then the resulting mixed signal beams and mixed idler beams are detected by two photodetectors $D_A$ and $D_B$. The photoelectric pulses from the two detectors are correlated in a coincidence counter C, providing the coincidence rate $R_{AB}$, which can be found to be proportional to

$$R_{AB} = \langle\psi|\hat{A}_A^{(-)}\hat{A}_B^{(-)}\hat{A}_B^{(+)}\hat{A}_A^{(+)}|\psi\rangle \sim 1 - \cos(\vartheta_1 - \vartheta_2 + \phi_1 - \phi_2) \,, \qquad (8.24)$$

providing the 100 % visibility, where $\vartheta_1 - \vartheta_2$ is the phase difference between the two pump waves, which can be determined from the measurement, $\phi_1$, $\phi_2$ are phases determined by the arrangement and the field is assumed in a state $|\psi\rangle$. We note that the description of the optical parametric down-conversion process in this case is similar to that given in Sec. 6.6. If a continuum of sub-frequency modes is taken into account in deriving (8.24) we obtain the field state $|\psi(t)\rangle$ in the form

$$\begin{aligned} |\psi(t)\rangle &= \exp\left[-\frac{i}{\hbar}\int_0^t \hat{H}_{\rm int}(t')\,dt'\right]|0\rangle \approx M|0\rangle_S|0\rangle_I \\ &+ \eta V\delta\omega t \sum_{\omega',\omega''}\phi(\omega',\omega'')\,{\rm sinc}\left[\frac{\omega'+\omega''-\omega}{2}t\right]\exp\left[i\frac{\omega'+\omega''-\omega}{2}t\right] \\ &\times |\omega'\rangle_S|\omega''\rangle_I\,, \end{aligned} \qquad (8.25)$$

where $M$ is a vacuum coefficient, $\delta\omega$ is the mode spacing, $\eta$ an efficiency coefficient, $V$ field amplitude, and $\phi(\omega',\omega'')$ is a spectral weighting function symmetric in $\omega'$, $\omega''$ and peaked at $\omega' = \omega'' = \omega/2$, which is approximately normalized, $\sum_{\omega'}|\phi(\omega',\omega-\omega')|^2\delta\omega = 1/2\pi$. Further it holds from the normalization of the state $|\psi(t)\rangle$ that $|M|^2+\eta^2|V|^2t = 1$. For a weak down conversion $|\eta Vt|^2 \ll 1$, so that $|M| \approx 1$. For the above arrangement we can write

$$\begin{aligned} |\psi_1\rangle_1 &= M_1|0\rangle_{S_1}|0\rangle_{I_1} + \eta_1 V_1|\omega_1\rangle_{S_1}|\omega_2\rangle_{I_1}\,, \\ |\psi_2\rangle_2 &= M_2|0\rangle_{S_2}|0\rangle_{I_2} + \eta_2 V_2|\omega_1\rangle_{S_2}|\omega_2\rangle_{I_2}\,, \\ \hat{A}_A^{(+)} &\sim \hat{a}_{S_1} + i\hat{a}_{S_2}\,, \\ \hat{A}_B^{(+)} &\sim \hat{a}_{I_1} + i\hat{a}_{I_2}\,. \end{aligned} \qquad (8.26)$$

It is possible to develop more realistic description of this experiment removing monochromaticity and including time resolution of the coincidence counter [Ou, Wang and Mandel (1989)]. Such a realistic description leads again to the coincidence interference (8.24) with the additional term $\omega\left(\delta\tau_S + \delta\tau_I\right)/2$ in the cosine function , where $\delta\tau_S = \tau_{S_1} - \tau_{S_2}$, $\delta\tau_I = \tau_{I_1} - \tau_{I_2}$, $\tau_{S_1}$, $\tau_{I_1}$, $\tau_{S_2}$, $\tau_{I_2}$ are propagation times from the crystals to the respective beam splitters. Therefore this process is described as an interference of two two-photon probability amplitudes for photon pairs produced from nonlinear crystals 1 and 2, with an essential role of the vacuum state reflected by the factor $M$, which is involved in a multiplication coefficient at (8.24). This quantum effect is really a result of the superposition of the vacuum state of the field and the two-photon down-conversion state. The dependence of the coincidence rate on the beam phase difference $\vartheta_1 - \vartheta_2$ and on the optical path difference $c\left(\delta\tau_S + \delta\tau_I\right)$ permits a test of "locality" violations. These conclusions have been confirmed in an experiment by Ou, Wang, Zou and Mandel (1990). The two-photon coincidence counting rate was measured as a function of the phase difference $\vartheta_1 - \vartheta_2$ between the two pump waves varied by displacement of the pumping beam splitters. They really observed cosinusoidal modulations, although not with the expected 100 % visibility, which was caused by various imperfections in the measurement.

Ou, Wang and Mandel (1990) suggested another experiment with parametric down conversions in two related nonlinear crystals. An idler photon emitted spontaneously from a first nonlinear crystal serves as stimulating idler input to a second nonlinear crystal. Detecting then the signals from the first and second crystal, one can show that the first idler photon can induce a stimulated down-conversion process, acting as an optical amplifier. The essence of their result is that the measured counting rate of the detector registering the signal beam from the second crystal is given by the expression

$$R_2 = |\eta_2 V_2|^2 \left[1 + \frac{1}{T_R}\int_0^{T_R} |\gamma_{12}(t_2 + \tau_1 - \tau_2)|^2 \, dt_2\right], \tag{8.27}$$

where $\eta_2$ is the conversion efficiency of the second crystal, $V_2$ the pumping field of the second crystal, $T_R$ the resolving time of the detector, $\gamma_{12}$ is a second-order degree of coherence related to the spectral functions of the beams, $\tau_1$ and $\tau_2$ are the propagating times between nonlinear crystals and detectors. The first term in (8.27) represents spontaneous emission from the second crystal, whereas the second term in (8.27) is due to stimulated emission from the second crystal induced by the idler beam from the first crystal. This result stimulated a number of interesting quantum experiments performed by the Mandel's group at the University of Rochester.

The above arrangement was experimentally realized by Zou, Wang and Mandel (1991a) and Wang, Zou and Mandel (1991a). The scheme of the experiment is shown in Fig. 8.23a. An argon-ion laser was used again of the wavelength 351.1 nm pumping two similar 25 mm-long crystals of $LiIO_3$ producing the idler beams $I_1$, $I_2$ at 632.8 nm wavelength and the signal beams $S_1$, $S_2$ at the 788.7 nm wavelength, which interfere at the beam splitter $BS_2$ and are registered by the detector $D_S$, whereas the idler beam $I_2$ is detected by the detector $D_I$. The interference filters $IF_S$ and $IF_I$ are centered at 788.7 and 632.8 nm, respectively and the coherence length of down-converted beams is about 0.3 mm. The neutral density filter placed between the nonlinear crystals $NL_1$ and $NL_2$ to the idler beam $I_1$ can be used to change the strength of the connection of both the nonlinear crystals. The authors observed the interference effect as a function of displacements of $BS_2$ in the counting rate of the signal detector $D_S$ when the transmittivity $|t|$ of the filter NF was 0.91 with the visibility of about 0.3, which was linearly decreasing with decreasing transmittivity and the fringes were missing for $|t| = 0$, when the idler beam $I_1$ was blocked (Fig. 8.23b). The visibility of the fringes is given by

$$C = \frac{2|f_1 f_2|\sqrt{\langle I_1\rangle\langle I_2\rangle}\,|\gamma_{12}|}{|f_1|^2\langle I_1\rangle + |f_2|^2\langle I_2\rangle}\,|t|, \tag{8.28}$$

where $\langle I_1\rangle$ and $\langle I_2\rangle$ are the mean intensities of the two pump beams, $|\gamma_{12}|$ is their degree of coherence at the two crystals and $|f_1|^2$ and $|f_2|^2$ are fractions of incident pump photons down-converted at the crystals. The coincidence

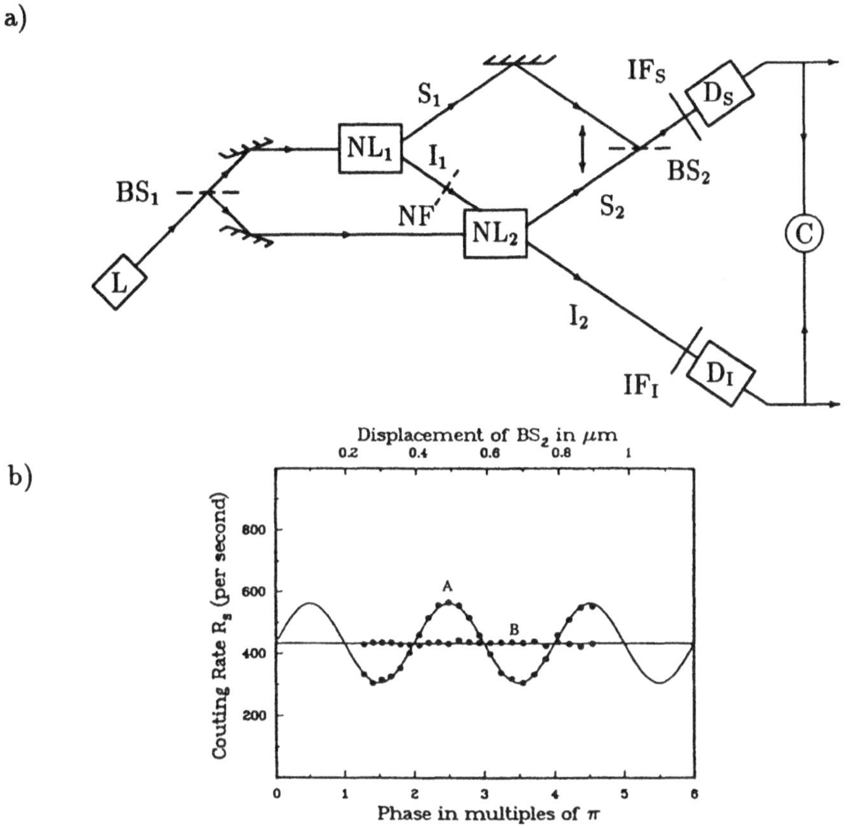


Figure 8.23: a) Scheme of the experiment by Zou, Wang and Mandel (1991a) for correlation of different signal beams; L is laser source, BS are beam splitters, NL nonlinear crystals, S and I signal and idler beams, IF interference filters, $D_S$ and $D_I$ detectors of signal and idler beams, C is coincidence counter, NF neutral density filter; b) Measured signal photon counting rate $R_S$ as a function of $BS_2$ displacement; $|t| = 0.91$ (curve A), $|t| = 0$ (curve B) [after X. Y. Zou, L. J. Wang and L. Mandel, *Phys. Rev. Lett.* **67** (1991) 318].

rate $R_{SI}$ exhibited modulation of the visibility about 0.53 for $|t| = 0.91$. The interpretation of this result is quantal. If the beam $I_1$ is blocked, then, in principle, the detector $D_I$ is able to determine, because the signal and idler photons are emitted together, from which signal a given signal photon comes to the signal detector $D_S$. Whenever a detection $D_S$ is accompanied by a simultaneous detection by $D_I$, the signal photon must arrive from $NL_2$, and whenever a detection by $D_S$ is not accompanied by a $D_I$ detection, the signal photon must arrive from $NL_1$. The consequence is that the possibility of this distinguishing between the signals $S_1$ and $S_2$ smoothes out the interference effect in the counting rate $R_S$, regardless of the auxiliary measurement by means of $D_I$ is actually performed or not. The experiment also demonstrates that induced emission need not accompany induced coherence. A similar experiment was performed by Herzog, Rarity, Weinfurter and Zeilinger (1993, 1994) using one nonlinear crystal, but pumping and down-converted beams were reflected back by mirrors creating pairs of undistinguishable beams being able to interfere. The change of the position of any mirror (in the signal, idler or pump beam) is changing the interference patterns in all beams, which demonstrates the "nonlocal" interference described in the single beams by the simple interference law $1 + \cos(\varphi - \varphi_S - \varphi_I)$, $\varphi$ being the pump phase, $\varphi_S$ the signal phase and $\varphi_I$ the idler phase. The effect represents the interference of two possible ways of photon emissions. Grayson, Zou, Branning, Torgerson and Mandel (1993) realized the experiment shown in Fig. 8.23a with an optical cavity providing a possible time delay for the idler beams. Whether the idler photon is to be delayed or not then determines whether the signal photon behaves like a wave exhibiting interference or it behaves like a particle. It is a kind of the delayed-choice experiment examing the interference pattern if conditions for the interference were created after a photon has chosen one of the interference paths. The result is again a manifestation of the entanglement of signal and idler photons. The above results demonstrate that the diagonal statistical operator reflects not only what is known about the physical system but also what is knowable in principle. Using polarization elements or additional interferometers inside the idler beams $I_1$ and $I_2$ in Fig. 8.23a Berry-Pancharatnam topological phase can be measured [Grayson, Torgerson and Barbosa (1994)] and "quantum erasers" can be created permitting to restore interference erasing distinguishability because the decision to measure wave-like or particle-like behaviour may be delayed until the detector irreversible process occurs [Kwiat, Steinberg and Chiao (1994)].

The above experiment was utilized by Zou, Grayson, Wang and Mandel (1992) to test again de Broglie pilot wave theory. It was shown that in the framework of that theory also "empty waves", which have lost all photons and cannot give rise to a photoelectric detection, are suppose to induce coherence effects [Selleri (1987, 1987a, 1988), Croca (1987), Croca, Garuccio and Selleri (1988), Croca, Garuccio, Lepore and Moreira (1990)]. Zou et al. placed a beam splitter $BS_I$ instead of the filter NF in Fig. 8.23a with the transmittivity $|t|^2 \approx 0.9$ and allowed the reflected idler beam to fall on another photocounting detector $D_{I_1}$. The cable length folloving $D_{I_1}$ was adjusted to make the idler $I_1$ pulse from $D_{I_1}$ to arrive at another coincidence counter between $D_{I_1}$ and $D_S$ simultaneously with the signal $S_1$ pulse from $D_S$. From the point of view of the pilot wave theory it follows that whenever $D_{I_1}$ detects a photon, the associated $I_1$ wave transmitted through $BS_I$ is "empty" of photons. If this empty wave induce coherence between the signal beams $S_1$ and $S_2$, the interference effect in the coincidence rate of detectors $D_S$ and $D_{I_1}$ should be observed. No interference effect is predicted by quantum theory, because in this case both detected photons originate in $NL_1$. The authors observed modulation in the counting rate $R_S$ with about 30 % visibility and in the counting rate $R_{SI}$ with higher visibility, as previously. However, no interference was observed in the coincidence rate $R_{SI_1}$ as a function of displacements of the beam splitter $BS_2$ in agreement with quantum theory and in disagreement with the empty wave theory.

As discussed in Sec. 6.1 if the detection time $T \gg \tau_c$, the interference effects can be observed in the frequency domain. Zou, Grayson and Mandel (1992) realized an experiment on this principle using basicly the arrangement shown in Fig. 8.23a again. The authors modified the original experiment inserting a glass plate in the $S_1$ beam inroducing an optical path of about 1.5 mm between the two arms. As this exceeds the coherence length of 0.3 mm, the interference effect in light falling on $D_S$ is lost, as was observed. However, insertion of a scanning Fabry-Perot interferometer with resolution of about 40 GHz before the detector $D_S$ reveals the spectral modulation according to the spectral second-order interference law (6.9). If the idler beam $I_1$ is blocked the interference disappears as a result of distinguishability of signal photons in the beams $S_1$ and $S_2$. Further the authors removed the glass plate from the $S_1$ beam and placed it inbetween the nonlinear crystals into the stimulating idler beam $I_1$, which caused the absence of the interference fringes in the intensity and the same spectral modulation as before, ever though neither of the two interfering signals $S_1$ and $S_2$ is delayed or directly disturbed by the insertion of the plate. Thus we have again nonclassical and "nonlocal" effect connected with the quantum correlated signal and idler photons. The absence of the interference in the total intensity measurement by the detector $D_S$ is again explained by the principle distinguishability of photon paths from $NL_1$ and $NL_2$ to $D_S$ if an efficient photodetector $D_I$ of the idler beam is applied. The spectral interference effect, however, cannot be smoothed out because it is measured with a high resolving interferometer whose passband is much narrower than $1/T$, which makes the time delay $T$ to be unresolvable. However, this interference disappears if the idler beam $I_1$ is blocked from the same reason. The authors recognized that the spectral modulation was a result of a superposition of one-photon states with different frequencies forming a non-stationary ensemble, and not a result of detection of a polychromatic photon.

Using the arrangement of Fig. 8.23a Zou, Grayson, Barbosa and Mandel (1993) demonstrated that the visibility of the interference of signal photons produced in the down-conversion process can be controlled by inserting differential time delays between idler photons, which is again a consequence of the quantum entanglement between signal and idler photons. Although two nonlinear crystals acting as parametric down converters are excited by two mutually coherent pump light beams, there is no mutual coherence between the generated signal beams, or idler beams. However, the interference effects can be observed between two signal beams when the two nonlinear crystals are connected by the idler beams, as discussed above. These authors inserted a time delay plate instead of the absorber in Fig. 8.23a. When $T$ exceeds the coherence time $\tau_c$ of the down-converted light, the interference effect at $D_S$ has to disappear. This is again the effect of principle distinguishability of the source of each photon using an auxiliary detector $D_I$ with high efficiency. This detector does not disturb the interference of the signal beams $S_1$ and $S_2$. If the detector $D_S$ registers a photon in coincidence with a photon at $D_I$, the photons were emitted at $NL_2$. If a detection by $D_S$ is followed by a $D_I$ detection after a delay $T$, then both the photons come from $NL_1$. This again eliminates any observable interference. That does not matter if the time delay is introduced to $I_1$ rather than to $S_1$, which is actually interfering because both the signal and idler photons are emitted simultaneously in an entangled state. In intermediate cases for $T < \tau_c$ the visibility will be reduced, but it will not be missing. The effect of the change of the state of the idler beam $I_1$ is changing the states of the signal beams $S_1$ and $S_2$ as a kind of "nonlocal" connection of physical system. Wang, Zou and Mandel (1992) analyzed to above arrangement of the experiment with a time-dependent modulator situated between the two down converters. It was found that the visibility of interference fringes is determined by the state of the modulator at one moment on the light cone backward from the detected photon, even though no photon may be passing the modulator at that time, which may have implications for delayed-choice experiments. These authors conclude that no explicit causal mechanism seems to be required either to destroy or to create coherence, because coherence is a complementary manifestation of indistinguishable photon paths. Auxiliary detectors are not necessary to operate in order to eliminate the interference effect, it is sufficient that the measurement can be made in principle. This suggests that the quantum state vector reflects not only what is known about the state but what is knowable in principle.

Ou, Wang, Zou and Mandel (1990a) suggested and Wang, Zou and Mandel (1991b) realized an experiment for interference of idler beams from two parametric down converters. They show that when two mutually coherent signal beams derived from a common strong laser beam are injected into two down converters to realize a stimulated process, the two idler beams are mutually coherent with the 100 % visibility. A scheme of the arrangement is shown in Fig. 8.24. The down converters consist of two similar 2.5 cm-long crystals of $LiIO_3$, which are pumped by light of the wavelength 351.1 nm from an argon-ion laser $L_1$. The beam splitter $BS_1$ directs the pump beam to the nonlinear crystals. Apertures $A_1$, $A_1'$ and $A_2$, $A_2'$ select directions of down-converted light corresponding to signal beams of the wavelength 632.8 nm. The idler beams have the wavelength 788.7 nm. An interference filter IF limits the

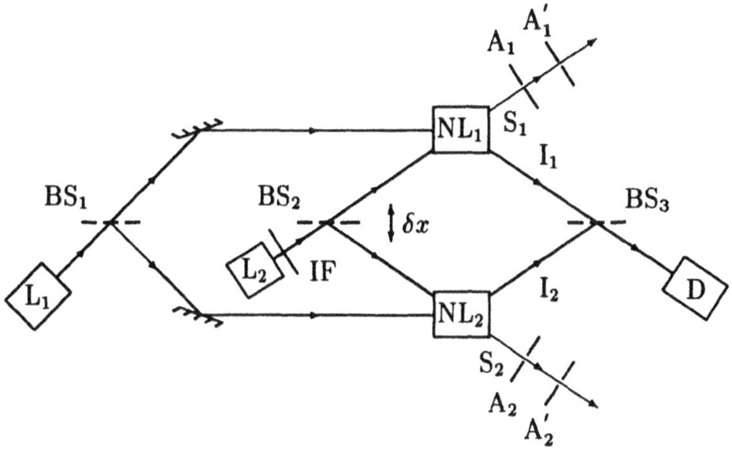


Figure 8.24: Scheme of the experiment with two nonlinear down converters and stimulated beams; L are lasers, BS beam splitters, IF is filter, NL are nonlinear crystals, D is detector, S and I are signal and idler beams, respectively, A and A' are apertures.

bandwidths $\Delta\omega$ of the signal photon beams to about $5 \times 10^{12}$ Hz. The light from a He-Ne laser $L_2$ oscillating at the 632.8 nm wavelength is divided by the beam splitter $BS_2$ providing inducing fields to the signal beams $S_1$ , $S_2$. Generated idler beams $I_1$ and $I_2$ are combined at the beam splitter $BS_3$ and the resulting field is detected by a detector D in a 5 s time interval. The interference pattern may be observed when changing displacement of the beam splitter $BS_2$. Typical photon counting rates were of the order of $10^4$ s$^{-1}$. It is obvious from eqs. (6.71) that for the spontaneous process the correlation of the signal and idler amplitudes are zero

$$\langle \hat{A}_S^{(-)} \hat{A}_I^{(+)} \rangle \approx \langle \hat{a}_S^\dagger \hat{a}_I \rangle = 0 \,, \tag{8.29}$$

because the initial amplitudes are zero. This means that the signal and idler beams have no phase relation regardless of the fact that they are strongly correlated in time. However, also in this case the correlation

$$\langle \hat{A}_S^{(+)} \hat{A}_I^{(+)} \rangle \approx \langle \hat{a}_S \hat{a}_I \rangle = \frac{i}{2} \exp(-i\omega t) \exp(i\varphi) \sinh(2gt) \neq 0 \,, \tag{8.30}$$

where $\omega$ is the pumping frequency, $\varphi$ pumping phase. In the case of the stimulated emission, the correlation $\langle \hat{A}_S^{(-)} \hat{A}_I^{(+)} \rangle$ is not zero, as follows from the expressions (6.71), indicating the case that the signal modes are stimulated ($\xi_{S_1}, \xi_{S_2} \neq 0$), whereas the idler modes start from the vacuum ($\xi_{I_1} = \xi_{I_2} = 0$). The interference pattern of the visibility of 0.7 was observed. The degree of coherence of the idler beams was increasing with increasing occupation number of photons per mode in the intensity increasing fields, i.e. when stimulated emission substantially exceeds spontaneous emission, generated fields are coherent.

Ou, Pereira, Kimble and Peng (1992) used an optical non-degenerate parametric amplification process for demonstration of quantum correlations for continuous dynamical variables having continuous spectra, such as field amplitudes, instead of spin or polarization variables. They have discussed the uncertainty relations and the EPR "paradox" in relation to observed reduction of vacuum fluctuations.

A similar experiment was performed by Sergienko, Shih and Rubin (1993) who used a KDP crystal for non-degenerate down conversion pumped by $Ar^+$ laser (351 nm) with a stimulated beam from a He-Ne laser (632.8 nm) and they experimentally investigated the idler beam using a Michelson interferometer. If the time difference in the interferometer is $\tau$, the mean vacuum intensity is denoted as $\langle I_{vac}\rangle$ and the mean intensity induced by the input signal in the idler beam is $\langle I_{ind}\rangle$, they have observed about 80 % visibility interference fringes for small $\tau$ and $\langle I_{vac}\rangle/\langle I_{ind}\rangle \ll 1$ even if the optical path difference in the interferometer was much larger than the coherence length of the spontaneous idler radiation; the interference fringe visibility was determined by the intensity of the coherent input signal. No interferences were observed with only spontaneous parametric radiation ($\langle I_{vac}\rangle/\langle I_{ind}\rangle \gg 1$).

In generating and detecting nonclassical light beams difficulties always arise related to occurrence of classical noise masking quantum effects, e.g. detector dark noise in a direct detection of photons from sources operating below threshold. It was discussed by Horowicz, Pinard and Reynaud (1987) for the process of generation of twin photons by means of the intracavity four-wave mixing that the best way to overcome these difficulties is to use cavities and light sources above threshold. This may provide intense macroscopic nonclassical light because if nonlinear crystal is inserted inside an optical cavity having mirrors with a high reflectivity for signal frequencies, the cavity favors its own eigenmodes and above the threshold the system oscillates and generates pairs of intense and quantum correlated light beams. However, as the photons are counted in pairs, the quantum noise in the difference of signal and idler intensities $I_1 - I_2$ of the non-degenerate process has to be reduced to obtain intense light beams with squeezed amplitude fluctuations.

Reynaud, Fabre and Giacobino (1987) analyzed the arrangement shown in Fig. 8.25 from this point of view. The optical parametric oscillator is pumped by light of 528 nm wavelength from a single-mode $Ar^+$ laser, stabilized in an external Fabry-Perot cavity. The nolinear 7 mm KTP crystal ($KTiOPO_4$) is inserted into a 3 cm-long cavity. The input mirror has a high transmittivity for the green light and the maximum reflectivity for the infrared radiation.

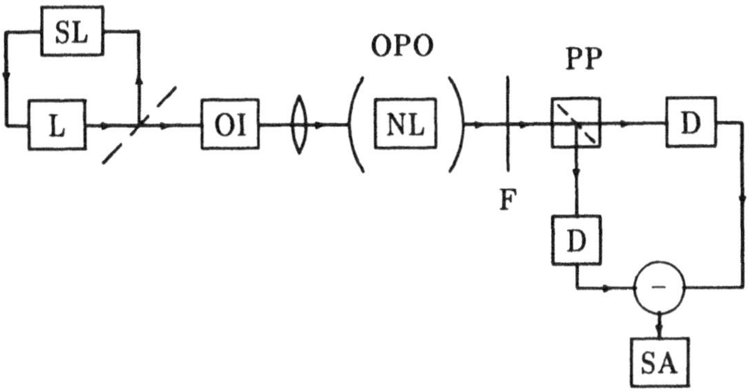


Figure 8.25: Scheme of the arrangement for down conversion in a resonator; L is laser source, SL stabilization loop, OI optical isolator, OPO optical parametric oscillator, NL nonlinear crystal, F filter, PP polarization prism, D are detectors, SA is spectral analyzer.

The output mirror has the 1 % transmittivity for the infrared radiation and also a high transmittivity for green light. An acousto-optic modulator served as an optical isolator preventing backreflected light of $Ar^+$ laser by the cavity. Above the pump threshold of about 100 mW, the system oscillates and two co-propagating cross-polarized beams with wavelengths approximately 1.06 $\mu$m are emitted. The filter shown blocks the pump beam and the two infrared beams are separated with the help of a polarizing beam splitter and are focused on two InGaAs photodiodes with the photoefficiency 0.9 at 1.06 $\mu$m wavelength. The two photocurrents are amplified and subtracted. The difference signal is monitored by means of a spectral analyzer. The authors performed the amplitude stability analysis along the lines discussed in Sec. 6.6. They analyzed the Fourier spectrum of noise,

$$S(\omega) = \int_{-\infty}^{+\infty} \langle \Delta I(t) \Delta I(t+\tau) \rangle_{\mathcal{N}} \exp(i\omega\tau)\, d\tau\,, \tag{8.31}$$

and obtained that above threshold the perfect squeezing is gained for any value of the ratio $I/I_0$ of the field intensity $I$ and the threshold intensity $I_0$. This means that there are no fluctuations in the difference of the amplitudes of the signal fields going out of the cavity. For the noise spectrum of the signal intensity difference they have obtained the simple formula

$$S(\omega) = S_0 \frac{\omega^2}{\omega^2 + \gamma^2} = S_0 \frac{1}{1 + \omega^{-2}\tau_c^{-2}}\,, \tag{8.32}$$

where $S_0$ is the shot noise and $\gamma$ the damping constant of the signal mode. Starting from the perfect quantum noise suppression at $\omega = 0$, one reaches the shot-noise limit for $\omega \gg \gamma$, i.e. for frequencies outside the cavity bandwidth. Perfect squeezing of the intensity difference $I_1 - I_2$ is degraded by intracavity losses and by the imperfect quantum efficiency of detectors. When monitoring the intensity $I_1$, we can use such a signal to modify $I_2$ for obtaining a Fock state. Reynaud (1987) discussed the generation of twin photon beams by a non-degenerate optical parametric oscillator (wavelengths $\lambda_1 = 1.048$ $\mu$m and $\lambda_2 = 1.067$ $\mu$m) and concluded that photon noise spectrum is suppressed at the frequencies which are multiples of the free spectral range. Heidmann, Horowicz, Reynaud, Giacobino, Fabre and Camy (1987) realized the experiment for observation of quantum noise reduction using twin beams generated by means of a two-mode optical parametric oscillator operating above threshold, which was able to generate high-intensity twin beams. They have observed the 30 % reduction of the signal intensities difference below the shot-noise limit over a broad range of frequencies. The analysis is similar to that given in Sec. 6.6.

Fabre, Giacobino, Heidmann and Reynaud (1989) investigated non-degenerate optical parametric oscillator from the point of view of generation of nonclassical light and reduction of quantum noise below the shot-noise limit. They analyzed the noise spectra (8.31) for the intensity difference and quadrature differences again along the lines of Sec. 6.6, including effects of detuning. They have found that the noise spectrum of the output signal is reduced to a minimum value at $\omega = 0$ and at the threshold, whereas the reduction is decreased for the pumping below the threshold in the resonant case. The quantum noise is further reduced for pumping above the threshold of oscillations. The similar behaviour can be found for the noise spectrum of the intensity difference in the unbalanced case. The authors also recognized that losses added to the optical system may lead to a large reduction of quantum noise provided that they are suitably adjusted. They have concluded that the noise reduction for the intensity difference of the signal and idler modes of non-degenerate parametric oscillator is rather insensitive to values of pump power, detuning and external noise for the balanced twin beams, whereas for slight unbalancing there is a strong peak of the noise spectrum for the intensity difference at low frequencies, which can be partly compensated by an external noise. The experiment with non-degenerate optical parametric oscillator above threshold was realized by Debuisschert, Reynaud, Heidmann, Giacobino and Fabre (1989) and Mertz, Debuisschert, Heidmann, Fabre and Giacobino (1991) who observed the quantum noise reduction 69 % in the intensity difference of the two sub-frequency beams at the frequency about 2 MHz and a curve of the noise reduction approaches the shot-noise limit for higher frequencies.

Some results for "nonlocal" diffraction were reported by Christanell, Weinfurter and Zeilinger (1993) and Christanell, Weinfurter, Erckert and Zeilinger

(1993). Using the photon coincidence measurements they were able to observe "nonlocal" diffraction effect in down-conwerted beams provided that a diffraction grating is placed in one of the downconverted beams. As a result of the energy conservation ($\omega = \omega_1 + \omega_2$) and momentum conservation ($\mathbf{k} = \mathbf{k}_1 + \mathbf{k}_2$) strong quantum correlations are arising between the sub-frequency beams. The diffraction of one of the beams is changing the quantum correlations, but these correlations are not generally cancelled by this interaction, which leads to quantum coincidences exhibiting "nonlocal" diffraction in both the beams caused by the grating in one of the beams. This has to be observed in coincidences of photons of both the beams, not in the single beams. The effect is a consequence of the existence of quantum correlations of both the beams and in this sense it is "nonlocal". Such an effect cannot arise if the interaction in one of the beams will cancel the quantum correlations.

Kwiat, Steinberg and Chiao (1992) were able to observe the effect of a "quantum eraser" investigating a revival of coherence in two-photon interference experiments.

# Chapter 9

# Conclusions

In this book, we have discussed contemporary stage of the development of quantum theory as the most important basis of modern physics from the point of view of methods used in quantum optics. It is worthwhile to mention that quantum optics originally representing an application of quantum electrodynamics and particle physics, was able to create new powerful mathematical tools, such as coherent-state technique, and experimental interference and coincidence methods which are so effective in investigating the fundamental concepts of quantum theory.

We have reviewed the basic elements of quantum theory together with the theory of quantum measurements, which is a particular feature of quantum description of physical systems. Quantum optics suggests new possibilities to perform quantum nondemolition measurements, quantum phase measurements and high precision measurements. As the most important theoretical tool the technique of coherent states can serve enabling us to examine cooperative phenomena in the process of interaction of a radiation field and a system of atoms. The coherent states can be introduced on quite general unified algebraic basis and can be adopted to discover various nonclassical features of such an interaction, including quantum squeezing of vacuum fluctuations, photon antibunching, sub-Poisson behaviour of optical fields, oscillations in photocount distributions of nonclassical fields, collapses and revivals of atomic motion, etc. The technique also makes it possible to develop effective Heisenberg-Langevin and Fokker-Planck approaches to quantum statistics of optical fields interacting with atoms and lossy systems.

Using classical interferometers with very low intensities so that "interferences of a photon with itself" can be observed, quantum optics created effective methods of research of quantum phenomena by means of single-photon interferences and coincidences. Particularly, when using the nonlinear process of frequency down conversion, producing pairs of photons in quantum strongly correlated entangled states, it was able to provide new information about the quantum interference, wave-particle duality and structure of a photon. All experiments performed till now making use of cascaded down converters, multichannel beam splitters and other devices of modern quantum optics, are consistent with the probabilistic interpretation of quantum theory and it can be concluded that the phenomenon of interference represents principal indistinguishability of photon paths, or more generally, of possibilities of realization of corresponding experiments, regardless of the fact whether a measurement is or is not performed to distinguish these possibilities.

The existence of optical effects having no classical analogues and of quantum entangled states permitted to verify strong violation of clasical inequalities and Bell's inequalities showing "nonlocal" properties of quantum theory as a consequence of strong quantum correlations. Owing to methods of quantum optics, many of those "non-local" effects can really be observed. Such results provide new interpretation of the Einstein-Podolsky-Rosen "paradox" and related questions. Such experiments were analyzed, including single-photon interferences, experiments with sub-Poisson and squeezed light, experiments with feedbacks, experiments using resonance fluorescence light, optical parametric process, four-wave mixing, phase conjugation, Kerr nonlinear media, micromasers, etc. Arrangements involving down-converted photon pairs proved to be the most fruitful tools in examining the quantum statistical properties of optical systems.

New perspective areas have been opened for further investigations of quantum phenomena in physical systems involving interferences of atoms and ions, which is a new field of research activities recording strong progress in these days.

We have originally planned to include a brief chapter on applications of coherent states in other branches of physics. However, interesting treatments of this subject are now available in various proceedings [Klauder and Skagerstam (1985), Han, Kim and Zachary (1992, 1993), Han, Kim and Man'ko (1993)]. Morover, this field is in so fast development now that we have decided to exclude such a treatment.

It was not possible to include all papers related to this subject here. However, we hope that just the most important ones have served as a basis of this monograph.

We believe that this book will contribute to deeper understanding of the meaning of quantum theory.

# References


Agarwal, G. S., 1973, Progress in Optics, Vol. 9, ed. E. Wolf, North-Holland, Amsterdam, p. 1.

Agarwal, G. S., 1993, *Phys. Rev.* **A 47**, 4608.

Agarwal, G. S. and P. Hariharan, 1993, *Opt. Commun.* **103**, 111.

Agarwal, G. S., D. Home and W. Schleich, 1992, *Phys. Lett.* **A 170**, 359.

Akhiezer, A. I. and V. B. Berestetsky, 1965, Quantum Electrodynamics, Interscience, New York.

Allen, L. and J. H. Eberly, 1975, Optical Resonance and Two-Level Atoms, J. Wiley, New York.

Alley, C. O. and T. H. Shih, 1988, *Phys. Rev. Lett.* **61**, 2921.

Ansari, N. A., 1993, *J. Mod. Opt.* **40**, 413.

Arnoldus, H. and T. F. George, 1991, *J. Mod. Opt.* **38**, 1429.

Arnoldus, H. and T. F. George, 1992, *Opt. Comm.* **88**, 127.

Arthurs, E. and M. S. Goodman, 1988, *Phys. Rev. Lett.* **60**, 2447.

Arthurs, E. and J. L. Kelly, Jr., 1965, *Bell. Syst. Tech. J.* **44**, 725.

Aschenwald, J., B. Dopfer, K. Mattle, H. Weinfurter and A. Zeilinger, 1993, private communication.

Aspect, A., 1976, *Phys. Rev.* **D 14**, 1944.

Aspect, A., 1984, The Wave-Particle Dualism, eds. S. Diner, D. Fargue, G. Lochak and F. Selleri, D. Reidel, Dordrecht, p. 377.

Aspect, A., J. Dalibard and G. Roger, 1982, *Phys. Rev. Lett.* **49**, 1804.

Aspect, A. and P. Grangier, 1991, International Trends in Optics, ed. J. W. Goodman, Acad. Press, Boston, p. 247.

Aspect, A., P. Grangier and G. Roger, 1981, *Phys. Rev. Lett.* **47**, 460.

Aspect, A., P. Grangier and G. Roger, 1982, *Phys. Rev. Lett.* **49**, 91.

Aspect, A., P. Grangier and G. Roger, 1989, *J. Optics* **20**, 119.

Badurek, G., H. Rauch and J. Summhammer, 1988, *Physica* **B 151**, 82.

Badurek, G., H. Rauch and D. Tuppinger, 1986, *Phys. Rev.* **A 34**, 2600.

Bajer, J. and P. Lisoněk, 1991, *J. Mod. Opt.* **38**, 719.

Bajer, J. and J. Peřina, 1991, *Opt. Comm.* **85**, 261.

Bajer, J. and J. Peřina, 1992, *Opt. Comm.* **92**, 99.

Ban, M., 1991, *J. Math. Phys.* **32**, 3077.

Ban, M., 1992, *J. Opt. Soc. Am.* **B 9**, 1189.

Bandilla, A. and H. Paul, 1969, *Ann. Physik* **23**, 323.

Barnett, S. M., P. Filipowicz, J. Javanainen, P. L. Knight and P. Meystre, 1986, Frontiers in Quantum Optics, eds. E. R. Pike and S. Sarkar, A. Hilger, Bristol, p. 485.

Barnett, S. M. and S. J. D. Phoenix, 1993, *J. Mod. Opt.* **40**, 1443.

Beck, M., D. T. Smithey, J. Cooper and M. G. Raymer, 1993, *Opt. Lett.* **18**, 1259.

Beck, M., D. T. Smithey and M. G. Raymer, 1993, *Phys. Rev.* **A 48**, 890.

Belinsky, A. V. and D. N. Klyshko, 1993, *Phys. Lett.* **A 176**, 415.

Belinsky, A. V. and D. N. Klyshko, 1993a, *Usp. fiz. nauk* **163**, No. 8, 1 (in Russian) and references therein.

Bell, J. S., 1987, Speakable and Unspeakable in Quantum Mechanics, Cambridge Univ. Press, Cambridge.

Bennett, C. H. and S. J. Wiesner, 1992, *Phys. Rev. Lett.* **69**, 2881.

Beran, M. and G. B. Parrent, Jr., 1964, Theory of Partial Coherence, Prentice Hall, Englewood Cliffs, New Jersey.

Berezin, F. A., 1966, The Method of Second Quantization, Academic Press, New York.

Berezin, F. A. and M. A. Shubin, 1991, The Schrödinger Equation, Kluwer, Dordrecht.

Berry, M. V., 1984, *Proc. Roy. Soc. (London)* **A 392**, 45.

Berry, M. V., 1990, *Physics Today*, 34.

Bhandari, R., 1992, *Phys. Rev. Lett.* **69**, 3720.

Bhandari, R. and J. Samuel, 1988, *Phys. Rev. Lett.* **60**, 1211.

Blank, J., P. Exner and M. Havlíček, 1993, Linear Operators in Quantum Mechanics, Carolinum, Prague.

Bochove, E. J., 1992, *J. Opt. Soc. Am.* **B 9**, 266.

Bondurant, R. S. and J. H. Shapiro, 1984, *Phys. Rev.* **D 30**, 2548.

Born, M. and E. Wolf, 1965, Principles of Optics, Pergamon, Oxford.

Braginsky, V. B. and F. Y. Khalili, 1992, Quantum Measurement, Cambridge University Press.

Braginsky, V. B. and Vorontsov Y. I., 1974, *Usp. Fiz. Nauk* **114**, 41.

Braginsky, V. B., Y. I. Vorontsov and K. S. Thorne, 1980, *Science* **209**, 547.

Braunstein, S. L., 1992, *Phys. Rev. Lett.* **69**, 3598.

Braunstein, S. L. and C. M. Caves, 1990, *Phys. Rev.* **A 42**, 4115.

Braunstein, S. L., A. S. Lane and C. M. Caves, 1992, *Phys. Rev. Lett.* **69**, 2153.

Brendel, J., E. Mohler and W. Martienssen, 1991, *Phys. Rev. Lett.* **66**, 1142.

Brendel, J., E. Mohler and W. Martienssen, 1992, *Europhys. Lett.* **20**, 575.

Broglie, L., de, 1978, Les Incertitudes d'Heisenberg et l'Interpretation Probabilistique de la Mécanique Ondulatoire, Gauthier-Villars, Paris.

Brown, Hanbury R. and R. Q. Twiss, 1956, *Nature* **177**, 27.

Brune, M., S. Haroche, J. M. Raimond, L. Davidovich and N. Zagury, 1992, *Phys. Rev.* **A 45**, 5193.

Burnham, D. C. and D. L. Weinberg, 1970, *Phys. Rev. Lett.* **25**, 84.

Bush, P., P. Kienzler, P. Lahti and P. Mittelstaedt, 1993, *Phys. Rev.* **A 47**, 4627.

Bužek, V., 1990, *J. Mod. Opt.* **37**, 1033.

Bužek, V. and I. Jex, 1989, *J. Mod. Opt.* **36**, 1427.

Bužek, V. and I. Jex, 1990, *Opt. Comm.* **78**, 425.

Bužek, V., H. Moya-Cessa, P. L. Knight and S. J. D. Phoenix, 1992, *Phys. Rev.* **A 45**, 8190.

Campos, R. A., 1994, *Phys. Lett.* **A 184**, 173.

Campos, R. A., B. E. A. Saleh and M. C. Teich, 1989, *Phys. Rev.* **A 40**, 1371.

Campos, R. A., B. E. A. Saleh and M. C. Teich, 1990, *Phys. Rev.* **A 42**, 4127.

Capasso, F. and M. C. Teich, 1986, *Phys. Rev. Lett.* **57**, 1417.

Carmichael, H. J., P. D. Drummond, D. F. Walls and P. Meystre, 1980, *Opt. Acta* **27**, 581.

Carmichael, H. J. and D. F. Walls, 1976, *J. Phys.* **B 9**, L43 and 1199.

Carnal, O. and J. Mlynek, 1991, *Phys. Rev. Lett.* **66**, 2689.

Caroll, J. E., 1986, *Opt. Acta* **33**, 909.

Carruthers, P. and M. M. Nieto, 1968, *Rev. Mod. Phys.* **40**, 411.

Caves, C. M., 1981, *Phys. Rev.* **D 23**, 1693.

Caves, C. M., 1982, *Phys. Rev.* **A 26**, 1817.

Caves, C. M., 1985, *Phys. Rev. Lett.* **54**, 2465.

Caves, C. M., K. S. Thorne, R. W. P. Drever, V. D. Sandberg and M. Zimmermann, 1980, *Rev. Mod. Phys.* **52**, 341.

Chiao, R. Y., A. Antaramian, K. M. Ganga, H. Jiao, S. R. Wilkinson and H. Nathel, 1988, *Phys. Rev. Lett.* **60**, 1214.

Chiao, R. Y., P. G. Kwiat and A. M. Steinberg, 1992, Workshop on Squeezed States and Uncertainty Relations, Univ. Maryland, eds. J. D. Han, Y. S. Kim and W. W. Zachary, NASA, p. 61.

Chiao, R. Y. and Y. S. Wu, 1986, *Phys. Rev. Lett.* **57**, 933.

Christanell, R., H. Weinfurter, B. Erckert and A. Zeilinger, 1993, Nonlocal effects in a PDC experiment by k-selection with a grating, to be published, Innsbruck Univ.

Christanell, R., H. Weinfurter and A. Zeilinger, 1993, Tech. Digest of EQEC'93, eds. P. De Natale, R. Meucci and S. Pelli, Vol. 2, p. 827.

Cini, M. and J.-M. Levy-Leblond, 1990, Quantum Theory without Reduction, A. Hilger, Bristol.

Clauser, J. F. and M. A. Horne, 1974, *Phys. Rev.* **D 10**, 526.

Clauser, J. F., M. A. Horne, A. Shimony and R. A. Holt, 1969, *Phys. Rev. Lett.* **23**, 880.

Cohen-Tannoudji, C., 1977, Frontiers in Laser Spectroscopy, eds. R. Balian, S. Haroche and S. Liberman, North-Holland, Amsterdam.

Cook, R. J., 1980, *Opt.Comm.* **35**, 347.

Cook, R. J., 1981, *Phys. Rev.* **A 23**, 1243.

Cresser, J. D., J. Häger, G. Leuchs, M. Rateike and H. Walther, 1982, Dissipative Systems in Quantum Optics, eds. R. Bonifacio and L. Lugiato, Springer, Berlin.

Croca, J. R., 1987, *Found. Phys.* **17**, 22.

Croca, J. R., A. Garuccio and F. Selleri, 1988, *Found. Phys. Lett.* **1**, 101.

Croca, J. R., A. Garuccio, V. L. Lepore and R. N. Moreira, 1990, *Found. Phys. Lett.* **3**, 557.

Dagenais, M. and L. Mandel, 1978, *Phys. Rev.* **A 18**, 2217.

Daniel, D. J. and G. J. Milburn, 1989, *Phys. Rev.* **A 39**, 461.

Davies, E. B., 1976, Quantum Theory of Open Systems, Academic Press, London.

Davies, E. B. and J. T. Lewis, 1970, *Commun. Math. Phys.* **17**, 239.

Debuisschert, T., S. Reynaud, A. Heidmann, E. Giacobino and C. Fabre, 1989, *Quant. Opt.* **1**, 3.

Dicke, R. H., 1954, *Phys. Rev.* **93**, 99.

Diedrich, F., J. Krause, G. Rempe, M. O. Scully and H. Walther, 1988, *IEEE J. Quant. Electr.* **24**, 1314.

Diedrich, F. and H. Walther, 1987, *Phys. Rev. Lett.* **58**, 203.

Dirac, P. A. M., 1927, *Proc. Roy. Soc. (London)* **A 114**, 243.

Dirac, P. A. M., 1958, The Principles of Quantum Mechanics, Clarendon, Oxford.

Drobný, G. and I. Jex, 1992, *Phys. Rev.* **A 45**, 1816.

Drobný, G., I. Jex and V. Bužek, 1993, *Phys. Rev.* **A 48**, 569.

Drummond, P. D. and S. J. Carter, 1987, *J. Opt. Soc. Am.* **B 4**, 1565.

Eberly, J. H., L. Mandel and E. Wolf, eds., 1989, Coherence and Quantum Optics 6, Plenum, New York.

Eberly, J. H., N. B. Narozhny and J. J. Sanchez-Mondragon, 1980, *Phys. Rev. Lett.* **44**, 1323.

Ehlotzky, F., ed., 1993, Fundamentals of Quantum Optics 3, Springer, Berlin.

Eichmann, U., J. C. Bergquist, J. J. Bollinger, J. M. Gilligan, W. M. Itano, D. J. Wineland and M. G. Raizen, 1993, *Phys. Rev. Lett.* **70**, 2359.

Einstein, A., B. Podolsky and N. Rosen, 1935, *Phys. Rev.* **47**, 777.

Eiselt, J. and H. Risken, 1990, *Phys. Rev.* **A 43**, 346.

Ekert, A. K., J. G. Rarity, P. R. Tapster and G. M. Palma, 1992, *Phys. Rev. Lett.* **69**, 1293.

Endo, T. and K. Toyoshima, 1992, *Opt. Comm.* **90**, 197.

Evans, M. and S. Kielich, eds., 1993, Modern Nonlinear Optics, Vol. 1–3, J. Wiley, New York.

Fabre, C., E. Giacobino, A. Heidmann and S. Reynaud, 1989, *J. Phys. (France)* **50**, 1209.

Filipovicz, P., P. Meystre, G. Rempe and H. Walther, 1985, *Opt. Acta* **32**, 1105.

Fougères, A., J. W. Noh, T. P. Grayson and L. Mandel, 1994, *Phys. Rev.* **A 49**, 530.

Fougères, A., J. R. Torgerson and L. Mandel, 1994, *Opt. Comm.* **105**, 199.

Franson, J. D., 1989, *Phys. Rev. Lett.* **62**, 2205.

Franson, J. D., 1991, *Phys. Rev. Lett.* **67**, 290.

Franson, J. D., 1991a, *Phys. Rev.* **A 44**, 4552.

Franson, J. D., 1993, *Phys. Rev.* **A 48**, 4610.

Freedman, S. J. and J. F. Clauser, 1972, *Phys. Rev. Lett.* **28**, 938.

Freyberger M. and W. Schleich, 1993, *Phys. Rev.* **A 47**, R30.

Friberg, S., C. K. Hong and L. Mandel, 1985, *Phys. Rev. Lett.* **54**, 2011.

Fry, E. S. and R. C. Thompson, 1976, *Phys. Rev. Lett.* **37**, 465.

Gaeta, A. L. and R. W. Boyd, 1988, *Phys. Rev. Lett.* **60**, 2618.

Gaeta, A. L. and R. W. Boyd, 1989, Coherence and Quantum Optics 6, eds. J. H. Eberly, L. Mandel and E. Wolf, Plenum, New York, p. 343.

Gardiner, C. W., 1991, Quantum Noise, Springer, Berlin.

Garuccio, A. and V. L. Lepore, 1992, *Phys. Rev. Lett.* **A 163**, 343.

Gähler, R. and A. Zeilinger, 1991, *Am. J. Phys.* **59**, 316.

Ghose, P., D. Horne and G. S. Agarwal, 1992, *Phys. Lett.* **A 168**, 95.

Ghosh, R. and L. Mandel, 1987, *Phys. Rev. Lett.* **59**, 1903.

Gibbs, H. M., 1985, Optical Bistability:Controlling Light with Light, Acad. Press, Orlando.

Glauber, R. J., 1963a, *Phys. Rev.* **130**, 2529.

Glauber, R. J., 1963b, *Phys. Rev.* **131**, 2766.

Glauber, R. J., 1986, Frontiers in Quantum Optics, eds. E. R. Pike and S. Sarkar, A. Hilger, Bristol, p. 534.

Golubev, Yu. M., 1987, *J. Exper. Theor. Phys. (USSR)* **93**, 463.

Gradshteyn, I. S. and I. M. Ryzhik, 1965, Table of Integrals, Series and Products, Academic Press, New York.

Graham, R., 1984, *Phys. Rev. Lett.* **52**, 117.

Grangier, P., M. J. Potasek and B. Yurke, 1988, *Phys. Rev.* **A 38**, 3132.

Grangier, P., G. Roger and A. Aspect, 1986, *Europhys. Lett.* **1**, 173.

Grangier, P., G. Roger, A. Aspect, A. Heidmann and S. Reynaud, 1986, *Phys. Rev. Lett.* **57**, 687.

Grayson, T. P., J. R. Torgerson and G. A. Barbosa, 1994, *Phys. Rev.* **A 49**, 626.

Grayson, T. P., X. Y. Zou, D. Branning, J. R. Torgerson and L. Mandel, 1993, *Phys. Rev.* **A 48**, 4793.

Greenberger, D. M., M. A. Horne, A. Shimony and A. Zeilinger, 1990, *Am. J. Phys.* **58**, 1131.

Greenberger, D. M., M. Horne and A. Zeilinger, 1989, Bell's Theorem, Quantum Theory, and Conceptions of the Universe, ed. M. Kafatos, Kluwer, Dordrecht, p. 73.

Greenberger, D. M., M. A. Horne and A. Zeilinger, 1993, *Phys. Today,* No. 8, 22.

Haake, F., 1973, Statistical Treatment of Open Systems by Generalized Master Equations, Quantum Statistics in Optics and Solid-State Physics, Springer Tracts in Modern Physics, Vol. 66, ed. G. Höhler, Springer, Berlin, p. 98.

Haake, F., L. M. Narducci and D. F. Walls, eds., 1986, Coherence, Cooperation and Fluctuations, Cambridge Univ. Press, Cambridge.

Haken, H., 1970, Handbuch der Physik, Vol. 25/2c, ed. S. Flügge, Springer, Berlin.

Hall, M. J., 1991, *Quant. Opt.* **3**, 7.

Han, D., Y. S. Kim and V. I. Man'ko, 1993, Second Internat. Workshop on Squeezed States and Uncertainty Relations, NASA, Maryland.

Han, D., Y. S. Kim and W. W. Zachary, 1992, Workshop on Squeezed States and Uncertainty Relations, NASA, Maryland.

Han, D., Y. S. Kim and W. W. Zachary, 1993, Workshop on Harmonic Oscillators, NASA, Maryland.

Hariharan, P., N. Brown and B. C. Sanders, 1993, *J. Mod. Opt.* **40**, 113.

Hariharan, P., N. Brown and B. C. Sanders, 1993a, *J. Mod. Opt.* 40, 1573.

Hariharan, P., I. Fujima, N. Brown and B. C. Sanders, 1993, *J. Mod. Opt.* **40**, 1477.

Hariharan, P., M. Roy, P. A. Robinson and J. W. O'Byrne, 1993, *J. Mod. Opt.* **40**, 871.

Haroche, S., 1992, Fundamental Systems in Quantum Optics, eds. J. Dalibard, J. M. Raimond and J. Zinn-Justin, North-Holland, Amsterdam, p. 767.

Haroche, S. and D. Kleppner, 1989, *Phys. Today* **42**, 24.

Haroche, S. and J. M. Raimond, 1985, *Adv. Atom. Molec. Phys.* **20**, 347.

Harvey, J. D. and D. F. Walls, 1986, Quantum Optics 4, Springer, Berlin.

Haus, H. A. and J. A. Mullen, 1962, *Phys. Rev.* **128**, 2407.

Heffner, H., 1962, *Proceedings of the IRE* **50**, 1604.

Heidmann, A., R. J. Horowicz, S. Reynaud, E. Giacobino, C. Fabre and G. Camy, 1987, *Phys. Rev. Lett.* **59**, 2555.

Heitler, W., 1954, The Quantum Theory of Radiation, 3rd ed., Oxford.

Helstrom, C. W., 1976, Quantum Detection and Estimation Theory, Academic Press, New York.

Herzog, T., J. Rarity, H. Weinfurter and A. Zeilinger, 1993, Tech. Digest of EQEC'93, eds. P. De Natale, R. Meucci and S. Pelli, Vol. 2, p. 823.

Herzog, T. J., J. G. Rarity, H. Weinfurter and A. Zeilinger, 1994, *Phys. Rev. Lett.* **72**, 629.

Hillery, M., 1987a, *Phys. Rev.* **A 36**, 3796.

Hillery, M., 1987b, *Opt. Comm.* **62**, 135.

Hillery, M., 1989, *Phys. Rev.* **A 40**, 3147.

Hillery, M., D. Yu and J. Bergou, 1992, Workshop on Squeezed States and Uncertainty Relations, eds. D. Han, Y. S. Kim and W. W. Zachary, NASA Conf. Publ. No. 3135, Maryland, p. 125.

Hirano, T. and M. Matsuoka, 1990, *Opt. Comm.* **15**, 1153.

Holevo, A. S., 1982, Probabilistic and Statistical Aspects of Quantum Theory, North-Holland, Amsterdam.

Holland, M. J. and K. Burnett, 1993, *Phys. Rev. Lett.* **71**, 1355.

Hollenhorst, J. N., 1979, *Phys. Rev.* **D 19**, 1669.

Hong, C. K. and L. Mandel, 1985, *Phys. Rev.* **A 32**, 974.

Hong, C. K. and L. Mandel, 1985a, *Phys. Rev.* **A 31**, 2409.

Hong, C. K., Z. Y. Ou and L. Mandel, 1987, *Phys. Rev. Lett.* **59**, 2044.

Horne, M. A., A. Shimony and A. Zeilinger, 1989, *Phys. Rev. Lett.* **62**, 2209.

Horowicz, R., M. Pinard and S. Reynaud, 1987, *Opt. Comm.* **61**, 142.

Hradil, Z., 1989, *Acta Phys. Pol.* **A 75**, 731.

Hradil, Z., 1990, *Phys. Rev.* **A 41**, 400.

Hradil, Z., 1991, *Phys. Rev.* **A 44**, 792.
Hradil, Z., 1992, *Quant. Optics* **4**, 93.
Hradil, Z., 1993, *Phys. Rev.* **A 47**, 2376.
Hradil, Z., 1993a, *Phys. Rev.* **A 47**, 4532.
Hradil, Z. and J. H. Shapiro, 1992, *Quant. Opt.* **4**, 31.
Huang, J. and P. Kumar, 1992, *Phys. Rev. Lett.* **68**, 2153.
Hussain, N., N. Imoto and R. Loudon, 1992, *Phys. Rev.* **A 45**, 1987.
Imoto, N., H. A. Haus and Y. Yamamoto, 1985, *Phys. Rev.* **A 32**, 2287.
Imoto, N., M. Ueda and T. Ogawa, 1990, *Phys. Rev.* **A 41**, 4127.
Jackiw, R., 1968, *J. Math. Phys.* **9**, 339.
Jaeger, G., M. A. Horne and A. Shimony, 1993, *Phys. Rev.* **A 48**, 1023.
Jakeman, E., E. R. Pike, P. N. Pusey and J. M. Vaugham, 1977, *J. Phys.* **A 10**, L257.
Jauch, J. M., 1968, Foundations of Quantum Mechanics, Addison-Wesley, Reading, Mass..
Jaynes, E. T. and F. W. Cummings, 1963, *Proc. IEEE* **51**, 89.
Jeffries, A. D., P. R. Saulson, R. E. Spero and M. E. Zucker, 1987, *Scientific Am.* **255**, 50.
Kampen, N. G. Van, 1981, Stochastic Processes in Physics and Chemistry, North-Holland, Amsterdam.
Kärtner, F. X. and H. A. Haus, 1993, *Phys. Rev.* **A 47**, 4585.
Keith, D. W., C. R. Ekstrom, Q. A. Turchette and D. E. Pritchard, 1991, *Phys. Rev. Lett.* **66**, 2693.
Kelley, P. L. and W. H. Kleiner, 1964, *Phys. Rev.* **136**, A316.
Kielich, S., 1981, Molecular Nonlinear Optics, Nauka, Moscow (in Russian).
Kiess, T. E., Y. H. Shih, A. V. Sergienko and C. O. Alley, 1993, *Phys. Rev. Lett.* **71**, 3893.
Kim, C. and P. Kumar, 1992, *J. Opt. Soc. Am.* **B 9**, 1379.
Kim, M. S. and V. Bužek, 1992a, *J. Mod. Opt.* **39**, 1909.
Kim, M. S. and V. Bužek, 1992b, *Phys. Rev.* **A 46**, 4239.
Kim, Y. S. and M. E. Noz, 1991, Phase Space Picture of Quantum Mechanics: Group Theoretical Approach, World Scientific, Singapore.
Kimble, H. J., 1992, Fundamental Systems in Quantum Optics, eds. J. Dalibard, J. M. Raimond and J. Zinn-Justin, North-Holland, Amsterdam, p. 545.
Kimble, H. J., M. Dagenais and L. Mandel, 1977, *Phys. Rev. Lett.* **39**, 691.
Kimble, H. J., M. Dagenais and L. Mandel, 1978, *Phys. Rev.* **A 18**, 201.
Kimble, H. J. and L. Mandel, 1976, *Phys. Rev.* **A 13**, 2123.
Kitagawa, M., N. Imoto and Y. Yamamoto, 1987, *Phys. Rev.* **A 35**, 5270.
Kitagawa, M. and Y. Yamamoto, 1986, *Phys. Rev.* **A 34**, 3974.
Klauder, J. R. and B.-S. Skagerstam, eds., 1985, Coherent States, World Scientific, Singapore.
Klauder, J. R. and E. C. G. Sudarshan, 1968, Fundamentals of Quantum Optics, W. A. Benjamin, New York.
Klyshko, D. N., 1986, Physical Fundamentals of Quantum Electronics, Nauka, Moscow (in Russian).
Klyshko, D. N., 1988, Photons and Nonlinear Optics, Gordon and Breach, New York.
Klyshko, D. N., 1993, *Phys. Lett.* **A 172**, 399.
Knight, P. L. and L. Allen, 1983, Concepts of Quantum Optics, Pergamon, Oxford.
Knight, P. L. and V. Bužek, 1995, Progress in Optics, Vol. 34, ed. E. Wolf, Elsevier, Amsterdam.
Knight, P. L. and P. M. Radmore, 1982, *Phys. Lett.* **A 90**, 342.
Knöll, L., W. Vogel and D. G. Welsch, 1986, *J. Opt. Soc. Am.* **B 3**, 1315.
Knöll, L. and D. G. Welsch, 1992, *Progr. Quant. Electr.* **16**, 135.
Koashi, M., K. Kono, T. Hirano and M. Matsuoka, 1993, *Phys. Rev. Lett.* **71**, 1164.

Kocher, C. A. and E. D. Cummins, 1967, *Phys. Rev. Lett.* **18**, 575.

Kodama, Y. and A. Hasegawa, 1992, Progress in Optics, Vol. 30, ed. E. Wolf, North-Holland, Amsterdam, p. 205.

Komrska, J., 1971, Advances in Electronics and Electron. Phys., Vol. 30, ed. L. Marton, Acad. Press, New York, p. 139.

Krause, J., M. O. Scully and H. Walther, 1987, *Phys. Rev.* **A 36**, 4547.

Krause, J., M. O. Scully, T. Walther and H. Walther, 1989, *Phys. Rev.* **A 39**, 1915.

Kürz, P., R. Paschotta, K. Fiedler and J. Mlynek, 1993, *Europhys. Lett.* **24**, 449.

Kwiat, P. G. and R. Y. Chiao, 1991, *Phys. Rev. Lett.* **66**, 588.

Kwiat, P. G., A. M. Steinberg and R. Y. Chiao, 1992, *Phys. Rev.* **A 45**, 7729.

Kwiat, P. G., A. M. Steinberg and R. Y. Chiao, 1993, *Phys. Rev.* **A 47**, R2472.

Kwiat, P. G., A. M. Steinberg and R. Y. Chiao, 1993a, *Phys. Rev.* **A 48**, 3999.

Kwiat, P. G., A. M. Steinberg and R. Y. Chiao, 1994, *Phys. Rev.* **A 49**, 61.

Kwiat, P. G., W. A. Vareka, C. K. Hong, H. Nathel and R. Y. Chiao, 1990, *Phys. Rev.* **A 41**, 2910.

Lai, M. and J.-C. Diels, 1992, *J. Opt. Soc. Am.* **B 9**, 2290.

Lai, Y. and H. A. Haus, 1989, *Quant. Opt.* **1**, 99.

Lamehi-Rachti, M. and W. Mitting, 1976, *Phys. Rev.* **D 14**, 2543.

Lane, A. S., S. L. Braunstein and C. M. Caves, 1993, *Phys. Rev.* **A 47**, 1667.

LaPorta, A., R. E. Slusher and B. Yurke, 1989, *Phys. Rev. Lett.* **62**, 28.

Larchuk, T. S., R. A. Campos, J. G. Rarity, P. R. Tapster, E. Jakeman, B. E. A. Saleh and M. C. Teich, 1993, *Phys. Rev. Lett.* **70**, 1603.

Lax, M., 1968, Fluctuations and Coherence Phenomena in Classical and Quantum Physics, Gordon and Breach, New York.

Lenstra, D., 1982, *Phys. Rev.* **A 26**, 3369.

Levenson, M. D. and R. M. Shelby, 1987, *J. Mod. Opt.* **34**, 775.

Levenson, M. D., R. M. Shelby, M. Reid and D. F. Walls, 1986, *Phys. Rev. Lett.* **57**, 2473.

Levy-Leblond, J. M., 1976, *Ann. Phys. (N.Y.)* **101**, 319.

Loudon, R., 1973, The Quantum Theory of Light, Clarendon, Oxford.

Loudon, R., 1989, *Opt. Comm.* **70**, 109.

Loudon, R., 1989a, Coherence and Quantum Optics 6, eds. J. H. Eberly, L. Mandel and E. Wolf, Plenum, New York, p. 703.

Loudon, R. and P. L. Knight, 1987, *J. Mod. Opt.* **34**, 709.

Louisell, W. H., 1973, Quantum Statistical Properties of Radiation, J. Wiley, New York.

Lukš, A. and V. Peřinová, 1991, *Czech. J. Phys.* **41**, 1205.

Lukš, A. and V. Peřinová, 1994, Presumable Solutions of Quantum Phase Problem and Their Flows, *Quant. Opt.*, in print.

Lukš, A., V. Peřinová and J. Křepelka, 1994, *J. Mod. Opt.*, in print.

Lyubomirsky, I., M. Shirasaki, F. X. Kärtner and H. A. Haus, 1993, *Quant. Optics* **5**, 241.

Machida, S. and Y. Yamamoto, 1986, *Opt. Comm.* **57**, 290.

Machida, S. and Y. Yamamoto, 1988, *Phys. Rev. Lett.* **60**, 792.

Machida, S. and Y. Yamamoto, 1989, *Opt. Lett.* **14**, 1045.

Machida, S., Y. Yamamoto and Y. Itaya, 1987, *Phys. Rev. Lett.* **58**, 1000.

Maeda, Mari W., P. Kumar and J. H. Shapiro, 1987, *Opt. Lett.* **12**, 161.

Maeda, Mari W., P. Kumar and J. H. Shapiro,1987a, *J. Opt. Soc. Am.* **B 4**, 1501.

Malkin, I. A. and V. I. Man'ko, 1979, Dynamical Symmetries and Coherent States of Quantum Systems, Nauka, Moscow (in Russian).

Mandel, L., 1962, *J. Opt. Soc. Am.* **52**, 1335.

Mandel, L., 1963, Progress in Optics, Vol. 2, ed. E. Wolf, North-Holland, Amsterdam, p. 181.

Mandel, L., 1966, *Phys. Rev.* **152**, 438.

Mandel, L., 1976, Progress in Optics, Vol. 13, ed. E. Wolf, North-Holland, Amsterdam, p. 27.

Mandel, L., 1979, *Opt. Lett.* **4**, 205.

Mandel, L., 1982, *Phys. Rev. Lett.* **49**, 136.

Mandel, L., 1983, *Phys. Rev.* **A 28**, 929.

Mandel, L., 1986, Frontiers in Quantum Optics, eds. E. R. Pike and S. Sarkar, A. Hilger, Bristol, p. 318.

Mandel, L., 1992, Workshop on Squeezed States and Uncertainty Relations, eds. D. Han, Y. S. Kim and W. W. Zachary, NASA Confer. Publ. 3135, p. 5.

Mandel, L. and E. Wolf, 1965, *Rev. Mod. Phys.* **37**, 231.

Mandel, L. and E. Wolf, eds., 1978, Coherence and Quantum Optics 4, Plenum, New York.

Mandel, L. and E. Wolf, eds., 1984, Coherence and Quantum Optics 5, Plenum, New York.

Mandel, L. and E. Wolf, eds., 1990, Coherence and Fluctuations of Light, SPIE, Bellingham, Washington.

Mandel, L. and E. Wolf, 1994/5, Optical Coherence and Quantum Optics, Cambridge Univ. Press, Cambridge.

Markov, M. A., ed., 1991, Squeezed and Correlated States of Quantum Systems, Nauka, Moscow (in Russian).

Marshall, T. W. and E. Santos, 1991, *J. Mod. Opt.* **38**, 1463.

Marte, M. A., H. Ritsch and D. F. Walls, 1988, *Phys. Rev. Lett.* **61**, 1093.

Martinez-Linares, J. and P. Garcia-Fernandez, 1994, *J. Mod. Opt.* **41**, 67.

Matteucci, G., 1990, *Am. J. Phys.* **58**, 1143.

Mattle, K. U., 1993, Zweiphotonen-Quantenstatistik am Dreifachstrahlteiler, Diplomaarbeit, Univ. Innsbruck.

McNeil, K. J. and D. F. Walls, 1974, *J. Phys.* **A 7**, 617.

Mermin, N. D., 1993, *Rev. Mod. Phys.* **65**, 803.

Mertz, J. C., T. Debuisschert, A. Heidmann, C. Fabre and E. Giacobino, 1991, *Opt. Lett.* **16**, 1234.

Mertz, J., A. Heidmann and C. Fabre, 1991, *Phys. Rev.* **A 44**, 3229.

Mertz, J. C., A. Heidmann, C. Fabre, E. Giacobino and S. Reynaud, 1990, *Phys. Rev. Lett.* **64**, 2897.

Meschede, D., H. Walther and G. Müller, 1985, *Phys. Rev. Lett.* **54**, 551.

Messiah, A., 1961, Quantum Mechanics I, II, North-Holland, Amsterdam.

Meystre, D., 1992, Progress in Optics, Vol. 30, ed. E. Wolf, North-Holland, Amsterdam, p. 261.

Meystre, P., A. Quattropani and H. P. Baltes, 1974, *Phys. Lett.* **A 49**, 85.

Meystre, P. and M. Sargent III, 1991, Elements of Quantum Optics, Springer, Berlin.

Milburn, G. J., 1991, *Phys. Rev.* **A 44**, 5401.

Milburn, G. J., M. D. Levenson, R. M. Shelby, S. H. Perlmutter, R. G. DeVoe and D. F. Walls, 1987, *J. Opt. Soc. Am.* **B 4**, 1476.

Milburn, G. J., D. F. Walls and M. D. Levenson, 1984, *J. Opt. Soc. Am.* **B 1**, 390.

Milonni, P. W., E. J. Bochove and R. J. Cook, 1989, *J. Opt. Soc. Am.* **B 6**, 1932.

Milonni, P. W., E. J. Bochove and R. J. Cook, 1989a, *Phys. Rev.* **A 40**, 4100.

Milonni, P. W. and S. Singh, 1991, Advances in Atomic, Molecular, and Optical Physics, Vol. 28, eds. D. Bates and B. Bederson, Acad. Press, New York, p. 75.

Milonni, P. W., M. L. Shih and J. R. Ackerhalt, eds., 1987, Chaos in Laser – Matter Interactions, World Scientific, Singapore.

Miranowicz, A. and S. Kielich, 1993, Modern Nonlinear Optics, Vol. 3, eds. M. W. Evans and S. Kielich, J. Wiley, New York.

Mišta, L. and J. Peřina, 1977, *Acta Phys. Pol.* **A 52**, 425.

Mišta, L. and J. Peřina, 1978, *Czech. J. Phys.* **B 28**, 392.

Mišta, L., V. Peřinová, J. Peřina and Z. Braunerová, 1977, *Acta Phys. Pol.* **A 51**, 739.

Mizobuchi, Y. and Y. Ohtaké, 1992, *Phys. Lett.* **A 168**, 1.

Mollow, B. R., 1968, *Phys. Rev.* **168**, 1896.

Morse, P. M. and H. Feshbach, 1953, Methods of Theoretical Physics, Vol. 1, McGraw-Hill, Amsterdam.

Muller, A., J. Buguet and N. Gisin, 1993, *Europhys. Lett.* **23**, 383.

Munro, W. J. and M. D. Reid, 1993, *Phys. Rev.* **A 47**, 4412.

Namiki, M. and S. Pascazio, 1993, *Phys. Lett.* **A 175**, 150.

Neumann, J., von, 1932, Mathematische Grundlagen der Quantenmechanik, Springer Verlag, Berlin.

Neumann, J., von, 1955, Mathematical Foundations of Quantum Mechanics, Princeton.

Neumann, R. and H. Haug, 1979, *Opt. Comm.* **31**, 267.

Newton, R. G., 1980, *Ann. Phys. (N.Y.)* **124**, 327.

Noh, J. W., A. Fougères and L. Mandel, 1992, *Phys. Rev.* **A 45**, 424.

Noh, J. W., A. Fougères and L. Mandel, 1992a, *Phys. Rev.* **A 46**, 2840.

Noh, J. W., A. Fougères and L. Mandel, 1993, *Phys. Rev. Lett.* **71**, 2579.

Orozco, J. A., M. G. Raizen, M. Xiao, R. J. Brecha and H. J. Kimble, 1987, *J. Opt. Soc. Am.* **B 4**, 1490.

Ou, Z. Y., C. K. Hong and L. Mandel, 1987, *Opt. Comm.* **63**, 118.

Ou, Z. Y. and L. Mandel, 1988, *Phys. Rev. Lett.* **61**, 50.

Ou, Z. Y. and L. Mandel, 1988a, *Phys. Rev. Lett.* **59**, 2046.

Ou, Z. Y. and L. Mandel, 1988b, *Phys. Rev. Lett.* **61**, 54.

Ou, Z. Y. and L. Mandel, 1990, *Quant. Opt.* **2**, 71.

Ou, Z. Y. and L. Mandel, 1990a, *J. Opt. Soc. Am.* **B 7**, 2127.

Ou, Z. Y., S. F. Pereira, H. J. Kimble and K. C. Peng, 1992, *Phys. Rev. Lett.* **68**, 3663.

Ou, Z. Y., L. J. Wang and L. Mandel, 1989, *Phys. Rev.* **A 40**, 1428.

Ou, Z. Y., L. J. Wang and L. Mandel, 1990, *J. Opt. Soc. Am.* **B 7**, 211.

Ou, Z. Y., L. J. Wang, X. Y. Zou and L. Mandel,1990, *Phys. Rev.* **A 41**, 566.

Ou, Z. Y., L. J. Wang, X. Y. Zou and L. Mandel, 1990a, *Phys. Rev.* **A 41**, 1597.

Ou, Z. Y., X. Y. Zou, L. J. Wang and L. Mandel, 1990, *Phys. Rev.* **A 42**, 2957.

Ou, Z. Y., X. Y. Zou, L. J. Wang and L. Mandel, 1990a, *Phys. Rev. Lett.* **65**, 321.

Paul, H., 1974, *Fortschr. Physik* **22**, 657.

Paul, H., 1982, *Rev. Mod. Phys.* **54**, 1061.

Paul, H., 1986, *Rev. Mod. Phys.* **58**, 209.

Pegg, D. T. and S. M. Barnett, 1989, *Phys. Rev.* **A 39**, 1665.

Perelomov, A. M., 1986, Generalized Coherent States and Their Applications, Springer, Berlin.

Peřina, J., 1985, Coherence of Light, D. Reidel, Dordrecht – Boston.

Peřina, J., 1991, Quantum Statistics of Linear and Nonlinear Optical Phenomena, Kluwer, Dordrecht – Boston.

Peřina, J., ed., 1991a, Photon Statistics and Coherence in Nonlinear Optics, SPIE, Bellingham, Washington.

Peřina, J., Jr., 1993, *Czech. J. Phys.* **43**, 615.

Peřina, J., Jr., 1993a, *J. Mod. Opt.* **40**, 2105.

Peřina, J. and J. Bajer, 1990, *Phys. Rev.* **A 41**, 516.

Peřina, J., J. Bajer, V. Peřinová and Z. Hradil, 1993, Photon Statistics of Nonclassical Fields, Modern Nonlinear Optics, Vol. 1, eds. M. W. Evans and S. Kielich, J. Wiley, New York, p. 405.

Peřina, J., R. Horák, Z. Hradil, C. Sibilia and M. Bertolotti, 1989, *J. Mod. Opt.* **36**, 571.

Peřina, J. and J. Křepelka, 1992, *J. Mod. Opt.* **39**, 2405.

Peřina, J., V. Peřinová and J. Koďousek, 1984, *Opt. Comm.* **49**, 210.

Peřina, J., V. Peřinová, C. Sibilia and M. Bertolotti, 1984, *Opt. Comm.* **49**, 285.

Peřinová, V., 1981, *Opt. Acta* **28**, 747.

Peřinová, V., J. Křepelka and J. Peřina, 1986, *Opt. Acta* **33**, 1263.

Peřinová, V. and A. Lukš, 1990, *Phys. Rev.* **A 41**, 478.

Peřinová, V. and A. Lukš, 1994, Progress in Optics, Vol. 33, ed. E. Wolf, Elsevier, Amsterdam.

Peřinová, V. and J. Peřina, 1981, *Opt. Acta* **28**, 769.

Pfleegor, R. L. and L. Mandel, 1967, *Phys. Rev.* **159**, 1084.

Pfleegor, R. L. and L. Mandel, 1968, *J. Opt. Soc. Am.* **58**, 946.

Phoenix, S. J. D. and S. M. Barnett, 1993, *J. Mod. Opt.* **40**, 979.

*Physics Today*, 1993.

Pike, E. R. and S. Sarkar, eds., 1986, Frontiers in Modern Optics, A. Hilger, Bristol.

Popescu, S., 1994, *Phys. Rev. Lett.* **72**, 797.

Radloff, W., 1968, *Phys. Lett.* **27 A**, 366.

Radloff, W., 1971, *Ann. Physik* **26**, 178.

Rai, S., 1992, *J. Opt. Soc. Am.* **B 9**, 590.

Rao, R. C., 1973, Linear Statistical Inference and Its Applications, J. Wiley, New York.

Rarity, J. G., J. Burnett, P. R. Tapster and R. Paschotta, 1993, *Europhys. Lett.* **22**, 95.

Rarity, J. G. and P. R. Tapster, 1989, *J. Opt. Soc. Am.* **B 6**, 1221.

Rarity, J. G. and P. R. Tapster, 1990, *Phys. Rev. Lett.* **64**, 2495.

Rarity, J. G. and P. R. Tapster, 1990a, *Phys. Rev.* **A 41**, 5139.

Rarity, J. G. and P. R. Tapster, 1992, *Phys. Rev.* **A 45**, 2052.

Rarity, J. G., P. R. Tapster and E. Jakeman, 1987, *Opt. Comm.* **62**, 201.

Rarity, J. G., P. R. Tapster, E. Jakeman, T. Larchuk, R. A. Campos, M. C. Teich and B. E. A. Saleh, 1990, *Phys. Rev. Lett.* **65**, 1348.

Rauch, H., 1993, *Phys. Lett.* **A 173**, 240.

Rauch, H. and J. P. Vigier, 1990, *Phys. Lett.* **A 151**, 269.

Raymer, M. G. and I. A. Walmsley, 1990, Progress in Optics, Vol. 28, ed. E. Wolf, North-Holland, Amsterdam, p. 181.

Raymer, M. G. and S. Yang, 1992, Workshop on Squeezed States and Uncertainty Relations, eds. D. Han, Y. S. Kim and W. W. Zachary, NASA, p. 91.

Reck, M. and A. Zeilinger, 1993, Quantum Interferometry, eds. F. De Martini and A. Zeilinger, World Scientific, Singapore.

Reid, M. D. and W. J. Munro, 1992, *Phys. Rev. Lett.* **69**, 997.

Reid, M. D. and D. F. Walls, 1986, *Phys. Rev.* **A 34**, 1260.

Rempe, G., W. Schleich, M. O. Scully and H. Walther, 1989, Proc. 3rd Int. Symp. Foundations of Quantum Mechanics, Tokyo, p. 294.

Rempe, G., F. Schmidt-Kaler and H. Walther, 1990, *Phys. Rev. Lett.* **64**, 2783.

Rempe, G., H. Walther and N. Klein, 1987, *Phys. Rev. Lett.* **58**, 353.

Reynaud, S., 1987, *Europhys. Lett.* **4**, 427.

Reynaud, S., C. Fabre and E. Giacobino, 1987, *J. Opt. Soc. Am.* **B 4**, 1520.

Reynaud, S., A. Heidmann, E. Giacobino and C. Fabre, 1992, Progress in Optics, Vol. 30, ed. E. Wolf, North-Holland, Amsterdam, p. 1.

Rice, D. A., C. F. Osborne and P. Lloyd, 1994, *Phys. Lett.* **A 186**, 21.
Risken, H., 1984, The Fokker-Planck Equation, Springer, Berlin.
Rosenbluh, M. and R. M. Shelby, 1991, *Phys. Rev. Lett.* **66**, 153.
Rubin, M. H. and Y. H. Shih, 1992, *Phys. Rev.* **A 45**, 8138.
Ruthen, R., 1992, *Scientific Am.* **262**, 90.
Saleh, B., 1978, Photoelectron Statistics, Springer, Berlin.
Saleh, B. E. A. and M. C. Teich, 1985, *Opt. Comm.* **52**, 429.
Saleh, B. E. A. and M. C. Teich, 1991, Fundamentals of Photonics, J. Wiley, New York.
Sánchez-Soto, L. L. and A. Luis, 1994, *Opt. Comm.* **105**, 84.
Santos, E., 1991, *Phys. Rev. Lett.* **66**, 1388.
Schleich, W. P. and S. M. Barnett, eds., 1993, *Phys. Scripta* **T 48**.
Schleich, W. and J. A. Wheeler, 1987, *Nature* **326**, 574.
Schleich, W. and J. A. Wheeler, 1987a, *J. Opt. Soc. Am.* **B 4**, 1715.
Schubert, M., K. E. Süsse, W. Vogel, D. G. Welsch and B. Wilhelmi, 1982, *Kvant. Elektr. (USSR)* **9**, 495.
Schubert, M. and B. Wilhelmi, 1986, Nonlinear Optics and Quantum Electronics, J. Wiley, New York.
Schumaker, B. L., 1984, *Opt. Lett.* **9**, 189.
Scully, M. O., B.-G. Englert and H. Walther, 1991, *Nature* **351**, 111.
Scully, M. O. and H. Walther, 1989, *Phys. Rev.* **A 39**, 5229.
Scully, M. O. and K. G. Whitney, 1972, Tools of Theoretical Quantum Optics, Progress in Optics, Vol. 10, ed. E. Wolf, North-Holland, Amsterdam, p. 89.
Selleri, F., 1987, *Phys. Lett.* **A 120**, 371.
Selleri, F., 1987a, *Found. Phys.* **17**, 739.
Selleri, F., 1988, *Phys. Lett.* **A 132**, 72.
Selleri, F., 1990, Quantum Paradoxes and Physical Reality, Kluwer, Dordrecht – Boston.
Selleri, F., ed., 1992, Wave-Particle Duality, Plenum, New York.
Sergienko, A. V., Y. H. Shih and M. H. Rubin, 1993, *J. Mod. Opt.* **40**, 1425.
Seward, S. F., P. R. Tapster, J. G. Walker and J. G. Rarity, 1991, *Quant. Opt.* **3**, 201.
Shapiro, J. H., 1985, *IEEE J. Quantum Electron.* **21**, 237.
Shapiro, J. H., G. Saplakoglu, S.-T. Ho, P. Kumar, B. E. A. Saleh and M. C. Teich, 1987, *J. Opt. Soc. Am.* **B 4**, 1604.
Shapiro, J. H. and S. R. Shepard, 1991, *Phys. Rev.* **A 43**, 3795.
Shapiro, J. H., S. R. Shepard and N. C. Wong, 1989, *Phys. Rev. Lett.* **62**, 2377.
Shapiro J. H. and S. M. Wagner, 1984, *IEEE J. Quantum Electron.* **20**, 803.
Shapiro, J. H., H. P. Yuen and J. A. Machado Mata, 1979, *IEEE Trans. Inf. Theory* **25**, 179.
She, C. Y. and H. Heffner, 1966, *Phys. Rev.* **152**, 1103.
Shelby, R. M., M. D. Levenson, S. H. Perlmutter, R. G. DeVoe and D. F. Walls, 1986, *Phys. Rev. Lett.* **57**, 691.
Shelby, R. M., M. D. Levenson, D. F. Walls, A. Aspect and G. J. Milburn, 1986, *Phys. Rev.* **A 33**, 4008.
Shih, Y. H. and M. H. Rubin, 1993, *Phys. Lett.* **A 182**, 16.
Shih, Y. H. and A. V. Sergienko, 1994, *Phys. Lett.* **A 186**, 29.
Shih, Y., A. V. Sergienko and M. H. Rubin, 1993, *Phys. Rev.* **A 47**, 1288.
Shih, Y., A. V. Sergienko and M. H. Rubin, 1993a, *Phys. Rev.* **A 48**, 4001.
Shore, B. W. and P. L. Knight, 1993, *J. Mod. Optics* **40**, 1195.
Short, R. and L. Mandel, 1983, *Phys. Rev. Lett.* **51**, 384.
Simon, R., H. J. Kimble and E. C. G. Sudarshan, 1988, *Phys. Rev. Lett.* **61**, 19.

Singer, F., M. J. Potasek, J. M. Fang and M. C. Teich, 1992, *Phys. Rev.* **A 46**, 4192.

Slusher, R. E., P. Grangier, A. LaPorta, B. Yurke and M. J. Potasek, 1987, *Phys. Rev. Lett.* **59**, 2566.

Slusher, R. E., L. W. Hollberg, B. Yurke, J. C. Mertz and J. F. Valley, 1985, *Phys. Rev. Lett.* **55**, 2409.

Slusher, R. E., B. Yurke, P. Grangier, A. LaPorta, D. F. Walls and M. Reid, 1987, *J. Opt. Soc. Am.* **B 4**, 1453.

Slusher, R. E., B. Yurke and J. Mertz, 1987, *J. Mod. Opt.* **34**, 761.

Smirnov, D. F. and A. S. Troshin, 1981, *J. Exper. Theor. Phys. (USSR)* **81**, 1597.

Smirnov, D. F. and A. S. Troshin, 1985, *Opt. Spektr. (USSR)* **59**, 1.

Smirnov, D. F. and A. S. Troshin, 1987, *Usp. Phys. Nauk (USSR)* **153**, 233.

Smithey, D. T., M. Beck, M. Belsley and M. G. Raymer, 1992, *Phys. Rev. Lett.* **69**, 2650.

Smithey, D. T., M. Beck, J. Cooper and M. G. Raymer, 1993, *Phys. Rev.* **A 48**, 3159.

Smithey, D. T., M. Beck, M. G. Raymer and A. Faridani, 1993, *Phys. Rev. Lett.* **70**, 1244.

Squires, E. J., 1992, *Phys. Lett.* **A 163**, 356.

Srinivas, M. D. and E. B. Davies, 1981, *Opt. Acta* **28**, 981.

Steinberg, A. M., P. G. Kwiat and R. Y. Chiao, 1992, *Phys. Rev. Lett.* **68**, 2421.

Stenholm, S., 1992, *Ann. Phys. (N.Y.)* **218**, 233.

Stoler, D., 1974, *Phys. Rev. Lett.* **33**, 1397.

Storey, P., M. Collett and D. Walls, 1992, *Phys. Rev. Lett.* **68**, 472.

Storey, P., M. Collett and D. Walls, 1993, *Phys. Rev.* **A 47**, 405.

Streater, R. F. and A. S. Wightman, 1964, PCT, Spin and Statistics and All That, W. A. Benjamin, New York.

Susskind, L. and J. Glogower, 1964, *Physics (N.Y.)* **1**, 49.

Swanson, R. C. and J. L. Carlsten, 1993, *Phys. Rev.* **A 47**, 2211.

Tan, S. M. and D. F. Walls, 1993, *Phys. Rev.* **A 47**, 4663.

Tan, S. M., D. F. Walls and M. J. Collett, 1991, *Phys. Rev. Lett.* **66**, 252.

Tapster, P. R., J. G. Rarity and J. S. Satchell, 1987, *Europhys. Lett.* **4**, 293.

Tapster, P. R., J. G. Rarity and J. S. Satchell, 1988, *Phys. Rev.* **A 37**, 2963.

Tapster, P. R., S. F. Seward and J. G. Rarity, 1991, *Phys. Rev.* **A 44**, 3266.

Taylor, G. I., 1909, *Proc. Cambridge Philos. Soc.* **15**, 114.

Teich, M. C., F. Capasso and B. E. A. Saleh, 1987, *J. Opt. Soc. Am.* **B 4**, 1663.

Teich, M. C. and B. E. A. Saleh, 1985, *J. Opt. Soc. Am.* **B 2**, 275.

Teich, M. C. and B. E. A. Saleh, 1988, Photon bunching and antibunching, Progress in Optics, Vol. 26, ed. E. Wolf, Elsevier, Amsterdam, p. 1.

Teich, M. C. and B. E. A. Saleh, 1989, *Quant. Opt.* **1**, 153.

Teich, M. C., B. E. A. Saleh and J. Peřina, 1984, *J. Opt. Soc. Am.* **B 1**, 366.

Teich, M. C., B. E. A. Saleh and D. Stoler, 1983, *Opt. Comm.* **46**, 244.

Titulaer, U. M. and R. J. Glauber, 1965, *Phys. Rev.* **140**, B676.

Titulaer, U. M. and R. J. Glauber, 1966, *Phys. Rev.* **145**, 1041.

Thirring, W., 1983, A Course in Mathematical Physics, Vol. 3: Quantum Mechanics of Atoms and Molecules, Springer, Berlin.

Tomita, A. and R. Y. Chiao, 1986, *Phys. Rev. Lett.* **57**, 937.

Townsend, P. D., J. G. Rarity and P. R. Tapster, 1993, *Electronics Lett.* **29**, 634.

Unruh, W. G., 1978, *Phys. Rev.* **D 18**, 1764.

Unruh, W. G., 1979, *Phys. Rev.* **D 19**, 2888.

Vilenkin, N. J., 1965, Special Functions and Theory of Group Representation, Moscow, Nauka (in Russian).

Vogel, K. and H. Risken, 1989, *Phys. Rev.* **A 40**, 2847.

Vogel, K. and W. Schleich, 1992, Fundamental Systems in Quantum Optics, eds. J. Dalibard, J. M. Raimond and J. Zinn-Justin, Elsevier, Amsterdam, p. 715.

Wagner, C., R. J. Brecha, A. Schenzle and H. Walther, 1993, *Phys. Rev.* **A 47**, 5068.

Walker, J. G., 1986, *J. Mod. Opt.* **33**, 213.

Walker, J. G., 1987, *J. Mod. Opt.* **34**, 813.

Walker, J. G. and E. Jakeman, 1985, *Opt. Acta* **32**, 1303.

Walker, N. G., 1987a, *J. Mod. Opt.* **34**, 15.

Walker, N. G. and J. E. Carroll, 1986, *Opt. Quantum Electron.* **18**, 355.

Walls, D. F., 1977, *Am. J. Phys.* **45**, 952.

Walls, D. F., 1993, Quantum Effects in Atomic Optics, p. 71.

Walls, D. F., M. J. Collett, E. P. Storey and S. M. Tan, 1993, *Austral. J. Phys.* **46**, 61.

Walther, H., 1993, Physics and Probability Essays in Honor of Edwin T. Jaynes, eds. W. T. Grandy, Jr. and P. W. Milonni, Cambridge.

Wang, L. J., X. Y. Zou and L. Mandel, 1991, *Phys. Rev. Lett.* **66**, 1111.

Wang, L. J., X. Y. Zou and L. Mandel, 1991a, *Phys. Rev.* **A 44**, 4614.

Wang, L. J., X. Y. Zou and L. Mandel, 1991b, *J. Opt. Soc. Am.* **B 8**, 978.

Wang, L. J., X. Y. Zou and L. Mandel, 1992, *J. Opt. Soc. Am.* **B 9**, 605.

Wheeler, J. A. and W. H. Zurek, eds., 1983, Quantum Theory and Measurement, Princeton Univ. Press, Princeton, New Jersey.

Wolf, E., 1983, *Opt. Lett.* **8**, 250.

Wolf, E., 1986, *J. Opt. Soc. Am.* **A 3**, 76 and 1920.

Wolf, E., 1987, *Phys. Rev. Lett.* **58**, 2646.

Wolf, E., 1991, International Trends in Optics, ed. J. W. Goodman, Acad. Press, Boston, p. 221.

Wu, L.-A., H. J. Kimble, J. L. Hall and H. Wu, 1986, *Phys. Rev. Lett.* **57**, 2520.

Wu, L.-A., M. Xiao and H. J. Kimble, 1987, *J. Opt. Soc. Am.* **B 4**, 1465.

Yamamoto, Y., ed., 1991, Coherence, Amplification, and Quantum Effects in Semiconductor Lasers, J. Wiley, New York.

Yamamoto, Y. and H. A. Haus, 1986, *Rev. Mod. Phys.* **58**, 1001.

Yamamoto, Y., N. Imoto and S. Machida, 1986, *Phys. Rev.* **A 33**, 3243.

Yamamoto, Y., S. Machida and O. Nilsson, 1986, *Phys. Rev.* **A 34**, 4025.

Yariv, A. and P. Yeh, 1984, Optical Waves in Crystals, J. Wiley, New York, Chap. 13.

Yuen, H. P., 1976, *Phys. Rev.* **A 13**, 2226.

Yuen, H. P., 1982, *Phys. Lett.* **91A**, 101.

Yuen, H. P., 1983, *Phys. Rev. Lett.* **51**, 719.

Yuen, H. P., 1992, Workshop on Squeezed States and Uncertainty Relations, eds. D. Han, Y. S. Kim and W. W. Zachary, NASA, Maryland, p. 13.

Yuen, H. P. and V. V.Chan, 1983, *Opt. Lett.* **8**, 177.

Yuen, H. P. and J. H. Shapiro, 1978, *IEEE Trans. Inf. Theory* **24**, 657.

Yuen, H. P. and J. H. Shapiro, 1980, *IEEE Trans. Inf. Theory* **26**, 78.

Yurke, B., 1985, *J. Opt. Soc. Am.* **B 2**, 732.

Yurke, B., P. Grangier, R. E. Slusher and M. J. Potasek, 1987, *Phys. Rev.* **A 35**, 3586.

Yurke, B., P. G. Kaminsky. R.E. Miller, E. A. Whittaker, A. D. Smith, A. H. Silver and R. W. Simon, 1988, *Phys. Rev. Lett.* **60**, 764.

Yurke, B., S. L. McCall and J. R. Klauder, 1986, *Phys. Rev.* **A 33**, 4033.

Yurke, B. and D. Stoler, 1992, *Phys. Rev. Lett.* **68**, 1251.

Zaheer, K. and M. S. Zubairy, 1991, Advances in Atomic, Molecular, and Optical Physics, Vol. 28, eds. D. Bates and B. Bederson, Acad. Press, New York, p. 143.

Zeilinger, A., H. J. Bernstein, D. M. Greenberger, M. A. Horne and M. Zukowski, 1993, Quantum Control and Measurement, eds. H. Ezawa and Y. Murayama, Elsevier, Amsterdam.

Zeilinger, A., M. Zukowski, M. A. Horne, H. J. Bernstein and D. M. Greenberger, 1993, Fundamental Aspects of Quantum Theory, eds. J. Arrandan and J. L. Safko, World Scientific, Singapore.

Zhang, W. M., D. H. Feng and R. Gilmore, 1990, *Rev. Mod. Phys.* **62**, 867.

Zou, X. Y., T. Grayson, G. A. Barbosa and L. Mandel, 1993, *Phys. Rev.* **A 47**, 2293.

Zou, X. Y., T. P. Grayson and L. Mandel, 1992, *Phys. Rev. Lett.* **69**, 3041.

Zou, X. Y., T. Grayson, L. J. Wang and L. Mandel, 1992, *Phys. Rev. Lett.* **68**, 3667.

Zou, X. Y., L. J. Wang and L. Mandel, 1991, *Opt. Comm.* **84**, 351.

Zou, X. Y., L. J. Wang and L. Mandel, 1991a, *Phys. Rev. Lett.* **67**, 318.

Żukowski, M., A. Zeilinger, M. A. Horne and A. K. Ekert, 1993, *Phys. Rev. Lett.* **71**, 4287.

Zurek, W. H., 1991, *Phys. Today*, No. 10, 36.

Zurek, W. H., S. Habib and J. P. Paz, 1993, *Phys. Rev. Lett.* **70**, 1187.

# Index

# Fundamental Theories of Physics

43. W.T. Grandy, Jr. and L.H. Schick (eds.): *Maximum-Entropy and Bayesian Methods.* Proceedings of the 10th International Workshop (Laramie, Wyoming, USA, 1990). 1991 ISBN 0-7923-1140-X
44. P.Pták and S. Pulmannová: *Orthomodular Structures as Quantum Logics.* Intrinsic Properties, State Space and Probabilistic Topics. 1991 ISBN 0-7923-1207-4
45. D. Hestenes and A. Weingartshofer (eds.): *The Electron.* New Theory and Experiment. 1991 ISBN 0-7923-1356-9
46. P.P.J.M. Schram: *Kinetic Theory of Gases and Plasmas.* 1991 ISBN 0-7923-1392-5
47. A. Micali, R. Boudet and J. Helmstetter (eds.): *Clifford Algebras and their Applications in Mathematical Physics.* 1992 ISBN 0-7923-1623-1
48. E. Prugovečki: *Quantum Geometry.* A Framework for Quantum General Relativity. 1992 ISBN 0-7923-1640-1
49. M.H. Mac Gregor: *The Enigmatic Electron.* 1992 ISBN 0-7923-1982-6
50. C.R. Smith, G.J. Erickson and P.O. Neudorfer (eds.): *Maximum Entropy and Bayesian Methods.* Proceedings of the 11th International Workshop (Seattle, 1991). 1993 ISBN 0-7923-2031-X
51. D.J. Hoekzema: *The Quantum Labyrinth.* 1993 ISBN 0-7923-2066-2
52. Z. Oziewicz, B. Jancewicz and A. Borowiec (eds.): *Spinors, Twistors, Clifford Algebras and Quantum Deformations.* Proceedings of the Second Max Born Symposium (Wrocław, Poland, 1992). 1993 ISBN 0-7923-2251-7
53. A. Mohammad-Djafari and G. Demoment (eds.): *Maximum Entropy and Bayesian Methods.* Proceedings of the 12th International Workshop (Paris, France, 1992). 1993 ISBN 0-7923-2280-0
54. M. Riesz: *Clifford Numbers and Spinors* with Riesz' Private Lectures to E. Folke Bolinder and a Historical Review by Pertti Lounesto. E.F. Bolinder and P. Lounesto (eds.). 1993 ISBN 0-7923-2299-1
55. F. Brackx, R. Delanghe and H. Serras (eds.): *Clifford Algebras and their Applications in Mathematical Physics.* Proceedings of the Third Conference (Deinze, 1993) 1993 ISBN 0-7923-2347-5
56. J.R. Fanchi: *Parametrized Relativistic Quantum Theory.* 1993 ISBN 0-7923-2376-9
57. A. Peres: *Quantum Theory: Concepts and Methods.* 1993 ISBN 0-7923-2549-4
58. P.L. Antonelli, R.S. Ingarden and M. Matsumoto: *The Theory of Sprays and Finsler Spaces with Applications in Physics and Biology.* 1993 ISBN 0-7923-2577-X
59. R. Miron and M. Anastasiei: *The Geometry of Lagrange Spaces: Theory and Applications.* 1994 ISBN 0-7923-2591-5
60. G. Adomian: *Solving Frontier Problems of Physics: The Decomposition Method.* 1994 ISBN 0-7923-2644-X
61 B.S. Kerner and V.V. Osipov: *Autosolitons.* A New Approach to Problems of Self-Organization and Turbulence. 1994 ISBN 0-7923-2816-7
62. A. Heidreder (ed.): *Maximum Entropy and Bayesian Methods.* 1994 ISBN 0-7923-2851-5
63. J. Peřina, Z. Hradil and B. Jurčo: *Quantum Optics and Fundamentals of Physics.* 1994 ISBN 0-7923-3000-5

KLUWER ACADEMIC PUBLISHERS – DORDRECHT / BOSTON / LONDON

GPSR Compliance

*The European Union's (EU) General Product Safety Regulation (GPSR) is a set of rules that requires consumer products to be safe and our obligations to ensure this.*

*If you have any concerns about our products, you can contact us on ProductSafety@springernature.com*

In case Publisher is established outside the EU, the EU authorized representative is:

Springer Nature Customer Service Center GmbH
Europaplatz 3
69115 Heidelberg, Germany

**Batch number: 10809439**

Printed by Printforce, the Netherlands